Probing Hadrons with Leptons

ETTORE MAJORANA INTERNATIONAL SCIENCE SERIES

Series Editor:
Antonino Zichichi
European Physical Society
Geneva, Switzerland

(PHYSICAL SCIENCES)

Probing Hadrons with Leptons

Edited by

Giuliano Preparata

C. E. R. N.
Geneva, Switzerland

and

Jean-Jacques Aubert

Particle Physics Laboratory
Annecy-le-Vieux, France

Plenum Press · New York and London

Library of Congress Cataloging in Publication Data

Seminar on Probing Hadrons with Leptons, 4th, Erice, Italy, 1979.
  Probing hadrons with leptons.

  (Ettore Majorana international science series: Physical sciences; v. 5)
  "Proceedings of the Fourth Seminar on Probing Hadrons with Leptons,
held in Erice, Sicily, Italy, in March, 1979."
  Includes index.
  1. Hadrons—Scattering—Congresses. 2. Leptons—Scattering—Congresses.
I. Preparata, Giuliano. II. Aubert, Jean-Jacques. III. Title. IV. Series.
QC793.5.H328S45 1979   539.7'216   80-12024
ISBN 0-306-40438-9

Proceedings of the Fourth Seminar on Probing Hadrons with Leptons, held in
Erice, Sicily, Italy, in March, 1979.

© 1980 Plenum Press, New York
A Division of Plenum Publishing Corporation
227 West 17th Street, New York, N.Y. 10011

Printed in the United States of America

FOREWORD

Physicists actively engaged in advanced research should be encouraged to discuss results and deepen their theoretical understanding of the data. It is practically impossible nowadays to achieve
the goal of the old times when small groups of scientists had the
privilege of debating their ideas and the details of their experiments in an informal and friendly way. Conferences are now too wide
in scientific coverage and, in consequence, there are often too many
participants.

The Highly Specialized Seminars of the Ettore Majorana Centre
for Scientific Culture are intended to provide such a forum for
scientists of outstanding reputation in their fields to exchange
information.

This volume deals with one of the most interesting topics in
subnuclear physics, and Professor Giuliano Preparata is one of the
world's leaders in the field. This volume contains the most recent
results on the study of deep inelastic phenomena, using neutrinos,
muons and electrons against nucleons, and (e^+e^-) interactions. It
represents our best knowledge of the field as given by some of the
most distinguished world experts in the theoretical and experimental
domain.

Antonino Zichichi, Director

Ettore Majorana Centre for
Scientific Culture

PREFACE

In March 1979, about 50 physicists gathered in Erice to participate in the Fourth Highly Specialized Seminar on "Probing Hadrons with Leptons".

In today's subnuclear physics, the hadronic structure at short distances, as revealed by highly inelastic collisions among leptons and hadrons plays a central role in our understanding of the basic forces of Nature. It is here that the relevance of the concepts of quarks and colour has found its most dramatic testing ground, and it is here that new intriguing effects keep emerging after ten years of momentous experimentation.

The most diverse aspects of deep inelastic physics were discussed in Erice in a very frank, friendly and productive fashion. Opposing views had the chance, quite uncharacteristically for the present situation, to be discussed and evaluated on their own merits and confronted with the known facts. As a result, we believe that the seminar was, for all those who participated, a very interesting and successful enterprise. It is in order to give the discussions of Erice a wider forum that the proceedings of the seminar are being published.

A special thanks goes to Kathie Hardy for skillful work in preparing this volume and to Sheila Navach for her organizing skill and constant efforts in arranging transportation to overcome a sudden strike.

Finally, we thank all those who have contributed and sacrificed some of their time in order to ensure that the material discussed at Erice would be made available to a wider audience.

Giuliano Preparata and Jean-Jacques Aubert
Geneva, November 1979

CONTENTS

FIRST RESULTS FROM MARK II AT SPEAR[*]

G. S. Abrams, M. S. Alam, C. A. Blocker, A. M. Boyarski,
M. Breidenbach, C. H. Broll, D. L. Burke, W. C. Carithers,
W. Chinowsky, M. W. Coles, S. Cooper, B. Couchman, W. E.
Dieterle, J. B. Dillon, J. Dorenbosch, J. M. Dorfan, M. W.
Eaton, G. J. Feldman, H. G. Fischer, M.E.B. Franklin,
G. Gidal, G. Goldhaber, G. Hanson, K. G. Hayes, T. Himel,
D. G. Hitlin, R. J. Hollebeek, W. R. Innes, J. A. Jaros,
P. Jenni, A. D. Johnson, J. A. Kadyk, A. J. Lankford,
R. R. Larsen, M. J. Longo, D. Lüke, V. Lüth, J. F. Martin,
R. E. Millikan, M. E. Nelson, C. Y. Pang, J. F. Patrick,
M. L. Perl, B. Richter, J. J. Russell, D. L. Scharre,
R. H. Schindler, R. F. Schwitters, S. R. Shannon, J. L.
Siegrist, J. Strait, H. Taureg, V. I. Telnov, M. Tonutti,
G. H. Trilling, E. N. Vella, R. A. Vidal, I. Videau
J. M. Weiss, H. Zaccone.

Presented by Peter Jenni

Stanford Linear Accelerator Center
Stanford University, Stanford, California 94305
                    and
Lawrence Berkeley Laboratory and Department of Physics
University of California, Berkeley, California 94720

[*] This work was supported primarily by the Department of Energy
under contract numbers DE-AC-03-76SF00515 and W-7405-ENG-48.

1

I.   INTRODUCTION

     In this talk I will report some of the first results from the
Mark II experiment at the Stanford Linear Accelerator Center e^+e^-
storage ring facility SPEAR. The first part will be devoted to a
description of the Mark II detector and of its performance. Then I
will show preliminary results on a measurement of the inclusive production of baryons (p, $\bar{p}$, Λ and $\bar{\Lambda}$) over the centre-of-mass energy
range from 4.5 to 6 GeV in a search for a threshold of charmed
baryon production. Data on decays of charmed D mesons have been
accumulated at the ψ'' (3770) resonance and first results will be
given in part three from an analysis of about one-third of the whole
data sample. Finally, we have observed, for the first time, evidence
for η' (958) production in two-photon collisions. We have measured th
radiative width of the η' through this process as will be discussed
in the last part of this talk.

II.   THE MARK II DETECTOR

     A schematic view of the Mark II detector is shown in Fig. 1.
A particle that moves outwards from the e^+e^- interaction region
first traverses the 0.15 mm thick stainless steel vacuum pipe and
two concentric 0.64 cm thick cylindrical scintillation counters. It
then enters the drift chamber[1] which contains 16 sense-wire layers
of radii 0.41 m to 1.45 m. The wire orientation is such as to provide the highest accuracy in the transverse direction: six of the

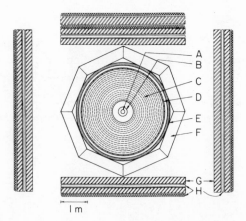

Fig. 1.   Schematic view of the Mark II detector. (A) vacuum chamber,
          (B) pipe counter, (C) drift chamber, (D) time-of-flight
          counters, (E) solenoid coil, (F) liquid argon shower
          counters, (G) iron absorber, (H) muon proportional tubes.

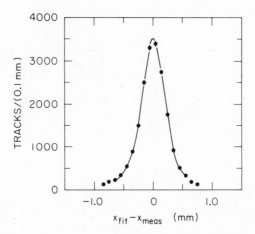

Fig. 2. Example of a distribution of the difference between fitted
and measured track positions in the drift chamber

layers have wires parallel to the beam axis and the other 10 are
at a small stereo angle (± 3°) with respect to the beam axis.
The average spatial resolution is about 210 μm per layer. Figure 2
shows an example of a distribution of the difference between fitted
and measured track positions. The magnetic field is 4.1 kG and the
momenta of charged particles are measured with an accuracy
$\delta p/p = \pm[(0.010p)^2 + (0.0145)^2]^{\frac{1}{2}}$ where p is the momentum in GeV/c.*
    The drift chamber is surrounded by a layer of 2.54 cm thick
scintillation counters. Each counter is viewed on both ends by
XP2230 photomultipliers. With the beam crossing reference signal
they provide an rms time-of-flight resolution of 300 ps for hadrons
and 270 ps for Bhabha scattered electrons (Figure 3).
    Next, the particle traverses 1.36 radiation lengths of solenoi-
dal coil and support material and enters one of the eight liquid
argon calorimeter modules[2] (LA). A module consists of a "massless"
trigger gap with 0.16 cm thick Al electrodes and two 0.8 cm thick
liquid argon gaps followed by 18 sampling layers with 0.2 cm thick
Pb ground planes alternating with 3.7 cm wide, 0.2 cm thick Pb readout strips and 0.3 cm liquid argon gaps. These shower counters are
about 14 radiation lengths deep. The read-out strips are parallel,
perpendicular and at 45° to the beam axis giving an rms angular
resolution of about 8 mrad both in azimuth and dip angle. The rms
energy resolution for electrons and photons at high energies
(≥ 0.5 GeV) has been measured to be $\delta E/E = 0.11/\sqrt{E}$ (E in GeV).
An example for Bhabha scattered electrons at the beam energy

---

* The momentum resolution is $\delta p/p = \pm[(0.005p)^2 + (0.0145)^2]^{\frac{1}{2}}$ when
tracks are constrained to pass through the known beam position.

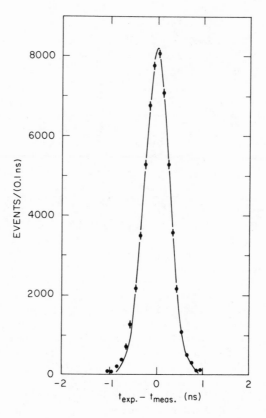

Fig. 3.   Expected time minus measured time from the TOF system
for a sample of Bhabha events at $E_{c.m.}$ = 4.16 GeV

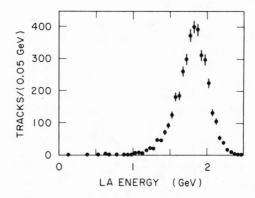

Fig. 4.   Energy distribution in the liquid argon calorimeter
modules for a sample of Bhabha scattered electrons
at $E_{c.m.}$ = 3.684 GeV

E = 1.842 GeV is shown in Figure 4. At lower energies the resolution
is worse $(0.13 / \sqrt{E})$ because of the increasing importance of the
energy loss in the coil material. The measured efficiency for photons within the geometrical acceptance of the LA detector is shown
in Figure 5 and agrees well with detailed electromagnetic shower
Monte Carlo calculations[3] (shown as curve in Figure 5) which are also
used to correct the measurements for the energy loss in the coil
material. The LA detector is also used for electron-pion separation.
Pion misidentification probabilities of less than 5% for electron
efficiencies above 75% are achieved for particle momenta greater than
500 MeV/c, and improve with higher momenta. Finally, two 23 cm thick
steel walls, each followed by one plane of proportional tubes, are
used for the detection of muons above $p \approx 700$ MeV/c.

The fraction of the full solid angle covered by the drift chamber and TOF counters is about 75%, by the LA modules 70% and by
the muon detection system, 55%. At small angles relative to the
beams there are additional shower counters (at one end a liquid argon
calorimeter and at the other end two planes of proportional chambers
each preceded by 1.1 cm of lead) which extend the solid angle coverage to 90% of 4π sr.

The detector is triggered with a two stage hardware trigger[4]
that selects, with efficiency $\geq 99\%$, all interactions that emit at
least one charged particle through the entire drift chamber and another particle through at least its first five layers. The luminosity
is measured with independent shower counters detecting Bhabha scattering at 22 mrad and checked against wide angle Bhabha events
observed in the detector. The systematic uncertainty in the luminosity is less than $\pm$ 6%.

In Figure 6 we show a particularly high multiplicity example
of a multiprong hadron event at $E_{CM} = 5.08$ GeV as reproduced by the
on-line event display. The event is shown in the projection into a

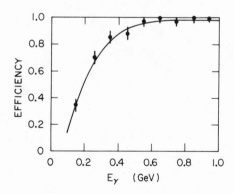

Fig. 5.   Detection efficiency for photons within the geometrical
          acceptance of the LA barrel modules. The curve is a
          Monte Carlo calculation.

plane perpendicular to the beam axis displaying the track measurements in the drift chamber, the time-of-flight counters and the liquid argon shower counters (with the corresponding energies in GeV).

III.   INCLUSIVE BARYON PRODUCTION

The inclusive production of baryons has been studied in a scan
over the centre-of-mass energy range from 4.5 to 6 GeV. At this
early stage of the analysis we present results for $\bar{p}$, Λ and $\bar{\Lambda}$
production, also including data from running at various fixed ener-
gies. The p and $\bar{p}$ identification has been done by the time-of-flight
measurement. The Λ ($\bar{\Lambda}$) have been observed by their $p\pi^-$ ($\bar{p}\pi^+$) decay
mode with a rms mass resolution of about 3 MeV/c^2.

The results, corrected for the acceptance, are given in Figure 7
as the ratio of the inclusive cross-section to the μ-pair production
cross-section. To avoid beam-gas backgrounds, only the $\bar{p}$ measurements have been used in the case of Figure 7(a); plotted is
$R(\bar{p}+p) = 2\sigma\bar{p}/\sigma_{\mu\mu}$. All the errors shown are only statistical;

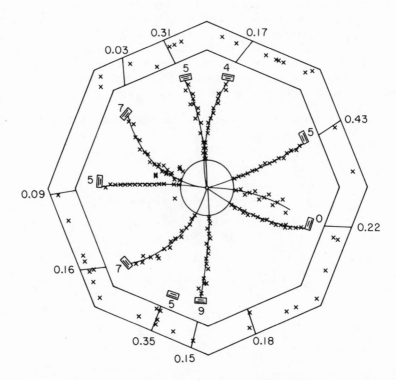

Fig. 6.   Example of a particularly high multiplicity multiprong
          hadron event at E$_{c.m.}$ = 5.08 GeV as produced by the on-line
          event display

the systematic uncertainty of $R(\bar{p} + p)$ and $R(\bar{\Lambda} + \Lambda)$ is estimated to be less than ± 20%. The measurements are consistent with previous experiments[5,6] and confirm in more detail the rise in the inclusive baryon production in $e^{+}e^{-}$ annihilation over the centre-of-mass energy range of about 4.5 to 5.5 GeV.

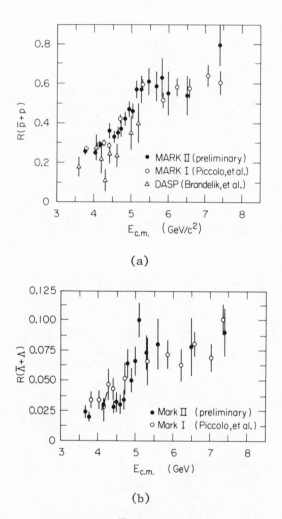

(a)

(b)

Fig. 7. Inclusive $\bar{p}$ and $\Lambda + \bar{\Lambda}$ production. (a) $R(\bar{p} + p) = 2\sigma(\bar{p})/\sigma_{\mu\mu}$ vs. c.m. energy. (b) $R(\bar{\Lambda} + \Lambda) = \sigma(\bar{\Lambda} + \Lambda)/\sigma_{\mu\mu}$ vs. c.m. energy. The Mark II results are preliminary, the systematic uncertainty for these measurements is estimated to be less than ± 20%. (Mark I is Reference 5; DASP is Reference 6.)

IV. CHARMED D MESON DECAYS

Various decay modes of the charmed D mesons are being studied with the data that have been accumulated at the ψ'' (3770) resonance[7]. The first results reported here are based on a total integrated luminosity of 770 nb^{-1} which represents slightly more than one quarter of the total data sample.

The charged particle identification has been done by the time-of-flight and liquid argon pulse height measurements and π^0's have been reconstructed in the liquid argon shower detectors. The total energy of a reconstructed D meson has been constrained to the known beam energy (because of ψ'' (3770) $\rightarrow$ D$\bar{\text{D}}$).

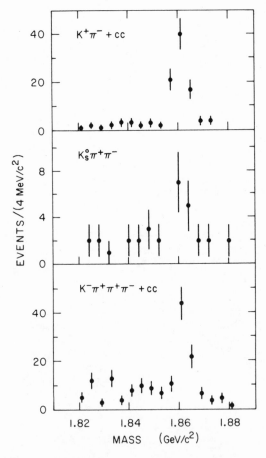

Fig. 8.   Invariant mass distribution for various D^0 decay modes

A further constraint is possible in the D decays involving π^0 or K_S^0 where the observed $\gamma\gamma$ or $\pi^+\pi^-$ final states have been fit to the known mass values respectively. The rms mass resolution achieved on the reconstructed D mesons is typically 2 - 3 MeV/c^2. Invariant mass distributions for different observed decay modes of the D^0, $\overline{D}^0$ and $D^\pm$ are shown in Figures 8 and 9. In Table 1 we summarize the branching ratios which have been obtained in this preliminary analysis. The errors given do include the uncertainty in the total **cross-section** measurement of the ψ'' (3770) resonance measured in a previous experiment[8]. However, a further estimated systematic uncertainty of ± 20% is not included. All values are in good agreement with the previous measurements of the Mark I collaboration[8,9].

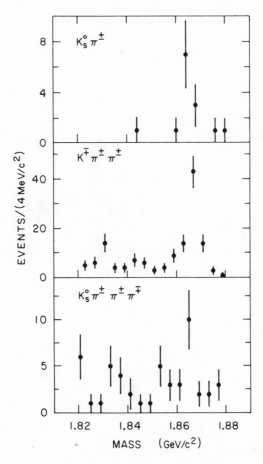

Fig. 9. Invariant mass distribution for various $D^\pm$ decay modes

Table 1.   Preliminary Measurement of Various
Branching Ratios of D mesons.

Decay Mode	Detection Efficiency	Number of Events	Branching Ratio
$D^0 \to K^\pm \pi^\mp$	0.44	74 ± 10	0.019 ± 0.005
$K^\pm \pi^\mp \pi^\mp \pi^\pm$	0.10	55 ± 10	0.061 ± 0.019
$K^\pm \pi^\mp \pi^0$	0.024	18 ± 5	0.083 ± 0.022
$K_S^0 \pi^+ \pi^-$	0.13	12 ± 4	0.010 ± 0.004
$D^\pm \to K^\mp \pi^\pm \pi^\pm$	0.29	65 ± 9	0.032 ± 0.008
$K_S^0 \pi^\pm$	0.24	10 ± 3	0.005 ± 0.003
$K_S^0 \pi^\pm \pi^\pm \pi^\mp$	0.04	12 ± 5	0.040 ± 0.023

V. EVIDENCE FOR η' (958) PRODUCTION IN TWO-PHOTON COLLISIONS

The observation of leptons and hadrons produced by two-photon interactions in electron-positron colliding beam experiments has been a challenge ever since the importance of the two-photon mechanism has been pointed out[10]. The basic diagram is shown in Figure 10. Lepton pairs produced by the two-photon process have been observed in several experiments[11,12]. The data on hadronic events are much more scarce as only a very few multi-hadron events have been seen[13] so far.

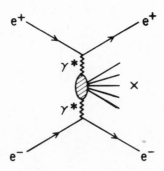

Fig. 10. Diagram for the two-photon production of the state X

We have observed and reported[14] evidence for η' (958) production in the reaction

$$e^+e^- \rightarrow e^+e^-\eta'$$ (1)

through the decay mode η' → ρ⁰γ resulting in a π⁺π⁻γ final state. The outgoing e⁺ and e⁻ were not detected.

Events having only two oppositely charged tracks coming from the interaction region and one photon detected in the LA modules were selected. The pions were identified by TOF measurement and by LA pulse heights (to eliminate electrons). A few kinematical cuts were applied to reduce possible backgrounds from one-photon e⁺e⁻ annihila- tions, where part of the final state particles remain undetected, from Bhabha scattered electrons with radiatively degraded initial states, and from lepton or pion pairs produced in two-photon inter- actions combined with noise-generated false photons. All the analy- sis cuts are listed in Reference 14; the two most important ones require that the transverse momentum of the π⁺π⁻γ state ($p_\perp$) be < 250 MeV/c and that the acoplanarity angle between the π⁺π⁻ pair and the γ momentum vectors defined with respect to the beam axis (Δφ) be < 0.5 rad (Δφ = 0 for back-to-back decays).

The π π γ mass distribution for the remaining events, given in Figure 11, shows a clear η' → π⁺π⁻γ signal. The η' mass resolution is dominated by the photon energy measurement and is consistent with the expectation.

The transverse momentum $p_\perp$, total energy E and angular distribu- tion cos θ (with respect to the beam axis) are shown in Figures 12a-c for all events (full histograms) and for the events lying in the η' mass region (shaded) defined as 800 < $m_{\pi\pi\gamma}$ < 1100 MeV/c². The η' events occur mainly at low $p_\perp$ in contrast with the background events.

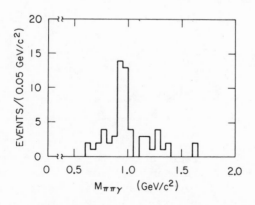

Fig. 11. π⁺π⁻γ invariant mass distribution

Their total energy peaks at low values thus excluding an interpretation of two-body production like η'γ where the γ is undetected. The
angular distribution is highly peaked in the forward and backward
directions. We also observe a flat rapidity (y) distribution, shown
in Figure 13, within the detector acceptance of about −0.6 < y < 0.6.
These kinematical features are those expected for η' production by
reaction (1) and are well reproduced by Monte Carlo generated two-
photon events, using the two-photon calculation of Reference 15.

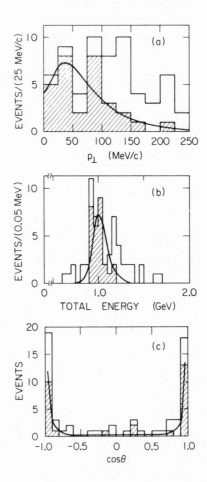

Fig. 12.   Kinematical distributions for (a) transverse momentum,
           (b) total energy, and (c) cosine of the production angle
           with respect to the beam axis. The full histograms contain
           all events; the events in the η' mass peak are shaded.
           The curve represents the Monte Carlo calculation assuming
           $e^+e^- \rightarrow e^+e^-\eta'$ normalized to the observed η' signal

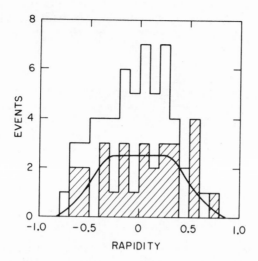

Fig. 13. Rapidity distribution for the $\pi^+\pi^-\gamma$ state
(Same definitions as for Figure 12)

The expected distributions, normalized to the same number of η'
events, are shown as solid curves in Figures 12 and 13.

The background from e^+e^- annihilation events has been studied
in multihadron events. The $\pi^+\pi^-\gamma$ mass combinations have been calculated independently of the existence of additional charged tracks or
photons, with all other criteria unchanged. No peaking in the mass
and energy distributions is seen, and the $p_\perp$ distribution peaks above
200 MeV/c. The correction for annihilation events is included in the
background subtraction using the adjacent mass regions which leaves
a total of 23 ± 6 η' events (see Table 2).

The **cross-section** for reaction (1) has been calculated using
the branching ratio $B(\eta' \to \rho^0\gamma) = 0.298 \pm 0.017$ and is also given in
Table 2. The cross section is directly proportional to the radiative
width $\Gamma_{\gamma\gamma}(\eta')$[16]. Using the two-photon cross-section calculation of
Reference (15), we determine $\Gamma_{\gamma\gamma}(\eta') = 5.9 \pm 1.6$ kev.[17] The error is
statistical only and does not include an estimated systematic uncer-
tainty of $\pm$ 20%. With the $B(\eta' \to \gamma\gamma) = 0.0197 \pm 0.0026$, the total
width can then be determined to be $\Gamma_{tot}(\eta') = 300 \pm 90$ KeV (or $\tau =$
$= (2.2 \pm 0.7) \times 10^{-21}$ sec). Our measurement of Γ_{tot} is in excellent
agreement with the only other available measurement (280 ± 100 keV)
recently reported by Reference 18.

There is considerable interest in a measurement of $\Gamma_{\gamma\gamma}(\eta')$[19-21].
Quark models with fractionally charged quarks and a small pseudoscalar
octet-singlet mixing angle lead, under the assumption of equal singlet
and octet decay constants[20], to the prediction $\Gamma_{\gamma\gamma}(\eta') \simeq 6$ keV [19].
This is in good agreement with our measurement. The data are also
in agreement with a recent relativistic quark model[21] calculation
which predicts $\Gamma_{\gamma\gamma}(\eta') = 7.3$ keV.

Table 2.   Summary of the Cross Section Calculation

E_b (GeV)	$\int \mathscr{L}dt$ (nb^{-1})	ϵ	$n_{\eta'}$	σ (nb)
2.21	798	0.017	5.1 ± 2.6	0.98 ± 0.50
2.25 – 2.50	2131	0.0224	4.3 ± 2.6	0.30 ± 0.18
2.50 – 3.00	1730	0.0217	10.3 ± 3.6	0.91 ± 0.32
3.70	984	0.0125	3.1 ± 2.2	0.84 ± 0.60

E_b is the beam energy, $\int \mathscr{L}dt$ the integrated luminosity, ϵ the detection efficiency [not including $B(\eta' \to \rho\gamma)$], $n_{\eta'}$ is the number of η' events (background subtracted) and σ is the observed cross-section. Errors shown are statistical only.

REFERENCES

1. W. Davies-White et al., Nucl. Instrum. Methods 160:227 (1979).
2. G.S. Abrams et al., IEEE Trans. on Nucl. Sci. NS-25:309 (1978).
3. R.L. Ford and W.R. Nelson, EGS Code, SLAC Report SLAC-210 (1978).
4. H. Brafman et al., IEEE Trans. on Nucl. Sci. NS-25:1 (1978).
5. M. Piccolo et al., Phys. Rev. Lett. 39:1503 (1977).
6. R. Brandelik et al., Nucl. Phys. B148:189 (1979).
7. P.A. Rapidis et al., Phys. Rev. Lett. 39:526 (1977);
    W. Bacino et al., Phys. Rev. Lett. 40:671 (1978).
8. I. Peruzzi et al., Phys. Rev. Lett. 39:1301 (1977).
9. D.L. Scharre et al., Phys. Rev. Lett. 40:74 (1978).
10. S.J. Brodsky, T. Kinoshita and H. Terazawa, Phys. Rev. Lett.
        25:972 (1970) and Phys. Rev. D4:1532 (1971);
    A. Jaccarini et al., Nuovo Cimento Lett. 4:933 (1970);
    N. Arteaga-Romera et al., Phys. Rev. D3:1569 (1971);
    V.N. Baier and V.S. Fadin, Nuovo Cimento Lett. 1:481 (1971).
11. V.E. Balakin et al., Phys. Lett. 34B:663 (1971);
    C. Bacci et al., Nuovo Cimento Lett. 3:709 (1972);
    G. Barbiellini et al., Phys. Rev. Lett. 32:385 (1974).
12. H.J. Besch et al., Phys. Lett. 81B:79 (1979).
13. S. Orito et al., Phys. Lett. 48B:380 (1974);
    L. Paoluzi et al., Nuovo Cimento Lett. 10:435 (1974).
14. G.S. Abrams et al., SLAC preprint SLAC-PUB-2231 (1979) (sub-
        mitted to Phys. Rev. Lett).
15. V.M. Budnev and I.F. Ginzburg, Phys. Lett. 37B:320 (1971);
    V.N. Baier and V.S. Fadin, Nuovo Cimento Lett. 1:481 (1971)
        and private communication.

16. F.E. Low, Phys. Rev. 120:582 (1960).
17. Upper limits on $\Gamma_{\gamma\gamma}(\eta')$ have been published from two-photon
     experiments in References 12 and 13.
18. D.M. Binnie et al., Imperial College London preprint
     IC/HENP/79/2 (1979).
19. S. Okubo in "Symmetries and Quark Models", ed. R. Chand,
     Gordon and Breach, New York (1970);
     H. Suura, T.F. Walsh and B-L Young, Nuovo Cimento Lett. 4:505
     (1972).
20. However, this assumption fails to describe $\Gamma(\eta' \to \pi^+\pi^-\gamma)/\Gamma(\eta' \to \gamma\gamma)$.
     See M.S. Chanowitz, Phys. Rev. Lett. 35:977 (1975) wherein
     an alternative analysis is proposed.
21   E. Etim and M. Greco, Nuovo Cimento 42:124 (1977).

FORMATION OF UPSILON MESONS AT DORIS

K.R. Schubert

Institut für Hochenergiephysik
Universität Heidelberg, Germany

In June 1977, Herb et al.[1] found a narrow state with mass
M = 9.5 GeV and width Γ < 0.5 GeV in the muon pair spectrum of the
reaction p(400 GeV/c) + nucleus → $\mu^+\mu^-$ + anything at FNAL. Soon
afterwards[2], the same group reported evidence for two or three states
in this mass region. Since the muon pair spectrum did not show any
structure between ψ'(3.7) and these new states, they were immediately
taken as the first experimental evidence for the "onium" states $b\bar{b}$
or $t\bar{t}$ of the fifth hadronic flavour. This evidence was highly wel-
come for incorporating CP violation into unified interaction theories[3].
Ten years ago it was found that $s\bar{d} \leftrightarrow \bar{s}d$ transitions show a small CP
violation and a small T violation, now it might become possible to
find a large CP and T violation in $b\bar{d} \leftrightarrow \bar{b}d$ transitions.

In September 1977, DESY decided to upgrade its electron-positron
storage ring DORIS from an upper energy limit of $\sqrt{s}$ ≈ 7 GeV to
$\sqrt{s}$ ≈ 10 GeV. Three experimental groups, DASP-2 [4], PLUTO[5] and NaI –
Lead Glass[6] aimed at finding the new states in e^+e^- annihilations
and at determining their resonance parameters. According to the
FNAL group, the Υ(9.46) state was expected[7] at $\sqrt{s}$ = (9458 ± 10 ±
± 100) MeV, where the central value is obtained by adjusting the
observed J/ψ mass to its world average; 10 MeV are statistics and
the systematical error of ± 100 MeV is due to alignment (10 MeV),
field integral (20 MeV), field shape (20 MeV) and muon energy loss
(50 MeV). The next state, Υ' (10.0) was expected Δm = (600 ± ?) MeV
higher.

I skip here the exciting work of transforming DORIS and turn to
the experiments immediately. Since the PLUTO results were presented
at this seminar by H. Meyer, I shall concentrate on the DASP-2 and

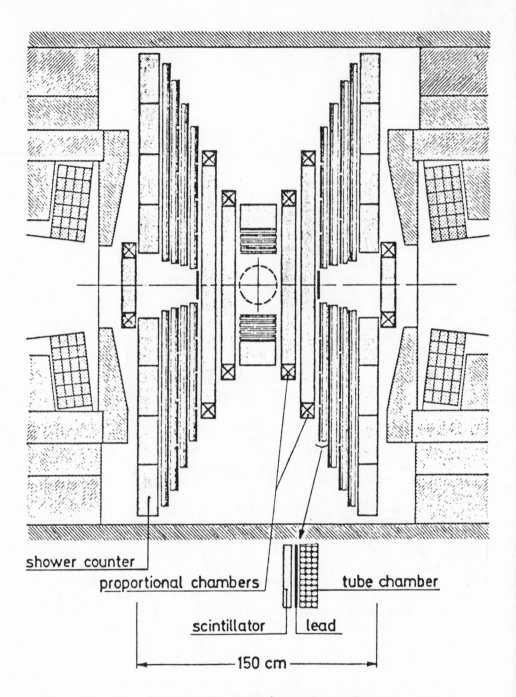

shower counter

proportional chambers   tube chamber

scintillator   lead

◄─────────────── 150 cm ───────────────►

Fig. 1 The DASP inner detector

the NaI – Lead Glass results. Figure 1 shows the DASP-2 inner detector with a solid angle of $\Delta\Omega \approx 0.50 * 4\pi$. Most of the DASP-2
data have been obtained with this non-magnetic part of the detector;
the two magnetic arms – extending to the left and right of Figure 1 –
cover only a solid angle of about $0.05 * 4\pi$. Figure 2 shows the
detector of the NaI – Lead Glass group; it is also non-magnetic and
covers a solid angle of $0.86 * 4\pi$.

The DASP-2 experiment was running in April/May 1978 on the T
and in July/August 1978 on the T'. During good runs, DORIS worked
with 15 mA/beam which corresponds to $\int \mathcal{L} dt \approx 10^{30}/cm^2/s$. We accumulated

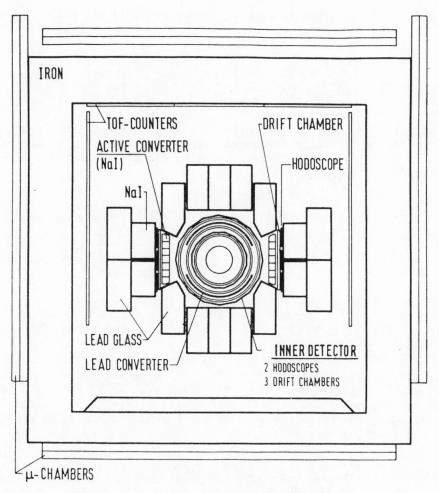

Fig. 2 The NaI – Lead Glass detector

$\int \mathcal{L} dt$ = 200/nb near the T, 300/nb on the T and 120/nb near and on the T'. The trigger was set-up for:

    a) $e^+e^- \to$ hadrons ($\geq$ three tracks or showers);

    b) $e^+e^- \to e^+e^-$;

    c) $e^+e^- \to \mu^+\mu^-$.

Reaction b) in the inner detector was used for monitoring together with small angle Bhabha scattering. For reaction a), $3 \cdot 10^7$ triggers were accumulated. The reconstruction criteria (energy/track < 3·5 GeV, total energy > 0.45 GeV, $\geq$ three tracks or showers reconstructed with a common vertex, not all tracks in one hemisphere) led to 3.10^3 events. The resulting cross-section for reaction a) is shown in Fig. 3. The corresponding results of NaI – Lead Glass are shown in Fig. 4. PLUTO observed only the T(9.46) state.

The visible cross-section for reaction a) is described by

$$\sigma_{had}^{(vis)} = \eta \left[R\sigma_{\mu\mu} + \frac{3\pi}{s} \frac{\Gamma_{ee} \Gamma_{had}}{(\sqrt{s} - M)^2 + \Gamma_{tot}^2/4} \right]$$

folded with Gaussian energy smearing and radiative corrections[8]. The radiative corrections suppress the peak height by a factor of about 1.5. Fits of the formula above to the data around the first T state yield:

M = (9.457 ± 0.010) GeV   (all three groups)

rms-width = (7.5 ± 1.0) MeV   as expected from the
                                   DORIS resolution

$\int \sigma_{had}^{(vis)} d\sqrt{s}$ = (337 ± 91) MeV nb   (DASP-2 value)

The last result requires a knowledge of the detector acceptance η. This can be obtained by Monte Carlo simulation or by knowing R outside the resonance and normalizing the data there. For reasons of consistence, the following results are based on the R value of PLUTO, R = 3.6 ± 0.5, $\tau\tau$ subtracted.

$$\int \sigma_{had}^{(vis)} d\sqrt{s} = \frac{6\pi^2}{M^2} \frac{\Gamma_{ee} \Gamma_{had}}{\Gamma_{tot}}$$

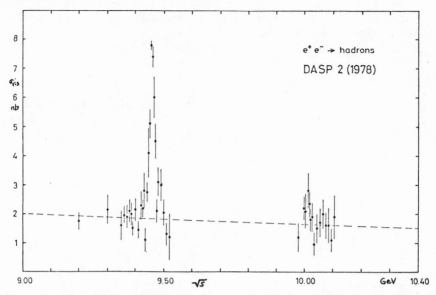

Fig. 3   The observed cross-section for e^+e^- → hadrons as measured by
DASP-2.   The dashed line is the off resonance expectation
for a constant value of R

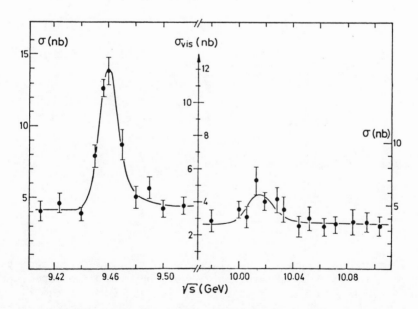

Fig. 4   The observed cross-section for e^+e^- → hadrons as measured
by NaI – Lead Glass

Since $\Gamma_{ee} \ll \Gamma_{tot}$, the integral yields approximately Γ_{ee}. With a determination of $B_{\mu\mu} = \Gamma_{\mu\mu}/\Gamma_{tot} = \Gamma_{ee}/\Gamma_{tot}$ one can determine Γ_{ee} better and, in addition, Γ_{tot}. All three groups have tried to measure $B_{\mu\mu}$ out of their limited amount of data. The DASP-2 group has used the time-of-flight system (TOF) of the inner detector to separate reaction c) from the main background of cosmic muons. Genuine events are selected by $\Delta TOF = (0 \pm 1)$ns, whereas cosmic muons have $\Delta TOF = 6$ns. On the Υ, the DASP-2 group finds $(86 \pm 13)\mu\mu$ events with a non-resonant expectation of (60 ± 14). The difference is due to $ee \to \Upsilon \to \mu\mu$. The results are as follows:

$$B_{\mu\mu} \text{ (DASP-2)} \qquad = \qquad (2.5 \pm 2.1)\%$$

$$B_{\mu\mu} \text{ (PLUTO)} \qquad = \qquad (2.2 \pm 2.0)\%$$

$$B_{\mu\mu} \text{ (NaI - Lead Glass)} \qquad = \qquad (1.0 \, {}^{+\,3.4}_{-\,1.0})\%$$

$$B_{\mu\mu} \text{ (average)} \qquad = \qquad (2.1 \pm 1.3)\%$$

This result gives a 6% correction to go from $\Gamma_{ee} \, \Gamma_{had}/\Gamma_{tot}$ to Γ_{ee}. With this correction, the three groups find:

$$\Gamma_{ee} \text{ (DASP-2)} \qquad = \qquad (1.39 \pm 0.37) \text{ keV}$$

$$\Gamma_{ee} \text{ (PLUTO)} \qquad = \qquad (1.33 \pm 0.14) \text{ keV}$$

$$\Gamma_{ee} \text{ (NaI - Lead Glass)} \qquad = \qquad (1.11 \pm 0.18) \text{ keV}$$

$$\Gamma_{ee} \text{ (average)} \qquad = \qquad (1.26 \pm 0.11) \text{ keV}$$

By combining the two results, one finds Γ_{tot}

$$\Gamma_{tot} = \Gamma_{ee}/B_{\mu\mu} = (60 \, {}^{+\,100}_{-\,23}) \text{ keV}$$

The errors correspond to one standard deviation. With two standard deviations, Γ_{tot} is larger than 27 keV, but has no upper bound. (Though 60 keV is the best estimate for the width of the Υ, its upper bound is still about 15 MeV from the DORIS energy resolution.)

For the T' meson, the following results were obtained:

M (DASP-2)	=	(10.012 ± 0.020) GeV
M (NaI - Lead Glass)	=	(10.020 ± 0.020) GeV
M (average)	=	(10.016 ± 0.020) GeV

The error is mainly the systematic DORIS calibration error. Since it is partially correlated with the systematic error on the $T(9.46)$ mass, one may quote: $\Delta M(T' - T) = (559 \pm 10)$ MeV. The electronic width, not corrected for finite Γ_{ee}/Γ_{tot}, is fitted to be:

Γ_{ee} (DASP-2)	=	(0.35 ± 0.14) keV
Γ_{ee} (NaI - Lead Glass)	=	(0.32 ± 0.13) keV
Γ_{ee} (T', average)	=	(0.33 ± 0.10) keV

The three results, Γ_{ee} (T), Γ_{ee}(T') and $B_{\mu\mu}$(T) allow a conclusion on the electric charge of the new quark with the fifth hadron flavour.

$$T, T' = b\bar{b} \; ; \; |Q(b)| = 1/3e_o$$

The first conclusion makes use of potential models for the quark-antiquark system. In the framework of these non-relativistic models, Γ_{ee} is given by the Van Royen-Weisskopf formula[9]:

$$\Gamma_{ee} = 16\pi\alpha^2 Q^2 |\psi(0)|^2/M^2$$

where M is the mass of the $q\bar{q}$ state, Q is the charge of quark q and $\psi(0)$ is the wave-function at the origin. Rosner et al.[10] have tried 20 different potentials and calculated bounds for Γ_{ee}(T) and Γ_{ee}(T'). Figure 5 shows their bi-plot of Γ_{ee}(T) vs Γ_{ee}(T') with the expected regions for $|Q(q)| = 1/3e_o$ and $2/3e_o$. The data point corresponds to the average values given in this report and excludes the quark charge $2/3e_o$ with more than three standard deviations.

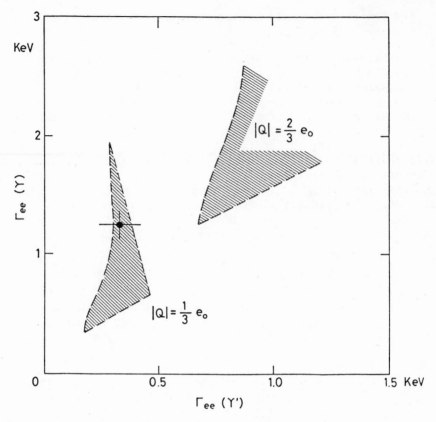

Fig. 5 Expectations for Γ_{ee} (T) and Γ_{ee} (T') with quark charges
       $1/3e_0$ and $2/3e_0$ (Ref. 10) and experimental results

A second conclusion uses $B_{\mu\mu}$(T) and is independent of the wave function at the origin:

$$\frac{\Gamma_{\mu\mu}}{\Gamma_{had}} = \frac{Q^2 \cdot 16\pi\alpha^2/M^2 \cdot |\psi(0)|^2}{K_{QCD} \cdot \alpha_s^6 \cdot |\psi(0)|^2}$$

where the denominator follows from T → three gluons → hadrons and K_{QCD} is a kinematical factor. Various authors[11] have calculated $B_{\mu\mu}$ in this QCD frame and find $B_{\mu\mu}$ around 3.5% for charge $1/3e_0$ and around 8.2% for charge $2/3e_0$. The average experimental result, $B_{\mu\mu}$ = (2.1 ± 1.3)%, excludes 8.2% with more than four standard deviations.

The last chapter of this report summarizes some results on the inclusive properties of the hadronic final state of the Υ decay. All three groups have presented results on multiplicities of final particles. Figure 6 shows the average observed multiplicity (not corrected for acceptance) for charged hadrons and photons of DASP-2. Off resonance we have $\langle n_{ch} + n_\gamma \rangle_{obs} = 5.2 \pm 0.1$, on resonance 5.65 ± 0.1. The on-resonance value has to be corrected for non-resonant background and for the "1γ" decay $\Upsilon \to \gamma^* \to$ hadrons which has the same multiplicity distribution as $e^+e^- \to \gamma^* \to$ hadrons off resonance. With this correction we find $\langle n_{ch} + n_\gamma \rangle_{obs}^{\Upsilon\ direct} =$ $= 5.95 \pm 0.17$, This corresponds to a rise of $(14 \pm 4)\%$ compared to the off-resonance value. PLUTO finds $\langle n_{ch} \rangle_{obs}^{off} = 4.9 \pm 0.1$ and $\langle n_{ch} \rangle_{obs}^{\Upsilon\ direct} = 5.9 \pm 0.1$, corresponding to a rise of $(20 \pm 3)\%$. Their solid angle is larger than ours at DASP, which may explain the small difference in the multiplicity rise. The NaI - Lead Glass group finds $\langle n_{ch} \rangle_{obs}^{off} = 6.4 \pm 0.2$ and $\langle n_{ch} \rangle_{obs}^{\Upsilon\ direct} = 7.3 \pm 0.2$, corresponding to a rise of $(14 \pm 5)\%$. Their solid angle is comparable to PLUTO's, but their values of $\langle n_{ch} \rangle$ contain a large fraction of beam-pipe-converted photons.

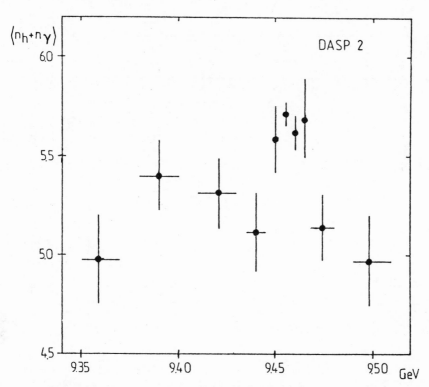

Fig. 6 Observed average multiplicities of charged particles plus converted photons in the DASP detector (not corrected for acceptance).

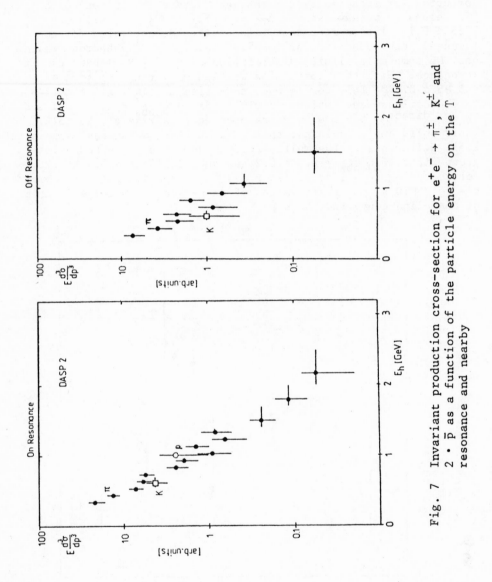

Fig. 7 Invariant production cross-section for $e^+ e^- \to \pi^{\pm}$, $K^{\pm}$ and $2 \cdot \bar{p}$ as a function of the particle energy on the Υ resonance and nearby

The conclusion on multiplicities is as follows. The observed average hadron multiplicity in the direct decay $T \rightarrow$ hadrons is only 14 to 20% higher than in $e^+e^- \rightarrow q\bar{q} \rightarrow$ hadrons at the same energy. The true rise, extrapolated to the full solid angle of 4π, may still be larger. A very preliminary PLUTO result quotes a true average multiplicity rise of (27 ± 9)% with rather large error. Simple three-gluon-jet expectations for $T_{direct} \rightarrow ggg \rightarrow$ hadrons gave a rise of 25 to 50%.

Figure 7 shows inclusive hadron spectra in the magnetic arms of DASP. The invariant production cross-section of $\pi^\pm$, $K^\pm$ and $2 \cdot \bar{p}$ is plotted against $E_h \sim \sqrt{m^2 + p_t^2}$; production angles are around $90°$. Above the momentum cut-off of 200 MeV/c, the pions show a purely exponential spectrum $Ed^3\sigma/dp^3 \propto \exp(-E/E_0)$ with E_0(on T) = (260 ± 25) MeV and E_0(off T) = (240 ± 40)MeV. There is no significant change from off to on resonance. The kaons on and off resonance and the protons on resonance follow roughly the well-known law that their production cross-section is the same as for pions at the same particle energy. The relative abundance of kaons is about the same on and off the T resonance.

Figure 8 shows the August 1978 results of DASP-2 on the topology changes of multi-hadron events off and on the T resonance[12]. These topology studies are motivated by the search for the decay of the T into three gluon jets as expected by QCD. The corresponding recent PLUTO results were presented by H. Meyer at this seminar. Since the DASP inner detector measures only directions of charged hadrons and not their momenta, we have tried to extract information on purely geometrical quantities. Analogous to the usual definitions, see, e.g., Ref. 13:

$$S = \frac{3}{2} \text{ Min } \frac{\sum p_t^2}{\sum p^2} \qquad\qquad \hat{S} = (\frac{4}{\pi} \text{ Min } \frac{\sum |p_t|}{\sum |p|})^2$$

$$T = \text{Max } \frac{\sum |p_e|}{\sum |p|} \qquad\qquad A = (2 \text{ Min } \frac{\sum |p_{out}|}{\sum |p|})^2$$

we can define "pseudo"-topological quantities using only the angles of the particles with a jet-axis or with the normal of a jet-plane:

$$PS = \text{Min } <\sin^2\delta> \qquad\qquad \hat{PS} = (\frac{4}{\pi} \text{ Min } <|\sin\delta|>)^2$$

$$PT = \text{Max } <|\cos\delta|> \qquad\qquad PA = (2 \text{ Min } <|\cos\delta|>)^2$$

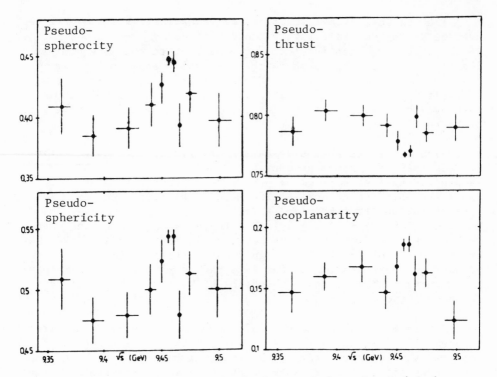

Fig. 8   The DASP-2 pseudotopological quantities of multi-hadron
         decays as a function of energy

As can be seen in Figure 8, all four quantities show a change when
passing through the T resonance. The topology changes in hadronic
final states are so pronounced that they are even visible in purely
geometrical quantities without using the momenta of the particles.
Are the changes due to three gluon jets or due to cascade decays or
just phase space? This is very difficult to answer, but different
Monte Carlo models can be compared to the data. The observed changes
in the DASP detector, including the multiplicity change mentioned
before, are in very good agreement with a description by two Field-
Feynman-jets outside the resonance and by three jets on the T:
$T \rightarrow ggg$, $g \rightarrow q\bar{q}$, $q,\bar{q} \rightarrow$ Field-Feynman-jets. In order to reach the
agreement, one has to assume a very special gluon fragmentation
$g \rightarrow q\bar{q}$. All the gluon momentum goes into one quark (antiquark) only,
whereas the antiquark (quark) is left with zero momentum.

     The conclusions on topologies as studied by DASP-2 are as
follows. The hadronic decay of the T is very different from the
two-quark-jet-like hadronic states near the T. It is compatible with
a three-jet decay of the T, but no direct evidence for three jets
has been seen. Analysis is being continued.

I thank J.J. Aubert and G. Preparata for their invitation to the seminar and for their hospitality.

REFERENCES

1. S.W. Herb et al., Phys. Rev. Lett. 39:252 (1977).
2. W.R. Innes et al., Phys. Rev. Lett. 39:1240 (1977).
3. M. Kobayashi and T. Maskawa, Progr. Theoret. Phys. 49:652 (1973).
4. C.W. Darden et al., Phys. Lett. 76B:246 (1978); Phys. Lett.
        78B:364 (1978); Phys. Lett. 80B:419 (1979).
5. Ch. Berger et al., Phys. Lett. 76B:243 (1978);
        Ch. Berger et al., DESY 79/19 (March 1979).
6. J.K. Bienlein et al., Phys. Lett. 78B:360 (1978).
7. C. Brown, W. Innes, and L. Ledermann, private communication.
8. J.D. Jackson and D.L. Scharre, Nucl. Instr. Meth. 128:13 (1975);
        M. Greco et al., Nucl. Phys. B101:234 (1975).
9. R. Van Royen and V.F. Weisskopf, Nuovo Cimento 50:617 (1967);
        Nuovo Cimento 51:583 (1967).
10. J.L. Rosner, C. Quigg and H.B. Thacker, FERMILAB-Pub-78/19-THY
        (Feb. 1978).
11. The quoted numbers are from J. Ellis et al., Nucl. Phys. B131:285
        (1977).
12. C.W. Darden et al., contributed paper to the Tokyo International
        Conference.
13. A. De Rújula et al., CERN TH. 2455 (Feb. 1978).

FIRST RESULTS OF THE MARK-J DETECTOR AT PETRA

A. Böhm

III. Physik. Institut, Technische Hochschule Aachen
Aachen, Germany

In this talk I shall report on the first results of the MARK-J detector at PETRA. The experiment is being done by a collaboration of Aachen-DESY-LAPP-MIT-NIKHEF and Peking. The names of the physicists participating in the experiment are listed in Ref. 1.

My talk will be divided into three parts: first, I will describe the MARK-J detector; second, the results of a test[1] of QED; and third, the first measurements[2] of the ratio

$$R = \frac{\sigma(e^+e^- \rightarrow \text{hadrons})}{\sigma(e^+e^- \rightarrow \mu^+\mu^-)}$$

1. MARK-J DETECTOR

1.1 Description of the detector

The detector has an approximately 4π solid angle and measures the direction and the momentum of μ, e, γ and the energy of charged and neutral hadrons. The physics objectives are:

a) to measure the interference effects between weak and electromagnetic interactions by studying the charge asymmetry in the reaction $e^+e^- \rightarrow \mu^+\mu^-$;

b) to measure the cross-section $e^+e^- \rightarrow$ hadrons;

c) to search for new heavy leptons and vector mesons by detecting decays into electrons and/or muons;

d) To study the various quantum electrodynamic processes.

The main aim of this experiment is the study of weak interaction effects in the reaction $e^+e^- \to \mu^+\mu^-$. As the cross-section for this reaction at high energies is very low, the observation of these effects will be limited in the first year of operation of PETRA by the luminosity. It should still be possible to measure the forward-backward asymmetry[3] which determines the product of the axial vector couplings of electrons and muons

$$A = \frac{F - B}{F + B} = \frac{3G_F}{2\sqrt{2}} \frac{m_Z^2}{2\pi\alpha} \cdot \frac{s}{s-m_Z^2} \cdot Re \ (g_A^*(e) \cdot g_A(\mu))$$

F means the number of scattered μ^- (μ^+) in the forward hemisphere with respect to the number of incoming electrons (positrons) and B the number of μ^- (μ^+) in the backward hemisphere. In the Salam-Weinberg theory there is μ-e universality and $g_A(e) = g_A(\mu) = -1/2$; therefore

$$A = \frac{3}{8} \frac{G_F m_Z^2}{\sqrt{2}} \cdot \frac{1}{2\pi\alpha} \frac{s}{s-m_Z^2}$$

For the centre-of-mass energy of $\sqrt{s}$ = 30 GeV, we expect an asymmetry of about -7%.

In order to measure such a small asymmetry, the detector has to be free of experimental asymmetries. We therefore built the detector in such a way that it can be rotated in the azimuthal angle ϕ around the e^+e^- beams by ± 90° and in the horizontal plane by a polar angle of $\Delta\theta$ = 180°.

Such rotations of the detector and the inversion of the magnetic field render it possible to eliminate most of the experimental asymmetries of the detector.

The side view of the Mark-J detector is shown in Fig. 1 and the end view in Fig. 2. The detector is designed to measure and distinguish muons, hadrons, electrons and neutral particles. Its acceptance has a solid angle of nearly 4π. In the polar angle θ, it covers angles between 9° and 171° and detects particles over the full range of the azimuthal angle ϕ.

In order to describe the detector it is best to follow the particles as they leave the intersection region. After passing the beam pipe, particles enter a ring of 32 lucite Cerenkov counters.

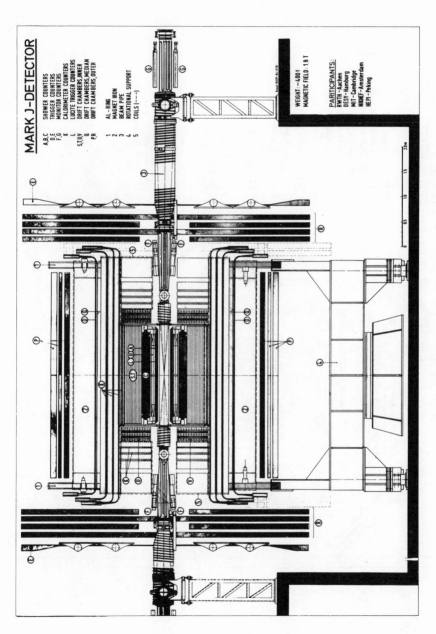

Fig. 1: Side view of the MARK-J detector.

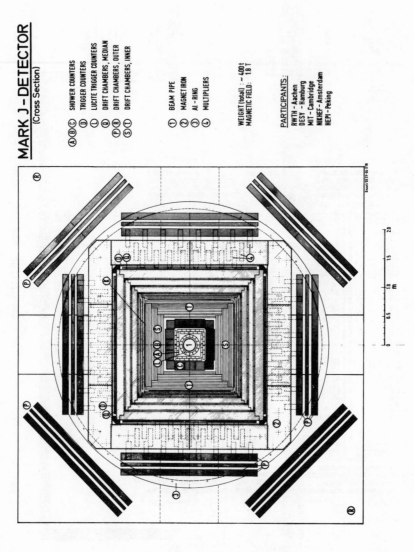

MARK J - DETECTOR
(Cross Section)

Ⓐ Ⓑ Ⓒ   SHOWER COUNTERS
Ⓓ   TRIGGER COUNTERS
Ⓛ   LUCITE TRIGGER COUNTERS
Ⓜ   DRIFT CHAMBERS, MEDIAN
Ⓟ Ⓡ   DRIFT CHAMBERS, OUTER
Ⓢ Ⓣ   DRIFT CHAMBERS, INNER

①   BEAM PIPE
②   MAGNET IRON
③   Al – RING
④   MULTIPLIERS

WEIGHT (total) : ~ 400 t
MAGNETIC FIELD : 1.8 T

PARTICIPANTS:
RWTH - Aachen
DESY - Hamburg
MIT - Cambridge
NIKHEF - Amsterdam
IHEP - Peking

Fig. 2:   End view of the MARK-J detector.

These counters are used to distinguish charged particles from neutral ones and are insensitive to synchrotron radiation. Then the particles enter the shower counters A, B and C, which consist of plates of one radiation length of lead with interjacent layers of five mm thick scintillator. The first two counters, A and B, are made of three layers of lead and scintillator and sample the maximum of the electromagnetic shower. Counter C contains 12 layers of lead and scintillator and has 12 radiation lengths.

All counters have one tube on each end and allow for locating the shower along the beam (z-direction) by two methods. The z-position of a shower can be measured by the time difference between the signals from the photomultiplier at each end. It can also be obtained from the ratio of pulse heights. Both methods give a similar resolution of about three cm for electromagnetic showers. For hadron showers, the spatial resolution is about a factor of three to four worse. The total thickness of the lead scintillator shower counter amounts to 18 radiation lengths, which is enough to absorb fully e and γ showers.

The remaining hadrons and muons traverse drift chamber modules S and T which measure the coordinates along the beam axis and vertically to the beam axis by a total number of 12 planes. The drift chambers have a spatial resolution of 400 μ per plane[4]. They are used to detect hadrons and to measure the direction of muons before they enter the magnetized iron.

The whole inner detector is surrounded by at least 90 cm of magnetized iron. This serves two purposes: it is a magnet for measuring the muon momentum and, it separates muons from hadrons by the absorption in the iron. Coils are placed around the iron in such a way as to obtain a toroidal shape for the field and to reach a field strength of B = 1.75 Tesla.

The momentum of the muon is determined by the drift chambers of the inner detector, two planes of drift chambers in the middle of the iron and eight to ten plances of large drift chambers outside the iron. The angular resolution of the drift chambers is adapted to the multiple scattering in the iron which limits the resolution to about 15% for high energy muons.

The iron of the spectrometer is divided into plates which have an increasing thickness of 3 × 2.5 cm, 2 × 10 cm, 15 cm and 45 cm. The gaps between the plates are filled with scintillator hodoscopes to measure the hadronic shower. For an evaluation of the hadronic energy, we also use the pulse height information from the lead scintillator sandwich in the inner detector. This calorimeter has been calibrated in a hadron and electron test beam at CERN. We obtain a resolution of $\sigma/E = 14\%/\sqrt{E}(GeV)$ for electrons and a resolution of $\sigma/E = 45\%$ to 30% for hadrons of 1 to 10 GeV. For most applications, we only want to measure the total hadronic energy, especially for the measurement of cross-section $e^+e^- \to$ hadrons. We therefore simulate the distribution of hadrons by a Monte Carlo program. When we take into account the measured resolution of the calorimeter, we obtain a resolution of 20% in the total energy of an hadronic event at $\sqrt{s} = 17$ GeV. As will be shown later, such a resolution is good enought to separate beam-beam events from a single beam background.

The scintillation counters D and E are used to trigger on single and multiple muon events and to reject cosmic ray and single beam background by requiring a coincidence with the beam bunch signal within 15 ns.

1.2 Status of the experiment

The major parts of the MARK-J detector, i.e., magnet, drift chambers and the central detector were installed in the August-September 1978 shutdown. After a short test and running-in period, the detector took data in December and January, when PETRA was running at $\sqrt{s} = 13$ GeV and 17 GeV at luminosities between 10^{29} cm^{2}s^{-1} and 10^{30} cm^2 s^{-1}. In the February shutdown we added the forward drift chamber (u, v) of the inner detector and the endcap magnets and calorimeters. In the next long shutdown, the large outer drift chamber in the forward direction will be installed. This will complete the MARK-J detector.

2.   FIRST RESULTS

Data have been taken at centre-of-mass energies $\sqrt{s} = 13$ GeV and 17 GeV triggering on the following reactions:

Bhabha scattering	$e^+e^- \to e^+e^-$
Muon pairs	$e^+e^- \to \mu^+\mu^-$
μ-hadron events	$e^+e^- \to \mu + $ hadrons
hadronic events	$e^+e^- \to$ hadrons

Here we present the first results on a test of QED in Bhabha scattering[1] and on the ratio R of cross-sections of $e^+e^- \to$ hadrons to $e^+e^- \to \mu^+\mu^-$ [2].

2.1 Test of QED

The study of Bhabha scattering serves two purposes. We can perform a test on QED and we can also use these events as a luminosity monitor.

Elastic electron-positron scattering can be selected by demanding two collinear particles which make electromagnetic showers in the lead scintillator sandwich with an energy equal to the beam energy. The energy of the electron and positron is obtained from the pulse height in the lead scintillator sandwichs A, B and C. We use an algorithm which was developed from the calibration of the counters in a test beam of energies between 0.5 GeV and 10 GeV.

The direction of the electron (positron) was determined from the position of the interaction region and the impact point at the counters. The azimuthal position of the impact point is given by the location of the finely segmented and staggered shower counters. The difference in time of the signals from the tubes at each end of a shower counter is a measure of the position along the counter. In the same way, the ratio of the pulse heights of each phototube depends on the position of the impact point along each counter. Both methods were calibrated in a test beam and gave a resolution of about three cm.

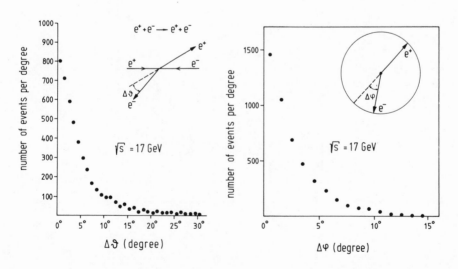

Fig. 3: The collinearity of the reaction $e^+e^- \rightarrow e^+e^-$. The angular difference of the outgoing particles is plotted for the polar angle θ and the azimuthal angle ϕ

In this way we can determine, by the shower counters, the polar angle θ and the azimuthal angle φ to better than 5°. To check the collinearity of the scattered electrons and positrons, we plot their angular defference in the polar angle and in the azimuthal angle (Fig. 3). Sharp peaks at Δθ = 0 and Δφ = 0 prove that there is very little background from inelastic events. Bhabha scattering events are identified by requiring two back-to-back showers, which are collinear to within 20° in θ and φ and have an energy greater than 8 GeV. The total number of e^+e^- events is 4660 at centre-of-mass energy, 17 GeV and 7193 at 13 GeV. The acceptance for Bhabha events is limited by the size of the first shower counter A, which subtended scattering angles from 12° to 168.

The results for $\sqrt{s}$ = 13 GeV and $\sqrt{s}$ = 17 GeV are shown in Fig. 4. Since the first order QED photon propagator produces a 1/s dependence in the $e^+e^- \to e^-e^+$ cross-section, the quantity s(dσ/dcosθ) vs θ is independent of s. This distribution is plotted for our data in Fig. 4 and shows excellent agreement with the QED predictions. To express analytically this agreement, we compare our data with the QED cross-section in the following form[5] (since charge is not distinguished here):

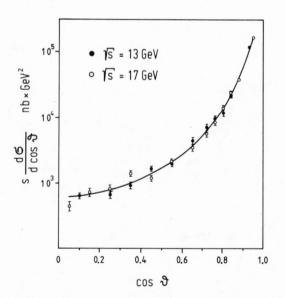

Fig. 4:   The data for $\sqrt{s}$ = 17 GeV and $\sqrt{s}$ = 13 GeV compared with the predictions from QED.

$$\frac{d\sigma}{d\cos\theta} = \frac{\pi\alpha^2}{s} \left[\frac{q'^4 + s^2}{q^4} \left| F_s \right|^2 + \frac{2q'^4}{q^2 s} \operatorname{Re}(F_s \cdot F_T^\ast) + \frac{q'^4 + q^4}{s^2} \left| F_T \right|^2 \right.$$

$$+ \frac{q^4 + s^2}{q'^4} \left| F_s' \right|^2 + \frac{2q^4}{q'^2 s} \operatorname{Re}(F_s' \cdot F_T^\ast) \qquad (1)$$

$$\left. + \frac{q'^4 + q^4}{s^2} \left| F_T \right|^2 \right] \quad \{1 + C(\theta)\}$$

where

$$F_s = 1 \bar{+} \frac{q^2}{q^2 - \Lambda_s^2}$$ is the form factor of the space-like photon

$$F_s' = 1 \bar{+} \frac{q'^2}{q'^2 - \Lambda_s^2}$$

$$F_T = 1 \bar{+} \frac{s}{s - \Lambda_T^2}$$ is the form factor of the time-like photon

$q'^2 = -s \cos^2(\theta/2)$, $q^2 = -s \sin^2(\theta/2)$, Λ is the cut-off parameter in the modified photon-propagator model and $C(\theta)$ is the radiative correction term as a function of θ.

The radiative correction to the $e^+ e^-$ elastic scattering process was calculated using a modified program from Berends[6] which includes the contribution of the heavy lepton τ loop and the hadronic vacuum polarization. The radiative corrections are +4.6% at $\theta = 14^\circ$ and +1.3% at $\theta = 90^\circ$ for $\sqrt{s} = 17$ GeV. They are slightly smaller at $\sqrt{s} = 13$ GeV.

Electron-positron pairs are generated according to Eq. (1) in a Monte Carlo program. Each electron is then traced through the detector. The effect of measured θ, ϕ resolutions are included. A χ^2 fit was made using this Monte Carlo generated angular distribution to both 13 GeV and 17 GeV data. The normalization was treated in two ways: a) the total number of Monte Carlo events in the region $0.9 < \cos\theta < 0.98$ was set equal to the total number of measured events in the same region; b) the minimum χ^2 fit to the entire data sample determined the normalization.

The two methods agree with each other to within 3% and give essentially the same result in the cut-off parameter Λ. The curve in Fig. 4 is the result of our fit to the data at both energies. The lower limits of Λ at 95% confidence level under various assumptions is shown in Table I.

In Table II our results are compared with measurements at lower energies. Note that there are different parametrizations for the form factors used. For simplicity, upper limits are only displayed for the case where a single cut-off parameter Λ was used for both the space-like and time-like photons.

Although earlier experiments give rather tight limits on the cut-off parameter, one should realize that we are testing QED at much larger energies and therefore at larger momentum transfer. The values of the cut-off parameter, especially that for the time-like photon, will improve substantially when we have enough events of the reaction $e^+e^- \rightarrow \mu^+\mu^-$ in order to include them in the test of QED. As apparent from the results of Augustin et al.[7], the limits increase drastically if both reactions are fitted simultaneously.

In summary, we can say that this test continues the previous work[7-10] on the test of QED up to the highest available energies.

Table I   Limits on Λ with 95% confidence level

Form factor	$1 - \dfrac{q^2}{q^2 - \Lambda_-^2}$	$1 + \dfrac{q^2}{q^2 - \Lambda_+^2}$
	Λ_- (GeV)	Λ_+ (GeV)
$1 \pm \dfrac{q^2}{q^2 - \Lambda_s^2}$ Λ_s	37	21
$1 \pm \dfrac{s}{s - \Lambda_T^2}$ Λ_T	22	21
$\Lambda_s = \Lambda_T$ Λ	38	26

Table II Comparison of our results on the cut-off parameter
with measurements at lower energies

	parametrization	$s\,(\mathrm{GeV}^2)$	$\Lambda_+\,(\mathrm{GeV})$	$\Lambda_-\,(\mathrm{GeV})$
Ref. 9 e^+e^-	$1 \pm \dfrac{q^2}{q^2 - \Lambda_\pm^2}$	27	22.8	14.4
Ref. 7 e^+e^- - - - - - e^+e^- and $\mu^+\mu^-$	$\left[1 \pm \dfrac{q^2}{\Lambda_\pm^2}\right]^{-1}$	9 and 14.4 and 23	15 - - - - 35	19 - - - - 47
Ref. 10 e^+e^-	$1 \pm \dfrac{q^2}{q^2 - \Lambda_\pm^2}$	49 54.8	38	33.8
This experiment e^+e^-	$1 \pm \dfrac{q^2}{q^2 - \Lambda_\pm^2}$	169 and 289	26	38

We find that Bhabha scattering indeed has a cross-section which
falls like $1/s$ up to $s = 289$ GeV^2 and shows the angular distribution as predicted by QED.

2.2 Measurement of the cross-section $e^+e^- \to$ hadrons

For this measurement we use mainly the inner detector and the
hadron calorimeter. As most of the events have two hadron jets
in opposite directions, we require for a trigger two A and two B
counters in coincidence with two A counters and one B counter in
the opposite quadrant of the inner detector. The signals of these
counters have to be in coincidence with the beam bunch signal within
15 ns. In addition, the drift chambers S or T have to have at least
three counts, allowing for the reconstruction of a line pointing

to the beam intersect. In this way, we suppress a large number of
triggers due to cosmic rays and single beam background and reduce
the trigger rate to, typically, $2s^{-1}$.

For each particle or group of particles emitted within a cone
of 20° which were not separable in the counters, the vector momentum
was computed from pulse height and counter position information
using the position of the interaction region.

To eliminate the bulk of the beam gas background, which is
mainly low energy and one-sided, we first required that the total
energy deposited in the calorimeter be greater than 5 GeV and that
the computed total p and total p_T each be less than 50% of the
observed energy. We further demanded at least one track in the
drift chambers pointing back to the interaction region to distinguish hadronic events from beam gas events. The interaction region
is defined by ee → $\mu^+\mu^-$ events to σ = 2 cm. To discriminate against
events of electromagnetic origin, such as ee → ee, ee → $\gamma\gamma\gamma$ and so
forth, we have accepted three types of events with different shower
properties:

   a) two narrow showers penetrating into the third and fourth layer
      of the hadron calorimeter counters (K). There is a total of
      33 radiation lengths from the intersection region. To discri-
      minate against e^+e^- and $\gamma\gamma$ final states, the total energy in
      the K calorimeter counters is required to be greater than 7%
      of the total shower energy in A + B + C;

   b) two broad showers penetrating into the first or second layer
      of the K calorimeter counters with energy greater than 1% of
      the total A, B and C shower energy;

   c) three or more broad showers with tracks in the drift chambers.

The energy spectrum of the events passing the cuts is shown in
Fig. 5. The energy spectrum of beam gas events, which are defined
by a chamber track pointing at least 15 cm from the intersection
point, is shown as the plain area in Fig. 5.

A comparison of those two spectra (which were taken simultaneously) shows that the real hadron events can be separated from
beam gas by requiring the total energy of the event to be larger
than 10 GeV for $\sqrt{s}$ = 17 GeV (the cut is 8 GeV for $\sqrt{s}$ = 13 GeV data).
This cut will also exclude most of the τ lepton decay from our sample.
The τ contribution is estimated to contribute 0.1 unit to R.

The acceptance for e^+e^- → hadrons via one photon annihilation
was computed using the Monte Carlo program. The program generates
two jet events according to the Feynman-Field ansatz[11] which

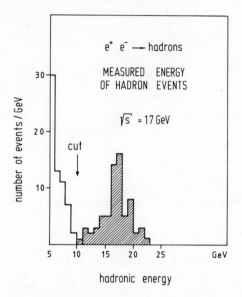

Fig. 5: Energy spectrum of the events passing the selection
criteria. The plain area is the energy spectrum of
the beam gas background.

includes only u, d and s quarks. Each final state track is then
traced through a simulated representation of the detector which
includes all the geometric details of the magnet (as shown in
Figs 1 and 2).

The resulting counter pulse heights and digitized chamber hits
are then subjected to a set of cuts approximating those of the
analysis program. Using these methods for Feynman-Field jet events
with average $p_\perp$ = 350 MeV, the acceptance as a fraction of 4π is
0.79 both at $\sqrt{s}$ = 17 GeV and at $\sqrt{s}$ = 13 GeV. Changing the average
P_T to 600 MeV increases these numbers by less than 5%, which is
included in the systematic error of our results. The two-photon
contribution has been calculated and found to be less than 5% in
this experiment. We have obtained the following results:

	$\sqrt{s}$ = 17 GeV	$\sqrt{s}$ = 13 GeV
$\int Ldt$ (nb^{-1})	60	53
events	68	98
R	4.9 ± 0.6	4.6 ± 0.5

where

$$R = \frac{\sigma \text{ hadrons}}{\sigma_{\mu\mu}} = \frac{\text{events}}{\int L dt} \cdot \frac{1}{\text{Acceptance}} \cdot \frac{1}{\sigma_{\mu\mu}}$$

We have assigned a rather conservative systematic error of ±0.7 on R at each energy for the following reasons:

 a) uncertainties due to models used in the Monte Carlo calcu-
    lation of acceptance (±

 b) the error in luminosity measurement (±6%);

 c) the possibility of confusion in event selection (±5%).

From the values of R we can compute the number of events $e^+e^- \rightarrow \mu^+\mu^-$. We expect 14 and 21 events at 17 GeV and 13 GeV, respectively. The numbers of μμ events found in a preliminary analysis are in agreement with the values calculated above. Unfortunatels, the statistical errors are still too large to permit a direct measurement of R from the simultaneous measurements of both reactions. When PETRA reaches higher luminosity, this method will lead to a measurement of R which is free of uncertainties due to the luminosity measurement.

The ratio of the two values of R is rather insensitive to systematic errors. We obtain

$$\frac{R (\sqrt{s} = 17 \text{ GeV})}{R (\sqrt{s} = 13 \text{ GeV})} = 1.07 \pm 0.17$$

This value implies that there is no large change in the hadronic cross-section of the e^+e^- interaction between the two c.m. energies.

We can compare our results to the values at lower energies[12-14] and to measurements done by the PLUTO group[15] and the TASSO group[16] at PETRA at the same energies. Figure 6 shows the data points above 5 GeV and the results of the SLAC group for energies up to 8 GeV as a shaded area. Also indicated is the trend of the data at lower energies and the position of the J/ψ and T resonances.

For energies slightly above the T resonance we will probably find a similar resonance region with large fluctuations of R as observed near the J/ψ resonance. At larger energies we can compare our measurement to the theoretical value[17,18] predicted by QCD. If there are N quark flavours with electric charges Q_i, the ratio R should be

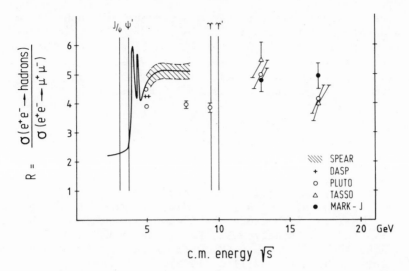

Fig. 6: The ratio R vs the c.m. energy $\sqrt{s}$. The trend of the lower
energy data[12],[13] is sketched by a solid line. The dashed
area indicates the results of the SLAC-LBL group[13] (heavy
lepton contribution not subtracted). Above 5 GeV are the
results of the DASP[14] and PLUTO[15] groups. Recent measurements with PETRA at energies $\sqrt{s}$ = 13 GeV and 17 GeV are
from the PLUTO[15], TASSO[16] and MARK-J[2] groups.

$$R = 3 \sum_{i=1}^{N} Q_i^2 \left(1 + \frac{\alpha_s}{\pi} + \ldots\right) \tag{2}$$

where α_s is the energy dependent strong coupling constant

$$\alpha_s = \frac{12\pi}{(33 - 2N) \ln \frac{s}{\Lambda^2}}$$

The parameter Λ determines the strength of the coupling constant
and is determined to $\Lambda \simeq 0.5$ GeV. If we use five quark flavours
u, d, s, c, b and assign an absolute value of charge $|Q_b| = 1/3$
to the b quark, we expect the value

$$R = \frac{11}{3} \left(1 + \frac{\alpha_s}{\pi}\right) \simeq 3.9$$

This value is compatible with our measurement. We cannot, however, exclude a charge of the b quark of $|Q_b| = 2/3$, for which expression (2) predicts a ration of $R \simeq 5.0$. But it can be expected that the statistical and systematical errors of this measurement will be substantially reduced in the near future, when the detector is completed and when PETRA runs at its full luminosity.

REFERENCES

1. D. Barber, U. Becker, H. Benda, A. Böhm, J.G. Branson, J. Bron,
        D. Buikman, J. Burger, C.C. Chang, M. Chen, C.P. Cheng,
        Y.S. Chu, R. Clare, P. Duinker, H. Fesefeldt, D. Fong,
        M. Fukushima, M.C. Ho, T.T. Hsu, R. Kadel, D. Luckey, C.M. Ma,
        G. Massaro, T. Matsuda, H. Newman, J. Paradiso, J.P. Revol,
        M. Rohde, H. Rykaczewski, K. Sinram, H.W. Tang, S.C.C. Ting,
        K.L. Tung, F. Vannuci, M. White, T.W. Wu, P.C. Yang, C.C. Yu,
        Phys. Rev. Lett. 42:1110 (1979).
2. D. Barber et al., Phys. Rev. Lett. 42:1113 (1979).
3. See for example, S. Berman, PEP Summer-Study 1974 and
        A. Benvenuti et al., PEP Summer-Studies 1975.
4. U. Becker et al., Nuclear Instrum. Methods, 128:593 (1975).
5. S.D. Drell, Ann. Phys. (New York) 4:75 (1958);
        T.D. Lee and G. Wick, Phys. Rev. D2:1033 (1970).
6. S.A. Berends et al., Phys. Lett. 63B:432 (1976) and private
        communication.
7. I.-E. Augustin et al., Phys. Rev. Lett. 34:223 (1975).
8. B. Borgia et al., Nuovo Cimento 3:115 (1972);
        V. Alles-Borelli et al., Nuovo Cimento 7A:330 (1972);
        M. Bernardini, Phys. Lett. 45B:510 (1973);
        H. Newman et al., Phys. Rev. Lett. 32:483 (1974).
9. B.L. Beron et al., Phys. Rev. Lett. 33:663 (1974).
10. L.H. O'Neill et al., Phys. Rev. Lett. 37:395 (1976).
11. R.D. Field and R.P. Feynman, Nucl. Phys. B136:1 (1978).
12. I. Burmeister et al., Phys. Lett. 66B:395 (1977);
        W. Braunschweig et al., Phys. Lett. 67B:243 (1977).
13. R.F. Schwitters, Proc. 1975 International Symposium on Lepton
        and Photon Interactions at High Energies, Stanford, p. 5.
14. R. Brandelik et al., Phys. Lett. 76B:361 (1978).
15. Ch. Berger et al., DESY preprint DESY 79/11 (1979) submitted
        to Phys. Lett.
16. H. Hartmann, talk at the spring meeting of the Deutsche
        Physikalische Gesellschaft, Bonn, March 1979 (unpublished).
17. T.W. Appelquist and H. Georgi, Phys. Rev. D8:4000 (1973).
18. A. Zee, Phys. Rev. D8:4038 (1973).

FINAL STATES IN e^+e^- ANNIHILATION

Mary K. Gaillard

LAPP, Annecy-le-Vieux, France
CERN, Geneva, Switzerland

INTRODUCTION

Most of this talk will be concerned with tests and applications of what is known as "perturbative QCD": the predictions of quantum chromodynamics in kinematic regimes where perturbation theory is believed to be applicable. I shall discuss tests of QCD in hadronic final states both in the continuum and in the decays of heavy quark bound states, and -- in the event that such tests prove affirmative -- the use of QCD predictions as a probe for locating new flavour thresholds. In addition, I shall comment on the possible influence of non-perturbative effects expected in QCD. I shall also point out some interesting weak interaction phenomena which may occur in the energy ranges accessible at CESR, PEP and PETRA.

1. PERTURBATIVE QCD: PHILOSOPHY

A problem which arises in perturbative QCD has long been known in QED: the emission of a soft or collinear gluon in some hadronic processes leads, in perturbative calculations, to a logarithmic infrared divergence. Recent theoretical activity in the field has been primarily concerned with finding predictions of the theory for which these singularities are absent. There are two general approaches to the problem:

a) the infrared singularities are absorbed into structure, fragmentation, or bound state wave functions which are hopefully universal, i.e., independent of the particular process studied. Then, while these universal functions cannot be calculated (although their momentum dependence can be) using perturbative methods, their universality allows one to relate measurements in different

processes via perturbation theory. For example, from measurements
of deep inelastic lepton-nucleon scattering,

$$\ell + N \rightarrow \ell' + X \qquad\qquad (1.1)$$

one can predict the behaviour of Drell-Yan lepton pair production,

$$N + N \rightarrow \ell^+ + \ell^- + X \qquad\qquad (1.2)$$

This property is known as "factorization", in the sense that the in-
calculable, infrared divergent function is a common removable factor, and has been demonstrated[1,2] to be correct in the (ultra-violet
or high energy) leading log approximation. We shall comment later
on the effects of non-perturbative phenomena on these results.

   b) the second approach, which we shall discuss in some detail,
involves a sum over final states so as to cancel out different infrared singular contributions. The idea is to look for variables
which are the same for physically indistinct processes, such as the
emission of a lone quark or of a quark accompanied by a zero energy
or collinear gluon. Variables which are <u>linear in momenta</u> are insensitive to soft or collinear gluon emission.

   As an example, consider e^+e^- annihilation into hadrons via
virtual photon exchange. Contributions to order α_s are shown in
Fig. 1. The radiative corrections to the $\gamma q\bar{q}$ vertex, e.g., Fig. 1b,

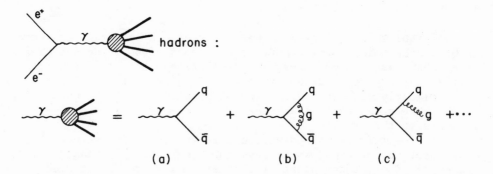

Fig. 1 $e^+e^- \rightarrow$ hadrons as described by perturbative QCD.

are logarithmically divergent for soft gluon exchange, as is the amplitude, e.g., Fig. 1c, for a gluon which is soft or collinear with one of the quarks. As in QED, these divergences cancel in the expression for the total cross-section to order α_s. In order to test the finer details of QCD, one must look for variables, more specific than the total cross-section, but with the same property of insensitivity to infrared singularities. The rule is that one may ask questions about the distribution of energy flow, but not about the nature or number of particles carrying the energy.

To lowest order in QCD perturbation theory, the annihilation process is given by the diagram of Fig. 1a, and the energy flow is confined to two back-to-back jets with no transverse component. In order α_s, the diagram of Fig. 1c, for example, contributes to events with non-vanishing transverse energy relative to the principle jet axis. A quantity which is calculable[3] at each order in perturbation theory is the probability $p(\varepsilon,\delta) = 1 - f(\varepsilon,\delta)$ that a fraction ε of the total energy will lie outside two back-to-back cones of angle 2δ centred on the jet axis. The function $p(\varepsilon,\delta)$ vanishes in lowest order and is free of infrared singularities in higher orders, since it receives no contributions from vertex corrections (no gluon emission) or from soft or collinear gluons as long as ε and δ are non-zero. Figure 2 shows[4] $p(\varepsilon,\delta)$ as a function of δ for different values of ε and for a centre-of-mass energy $Q^2 = 40$ GeV^2 (the energy dependence is very slight), calculated to lowest order in the (running) coupling constant $\alpha_s(Q^2)$.

The prediction of Fig. 2 can in principle be directly compared with the data. However, there are two caveats:

a) For very small values of (ε,δ), $p(\varepsilon,\delta)$ blows up:

$$ p(\varepsilon,\delta) \underset{\varepsilon,\delta\to 0}{\sim} \sum_n \left[\ln\varepsilon \, \ln\delta \, \frac{\alpha_s(Q^2)}{\pi} \right]^n \tag{1.3} $$

The perturbation expansion in α_s is obviously useful only when it converges sufficiently rapidly, i.e., when

$$ |\ln\varepsilon \, \ln\delta| \ll \left[\frac{\alpha_s(Q^2)}{\pi} \right]^{-1} \sim \frac{33 - 2N_g}{12} \ln Q^2/\Lambda^2 \tag{1.4} $$

where $\Lambda \simeq (300-500)$ MeV is the usual parameter for defining the coupling strength, and N_f is the number of quark flavours with production threshold $\lesssim Q^2$. The divergence for small ε, δ reflects the increasing importance of multigluon emission with small transverse energy as one demands a more two-jet-like configuration.

     b) Quark confinement does not follow from perturbation theory. This means that the "hadronization" process by which quarks and gluons combine to form hadrons, and which will also contribute to transverse energy flow, cannot be described by perturbative QCD.

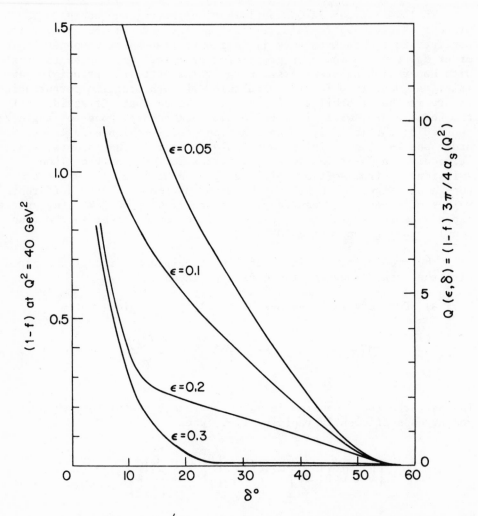

Fig. 2   The probability[4] $(1 - f)$ that a fraction ε of the energy in e^+e^- hadrons will lie outside a cone of half angle δ. The scale on the right has the coupling constant (and therefore Q^2) dependence removed.

The hope is that the transverse energy associated with hadronization is characterized by some constant, for example, a confinement radius,

$$E_T \propto R^{-1}_{confinement} \qquad (1.5)$$

so that its contribution will become unimportant at high energies.

For example, dimensionless quantities like the mean values of ε or δ (with the other one fixed) are governed by the ratio $<E_T>/Q$. For hadronization effects we expect (or at least hope) that

$$\left[<E_T>/Q \right]_{had} \sim \frac{1}{Q R_{conf.}} \qquad (1.6)$$

while, since there is no explicit dimensional parameter in QCD perturbation theory, we necessarily have:

$$<E_T>_{pert.} \propto \frac{\alpha_s(Q^2)}{\pi} Q \qquad (1.7)$$

so that

$$\left[<E_T>/Q \right]_{pert.} \sim \frac{\alpha_s(Q^2)}{\pi} \sim \frac{1}{\ln(Q^2/\Lambda^2)}. \qquad (1.8)$$

At sufficiently high momentum transfer, the transverse energy contributions which can be calculated perturbatively will then dominate over hadronization effects. At the present stage of understanding, the latter can only be included phenomenologically by parametrizing the data at lower energies where they are believed to account for most of the transverse energy.

The moral is that in applying QCD perturbation theory, or in attempting to confront it with the data, one must:

a)   choose observables which depend only on energy flow;

b)   study these observables for values sufficiently far from a zero width two-jet configuration that the perturbation expansion converges rapidly; and

c)   choose a kinematic region where hadronization effects are
damped relative to calculable effects.

Figure 3 shows data[5] for a quantity closely related to that of
Fig. 2, namely the average fraction of energy flow outside a cone
of fixed half angle δ. The dashed line is the QCD prediction[6], and
comparison with the data indicates that the last criterion is not
satisfied at present energies, at least for this variable. We mention later a possibility for improving the perturbation expansion
so as to include more of the total contribution into the "calculable"
part.

2.   OTHER VARIABLES: LOOKING FOR GLUON JETS[7]

To the extent that non-calculable hadronization effects can
be removed, comparison of data with predictions like those of Fig. 2
provide tests of QCD. Specifically, they provide a measure of the
energy spread due to gluon bremsstrahlung by quarks. However, just
as the observation of two-jet events, with limited transverse momentum and a $1 + \cos^2\theta$ distribution characteristic of spin 1/2
particles, has given impressive support to the physical reality of
quarks, one would like more specific evidence for gluon jets with
properties characteristic of spin 1 objects. I shall describe some
of the variables proposed for this purpose.

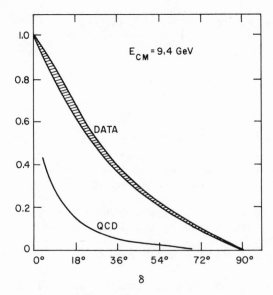

Fig. 3   Comparison of data[5] with QCD predictions[6] for the average
energy fraction outside a cone of half angle δ.

As discussed above, these variables must be linear in the momenta of the final state hadrons; this criterion assures insensitivity both to the infrared divergences of perturbation theory and to the details of hadronization to the extent that the transverse momenta of quark and gluon fragments can be neglected. Two commonly used variables are thrust and spherocity. Thrust is defined by[8]:

$$ T = 2 \, max \left\{ \frac{\sum_n p_{\shortparallel}}{\sum p} \right\} , \frac{1}{2} \leqslant T \leqslant 1 , \qquad (2.1) $$

where p is the momentum of a final state hadron and $p_{\shortparallel}$ its momentum component parallel to the axis which maximizes the quantity in brackets. The notation $\sum_n$ means that the sum is over particles in the forward hemisphere only, i.e., those with $p_{\shortparallel} > 0$. Spherocity is defined by[9]

$$ S = \left(\frac{4}{\pi}\right)^2 min \left\{ \frac{\sum p_T}{\sum p} \right\} , 0 \leqslant S \leqslant 1 \qquad (2.2) $$

where p_T is the transverse momentum relative to the axis which minimizes the quantity in brackets. In order $(\alpha_s)^0$, the only contribution to e+e− annihilation into hadrons is that of Fig. 1a, giving a collinear final state: T = 1, S = 0. In order α_s there can be contributions like that of Fig. 1c. Defining the fractional energies of the final state quanta by

$$ x_i = \frac{2E_i}{Q} \qquad i = 1,2,3 \quad , \quad \sum x_i = 2 \qquad (2.3) $$

we get for a three body final state:

$$ T = max(x_i) \quad , \quad S = \frac{64}{\pi^2} \prod_{i=1}^{3} (1 - x_i) \qquad (2.4) $$

with the kinematical constraints

$$ \frac{2}{3} \leqslant T \leqslant 1 \quad , \quad \frac{64}{\pi^2} \frac{(1-T)^2(2T-1)}{T^2} \leqslant S \leqslant \frac{16}{\pi^2} (1-T) \qquad (2.5) $$

The variables S and T are correlated but independent, and their distributions provide tests of perturbative QCD. They are derivable from the matrix element for gluon bremsstrahlung[7,10]

$$\frac{1}{\sigma} \frac{d^2\sigma}{dx_1 dx_2} = \frac{2\alpha_s(Q^2)}{3\pi} \frac{(x_1^2 + x_2^2)}{(1-x_1)(1-x_2)} + 0\left(\alpha_s^2\right) \tag{2.6}$$

where x_1 and x_2 are defined to be the quark and antiquark fractional momenta. The expression (2.6) diverges when a two-jet like configuration is approached,

$$1-T, \; S \to 0$$

where the perturbation expansion fails. It can therefore be applied only for

$$1-T, S > \Delta_{QCD} \gg \alpha_s(Q^2) \sim \frac{1}{\ln Q^2} \tag{2.7}$$

In addition, this formalism is useful only in a region where hadronization effects are not dominant; these are governed by the intrinsic p_T spread of a jet and its multiplicity n. Thus we require:

$$1-T, S > \Delta_{NP} \sim n(Q^2) \frac{p_T}{Q^2} \sim \frac{\ln Q^2}{Q^2} \tag{2.8}$$

and we expect that at asymptotic energies the dominant limitation will be the one intrinsic to perturbation theory. In Fig. 4 we show[11] calculated thrust distributions for $e^+e^- \to$ hadrons at three energies. The curves labelled $(\bar{q})_{NP}$ represent an estimate of the thrust distribution arising from hadronization effects for a two jet event. For this a simple fragmentation model was used which roughly reproduces the single-particle p_T and z distributions and the (Q^2 dependent) multiplicity measured at SLAC and DORIS. The dashed curve is the convolution of the gluon bremsstrahlung contribution (q$\bar{q}$g) calculated in perturbation theory, and the (q$\bar{q}$)$_{NP}$ curve; this takes into account hadronization of the three quanta final state. Figure 5 shows[11] the thrust distribution for the decal

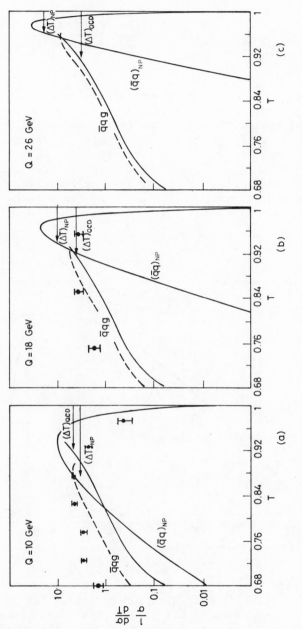

Fig. 4   Thrust distributions[11] for $e^+e^- \to$ hadrons at three centre-of-mass energies in order α_s in perturbative QCD. The solid (dashed) lines show hard gluon bremsstrahlung ($q\bar{q}g$) without (with) hadronization effects, and are to be compared with a two-jet model $(q\bar{q})_{NP}$ and data[13] at 9.4 GeV, (a), and 17 GeV, (b) (resolution smearing not unfolded[14]).

$$\Upsilon\,(9.4) \to 3g$$

derived from the decay matrix element[10]:

$$\frac{1}{\Gamma}\frac{d^2\Gamma}{dx_1 dx_2} = \frac{1}{\pi^2 - 9}\left\{\left(\frac{1-x_1}{x_2 x_3}\right)^2 + \left(\frac{1-x_2}{x_1 x_3}\right)^2 + \left(\frac{1-x_3}{x_1 x_2}\right)^2\right\} \qquad (2.9)$$

where again the dashed line includes hadronization effects. For this a gluon fragmentation model was used similar to the one for quarks except that the multiplicity was taken somewhat higher: the ratio of multiplicities in gluon and quark jets was assumed to approach 9/4 at asymptotic energies. This is based on the conjecture that multiplicity of a fragmenting quantum is proportional to its colour charge, and implies that the intrinsic spread, cf. Eq. (2.8), of a gluon jet will be greater than for a quark jet.

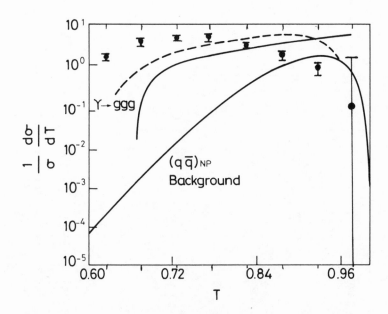

Fig. 5 Thrust distributions[11] on resonance for $T \to 3g$ with (dashed) and without (solid) hadronization effects. The two-jet $(q\bar{q})_{NP}$ background was normalized to 1/6 of the three gluon contribution. The data[13] points are with the two-jet contributions subtracted out (resolution smearing not unfolded[14]

This expectation is also supported by perturbative calculations which show[12] that for fixed ε, a fraction $1 - \varepsilon$ of the total energy lies within a cone of half angle δ with δ_g for a gluon jet and δ_q for a quark jet related by (for ε, $\delta \ll 1$)

$$\ln \delta_q / \ln \delta_g = \frac{9}{4} \qquad (2.10)$$

so that for a kinematic regime where predominantly perturbative effects dominate jet spread, we expect $\delta_g \gg \delta_q$. This result led to the conclusion that gluon jets might not be separable at present energies; however, Fig. 3 indicates that the relation (2.10) is not applicable at present energies, so no quantitative conclusion can be drawn. The curve labelled $(\bar{q}q)_{NP}$ background is the expected thrust distribution from the decay $\Upsilon \rightarrow \gamma \rightarrow \bar{q}q$ and from the direct process $e^+e^- \rightarrow q\bar{q}$ at $\sqrt{s} = m_\Upsilon$. The relative importance of $q\bar{q}$ to $3g$ final states was taken to be $1/6$, corresponding to the assumptions $\Gamma(\Upsilon \rightarrow ee) = 1.3$ keV and $\alpha_s(m_\Upsilon) = 0.26$. Data points[13] in Figs. 4 and 5 are from the PLUTO group at DESY. The continuum points are at 9.4 GeV and 17 GeV, respectively. Smearing due to measurement errors has not been unfolded in the data points, but this is expected to give a small effect since data from different experiments are in good agreement[14]. The Υ data points are with the $e^+e^- \rightarrow q\bar{q}$ and $\Upsilon \rightarrow \gamma \rightarrow q\bar{q}$ two jet background contributions subtracted.

Spherocity distributions show similar characteristics, but data analysis turns out to be less manageable[14,15] in terms of spherocity. Experimenters prefer instead to use sphericity, defined by

$$\hat{S} = \frac{3}{2} \min \frac{\Sigma |p_T|^2}{\Sigma |p|^2} \qquad (2.11)$$

but this is not a good variable for testing QCD perturbation theory since it is not identical for different collinear configurations of the same total momentum; it is infrared divergent in perturbation theory and depends on the details of fragmentation even in the zero transverse energy limit. Since thrust seems to be acceptable to both theorists and experimenters, we have shown these distributions only.

Once the thrust axis has been determined, its angular distribution can be measured as a further test of the theory. A broad thrust distribution is really only indicative of a multi-quanta final state; the angular distribution of its axis provides a precise test of the spins of quarks and gluons. Figure 6 shows data points from PLUTO[16] for a) $e^+e^- \rightarrow \gamma \rightarrow$ hadrons and for b) $\Upsilon \rightarrow$ hadrons with

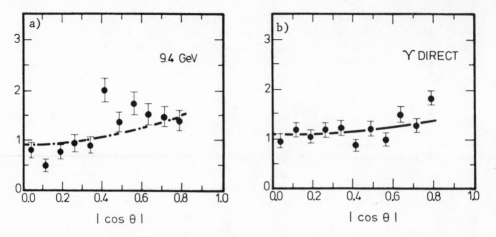

Fig. 6   Experimental[16] sphericity axis distributions and theoretical[17] thrust axis distributions in the continuum (a) and on resonance (b) with $e^+e^- \to q\bar{q}$ contributions subtracted.

the contribution from $T \to \gamma \to$ hadrons subtracted. The solid lines are QCD predictions[17] for a) $e^+e^- \to \gamma \to q\bar{q}$ and b) $T \to 3$ gluons. The prediction for $T \to 3$ scalar gluons would give the opposite slope and appears to be ruled out[14].

We can now go further and try to display more explicitly the three jet character of three quanta final states. If we select an event sample in $e^+e^- \to$ hadrons with

$$1-T \gg (\Delta T)_{NP}$$

we expect this sample to be dominated by $q\bar{q}$-gluon final states. Similarly, T decay final states will be dominated by three gluons if the $T \to \gamma \to q\bar{q}$ contribution is subtracted. What we would like to do is make the three quanta structure apparent by measuring the energy flow in the plane of the quanta. This amounts to a statistical reconstruction of jets. We define the "pointing vector"[11]:

$$P(\theta) = p(\theta)\ \frac{d\sigma(\theta)}{d\theta} \qquad\qquad (2.12)$$

where $p(\theta)$ is the total momentum in the element $d\theta$ around θ, and θ is the angle in the event plane relative to the thrust (most energetic jet) axis, with the angular direction defined so that the second most energetic jet has $\theta < 180^\circ$. Consider first the idealized 3-quanta final state. For fixed thrust, and with the thrust axis aligned in the $\theta = 0$ direction, the remaining two momentum vectors will lie within fixed kinematic limits. How they vary in length and in angle within those limits depends on the matrix element. The pointing vector is plotted[11] in Figs. 7c to 7e for $\gamma \to q\bar{q}g$ and in Figs. 8c to 8e for $T \to 3$ gluons. The relative influence of kinematics and dynamics can be estimated by comparing these two cases. In Figs. 7 and 8, f to h, we show "smeared" pointing vectors, obtained from the idealized ones using the fragmentation models discussed above. We see that for sufficiently low thrust the three statistically reconstructed jets are well-separated even at the T mass. However, in both cases the thrust distribution is peaked at $T = 1$, as shown in Figs. 7a and 8a, so a statistically significant sample at low thrust requires a very large event sample.

The pointing vector analysis for T decay final states has been reported at this meeting by H. Meyer[13], and three jet patterns very much like those of Figs. 8f to 8h appear in the data. Unfortunately, the statistical reconstruction procedure involves an alignment of events which automatically introduces a bias in favour of a three jet structure, as illustrated by the Monte Carlo pointing vector[13] for a phase space model. For this reason the present data, while indicative of a three gluon final state, particularly when taken together with the thrust angular distribution, average thrust, etc., are not conclusive.

The variables S and T are averaged quantities which contain the minimal information on the multi-jet nature of final states. In contrast, the pointing vector potentially contains the maximal information as it reflects most faithfully the underlying quanta. Because of the problems of bias, and of the possibility mentioned above that gluon jets may be too wide to be separable, various authors have suggested intermediate analyses.

For example, the average values of thrust and spherocity can be regarded as first moments of the final state fractional momenta. One can also construct[18] higher moments for which collinear momenta are added linearly, and which would be calculable in perturbative QCD. The more moments measured, the more information is extracted; measurement of a complete set of moments is, of course, equivalent to reconstruction of the energy flow diagram. Brandt and Dahmen[15] have proposed a variable which they call triplicity, T_3, which is the maximum of the sum of fractional energy flowing along a set of three axes, and has $T_3 = 1$ for an ideal three jet event. This

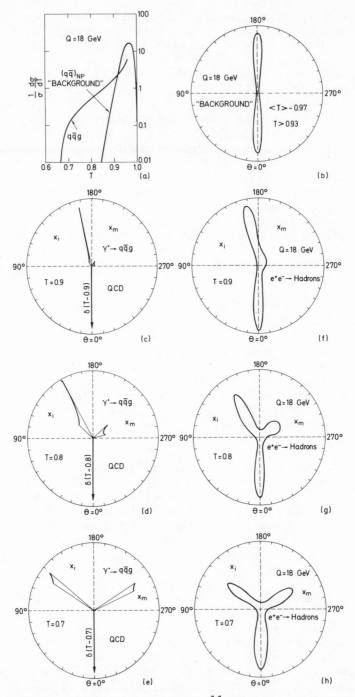

Fig. 7 Pointing vector distributions[11] for $q\bar{q}g$ final states with (f – h) and without (c – e) hadronization effects, and for $q\bar{q}$ background (b).

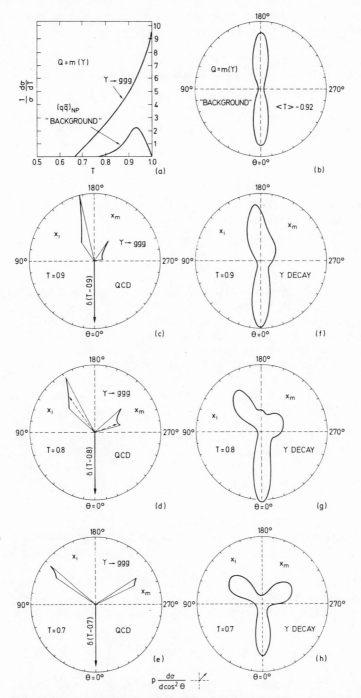

Fig. 8 Pointing vector distributions[11] for $\Upsilon \to 3g$ calculated in
        QCD (c − e), and including hadronization and a $q\bar{q}$ background
        with $\sigma(q\bar{q})/\sigma(3g) = 1/6$ (a, f − h).

is a generalization of thrust and sphericity, which measure two-jettiness, to a measure of three-jettiness.

Other approaches make use of the fact that a three quanta final state defines two axes which have calculable angular correlations. Examples are:

a) The jet boost[11,19]. For a three jet system of fixed thrust, the mass of the system recoiling against the most energetic jet is (in the zero mass jet approximation; measured jet masses are typically $0(1 \text{ GeV})$ [14]):

$$m_x = Q\sqrt{1 \cdot T}$$ (2.13)

The final state is transformed to the rest frame of the recoil system by the Lorentz transformation

$$ch\,\xi = \frac{2-T}{2\sqrt{1-T}}$$ (2.14)

In this frame there are only two axes of momentum flow, and their angular correlation is calculable. Figure 9 shows the cross-section shape as a function of their angular separation $\tilde{\theta}$ in the recoil rest frame for three final states ($q\bar{q}g$, $q\bar{q}$ + scalar gluon, 3 gluons) and for two values of thrust. The angle $\tilde{\theta}$ can be expressed in terms of thrust and spherocity by

$$\sin\tilde{\theta} = \frac{\pi\sqrt{S}}{4\sqrt{1 \cdot T}}$$ (2.15)

The requirement that T is the largest momentum in the rest frame determines the kinematic limits (Eq. (2.4)):

$$4(1-T)(2T-1)/T^2 \leq \sin^2\tilde{\theta} \leq 1$$ (2.16)

b) Azimuthal asymmetry[20]. Pi et al., define a "coplanarity axis" $\hat{C}$ as the axis which minimizes the transverse momentum in the plane perpendicular to the thrust axis $\hat{T}$; this defines the normal to the plane of the final state quanta. Defining the polar coordinates θ and ϕ by $\cos\theta = \hat{k} \cdot \hat{T}$, $\sin\theta \cos\phi = \hat{C} \cdot \hat{k}$, where $\hat{k}$ is the beam axis, the azimuthal asymmetry parameter B is defined by

$$\frac{2\pi}{\sigma}\frac{d\sigma}{d\phi} = 1 + B\cos2\phi \quad ,$$ (2.17)

and is identically zero for two jet events. Figure 10 shows[21] B as a function of energy, neglecting fragmentation effects. In this approximation there is a sharp peak on resonance, the

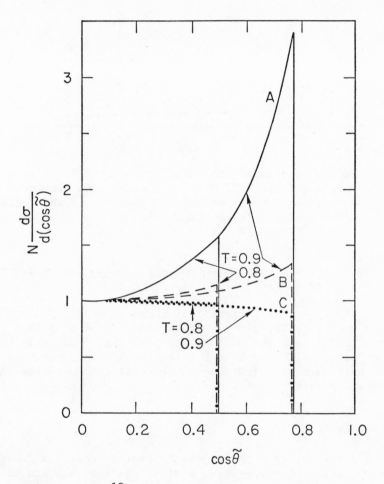

Fig. 9   Distributions[19] of the jet angular separation in the rest
frame of the system recoiling against the thrust axis for
$e^+e^- \to q\bar{q}g$ (A), $q\bar{q}$ + scalar gluon (B), ggg (C).

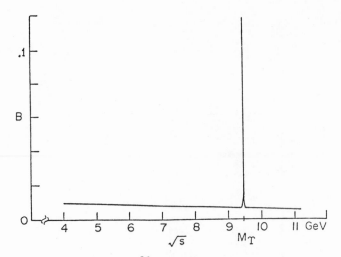

Fig. 10 Azimuthal asymmetry[21] B defined by Eq. (2.17) in the continuum and on the $\Upsilon(9.4)$ resonance. The continuum contribution was calculated using $\alpha_s(m_\Upsilon) = 0.15$ and hadronization
effects have not been included.

contribution for the continuum coming from $0(\alpha_s)$ gluon bremsstrahlung contributions.

There are many variations[22] of these proposals which are all
based on the same principle: the determination of energy flow or
"antenna" patterns. Which tests are most useful can only be determined a posteriori. At present energies the main problem seems to
be the separation of calculable QCD effects from uncontrollable
fragmentation effects.

3. NEW FLAVOURS

While the evidence for 3 quanta final states is not yet conclusive, the use of energy flow analysis already appears to be a
promising probe for new physics. In the continuum region in e^+e^-
annihilation the mean values $<s>$ and $<1 - T>$ are $0(\alpha_s)$ in perturbation theory as they arise from gluon bremsstrahlung corrections
to the dominant $q\bar{q}$ pair creation process. On a 1^{--} resonance, they
are $0(1)$ in perturbation theory since the lowest allowed direct
decay mode into hadrons is a three quanta state. They are also
$0(1)$ for pair creation of heavy naked flavour states because the
heavy hadrons will have little momentum and each will decay predominantly into three quanta with roughly isotropic distributions
via the weak V-A transitions

$$\left. \begin{array}{l} Q \rightarrow q q \bar{q} \\ Q \rightarrow q \ell \bar{\nu}_e \end{array} \right\} \tag{3.1}$$

As the energy increases, these events will become increasingly two jet-like, with their contributions giving $\langle S \rangle$, $\langle 1 - T \rangle = 0(\alpha_s)$ well above threshold. As a result, plots of $\langle S \rangle$ and $\langle 1 - T \rangle$ as a function of centre-of-mass energy should reflect the shape of plots of the total cross-section ratio

$$R = \frac{\sigma(e^+e^- \rightarrow hadrons)}{\sigma(e^+e^- \rightarrow \mu^+\mu^-)} \tag{3.2}$$

except that new structures will be relatively enhanced. Figure 11 shows a plot[11] of $\langle 1 - T \rangle$ with the T resonance and $B\bar{B}$ threshold contributions superimposed on a gluon bremsstrahlung background, labelled QCD. The curve labelled NP is the contribution from fragmentation in the continuum, using the simple model described above.

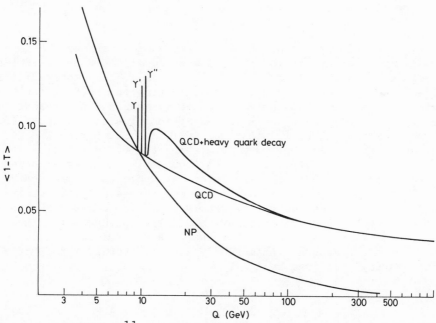

Fig. 11   Calculated[11] energy dependence of $\langle 1-T \rangle$ from gluon bremsstrahlung (QCD), $1^{--} \rightarrow$ 3g resonance decay, heavy quark production and two-jet hadronization (NP).

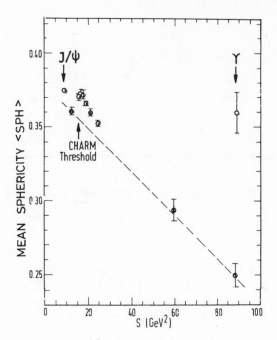

Fig. 12 Energy dependence[13] of $\langle \mathcal{S} \rangle$ showing resonance and charm
         threshold structure.

The theoretical curves for $\langle S \rangle$ have very similar shapes. Figure 12
shows[13] sphericity distributions from PLUTO at DORIS for energies
through charm threshold and up to the Υ mass. While sphericity is
not a "good" variable theoretically, it turns out to be closely
correlated to the less amenable spherocity variable at present
energies[14], and indeed, shows the expected structure. Figure 13
shows[13] combined PLUTO results from DORIS and PETRA for $\langle 1 - T \rangle$.
The open circles are continuum points and the closed circles are
the $(c\bar{c})$ 1^{--} states at 3.1, 3.7 and 4.03 GeV. The square surround-
ing the Υ point (triangle) is a three jet Monte Carlo prediction
for $\langle 1 - T \rangle$ on resonance (the $\Upsilon \rightarrow \gamma \rightarrow q\bar{q}$ process has been subtracted out), and the dashed curves are the combined QCD and NP curves
of Fig. 11, with and without heavy quark ($B\bar{B}$) production. These
results indicate that energy flow analysis may indeed be a useful
probe for new flavours. Analogously, one can consider the variable
"acoplanarity", defined by[11]

$$ A = 4 \min \left| \frac{\Sigma P_{out}}{\Sigma P} \right| , \quad 0 \leq A \leq 1 \qquad (3.3) $$

where P_{out} is the momentum component transverse to the plane which
minimizes (3.3). The upper and lower bounds on A correspond to a

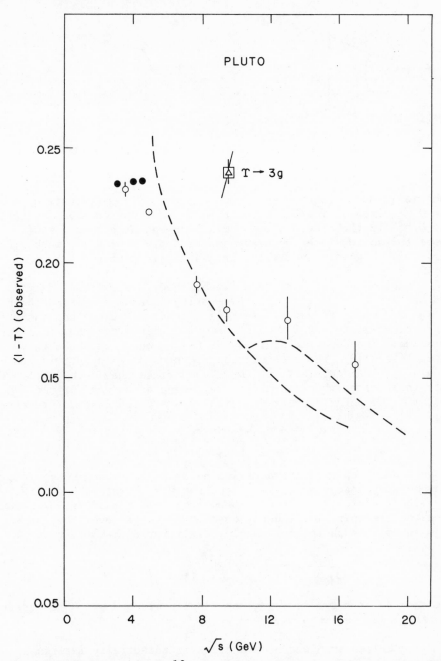

Fig. 13   Energy dependence[13] of <1−T> showing resonance (dark circles
          and triangle) structure and an indication of B$\bar{\text{B}}$ threshold.
          The dashed curve combines the NP and QCD + heavy quark
          curves of Fig. 11.   The square is a Monte Carlo calculation
          for T → 3g (T → q$\bar{\text{q}}$ has been subtracted).

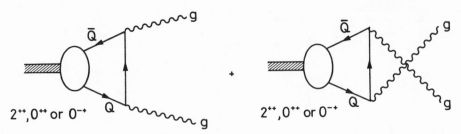

Fig. 14   Lowest order hadronic decays of 2^{++}, 0^{++} and 0^{-+} states.

spherical and a planar event, respectively. In perturbation theory
$A \neq 0$ requires at least four quanta in the final state, so we get
$A = 0(\alpha_s^2)$ in the continuum, $A = 0(\alpha_s)$ on resonance and $A = 0(1)$
above threshold since two heavy quarks decaying via the weak processes (3.1) give effectively six quanta final states.

Once new states have been established, energy flow analysis
can be used to determine the spin-parity quantum numbers of both
parent and decay products. We have already discussed the three
gluon final states of 1^{--} onia for which there is by now supporting
evidence. The decay mechanisms for C-even onia are shown in
Figs. 14 and 15. The even angular momentum onia may provide particularly clean cut tests of the theory since they decay into two
jet final states which should be well-separated even for the Υ
system. The decay angular distributions for the cascade processes

$$1^{--} \rightarrow 0 + \gamma \qquad (3.4)$$
$$\quad\hookrightarrow gg$$

depend on the spin of the decaying onium and of the gluons[11,23].
The photon-jet angular correlations for $0 = 2^{++}$ are shown[11] in
Fig. 16 (a-c) for fixed photon-beam angle and for different assumptions on the spin of the decay quanta. Averaged over the photon-
beam angle, the photon-jet angular distribution is of the form:

$$\frac{d\sigma}{d\cos\theta_{\gamma j}} = 1 + \alpha \cos^2\theta_{\gamma j} \qquad (3.5)$$

with $\alpha > 0$ for a two vector gluon final state and $\alpha < 0$ for $q\bar{q}$ or
scalar gluon final states.

The C-even spin-one p wave state is potentially interesting
as a probe of the bound state structure. In lowest order in QCD
perturbation theory there are three decay mechanisms, each of which
diverges when a final state gluon becomes soft. The divergences
in Figs. 15b and c cancel, leaving 15a as the dominant contribution,
the preferred configuration corresponding to a near two quark jet

event, but with a slight acollinearity arising from the soft gluon.
However, the divergence at vanishing gluon momentum must necessarily also be cancelled; how this comes about depends on the details of the bound state wave function. The picture just described
is valid for a Coulomb-like system for which the effective lower
cut-off on the gluon momentum is of order $\alpha_s M(\bar{Q}Q)$. To the extent
that it is correct, the angular distribution of the quasi-two-jet
axis can be measured and is predicted to have the opposite slope
of that corresponding to a direct 1^{++} $(q\bar{q})$ coupling as shown[11]
in Fig. 16 (d-f).

4. IMPROVING THE THEORY: LOOKING FOR GLUON BREMSSTRAHLUNG

The first application of the scaling parton model to deep inelastic scattering was justified as a good approximation within
QCD by the use of renormalization group equations (RGE) and a
light cone (operator product) expansion (OPE) This result was
later reproduced[24] in terms of a sum over a certain class of
Feynman diagrams, a framework which is more easily generalized to

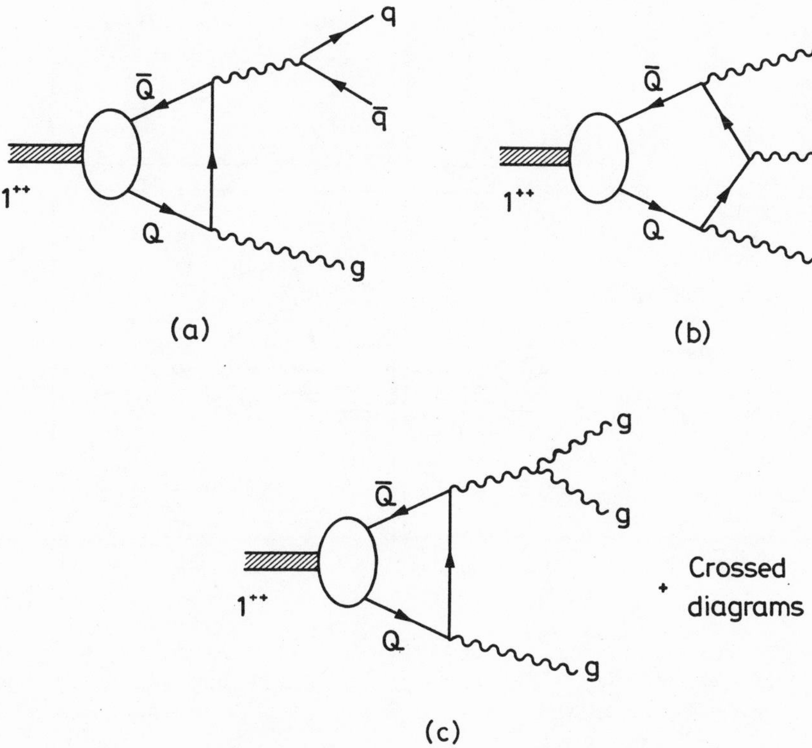

Fig. 15   Lowest order hadronic decays of a 1^{++} state.

Jet angular distributions:

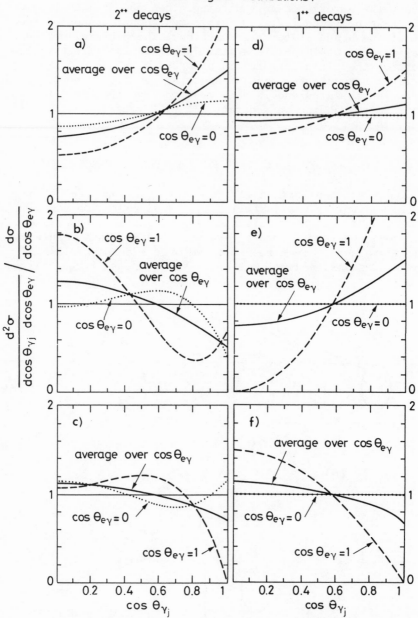

Fig. 16 Photon-jet angular correlations[11] for the decay chain
$1^{--} \rightarrow J^{++} + \gamma$, $J^{++} \rightarrow$ hadrons for different values of the
beam-photon angle: a) $2^{++} \rightarrow gg$, b) $2^{++} \rightarrow$ 2 scalar gluons
or $(q\bar{q})$ with helicity zero, c) $2^{++} \rightarrow (q\bar{q})$ with helicity 1,
d) $1^{++} \rightarrow q\bar{q}$ + soft gluon, e) $1^{++} \rightarrow$ spin zero gluons,
f) $1^{++} \rightarrow q\bar{q}$ (direct coupling).

other processes, and in particular, to e^+e^- annihilation. To leading order in $\ln Q^2$ the total cross-section for hard processes is
determined by a near collinear configuration, Fig. 17a, where no
"branch" of the "tree" is far off mass shell: $p_i^2 \ll Q^2$. The next
to leading order[24,2] in $\ln Q^2$ corresponds to one "branch" far off
shell, i.e., p_i^2/Q^2 finite, all other $p_i^2 \ll Q^2$. Since $\alpha_s(Q^2) \propto$
$\propto 1/\ln Q^2$, the subdominant process is related to the dominant one
by a perturbative expansion in the "effective" coupling constant
$\alpha_s(Q^2)$. Just as the parton model diagram of Fig. la gives a good
approximation to the total cross-section, i.e., to Fig 18a, the
diagram of Fig 1c is a good approximation to the cross-section for
highly acollinear events, arising from the subdominant diagram of
Fig. 17b. However, just as the sum of leading logs improves the
calculation of the total cross-section by taking into account
scaling violations, it is expected that a sum over next-to-leading
logs should improve the calculation of deviations from two jet
configurations. Part of the problem is purely kinematic; one is
trying to compare calculations based on two and three quanta processes with multiparticle physical processes. Since the values of
<T> and <1 − S> are highly correlated[14] with multiplicity, it is
hoped[25] that a resummation, including multi-(soft of collinear)
quanta will more closely resemble the physical process. Then it
may be that a substantial part of the contribution which, with
present techniques we are necessarily calling "uncalculable", can
be incorporated instead into the "calculable" contribution. We
will, of course, be unable to account entirely for hadronization
effects until we understand the confining mechanism which requires
the quanta to bind in forming the physical final states.

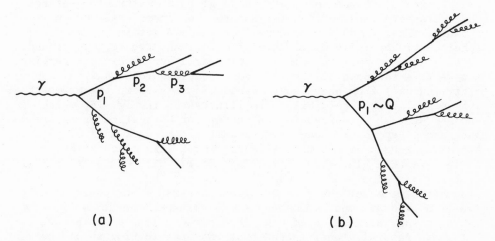

(a)   (b)

Fig. 17 Dominant (a) contribution to $e^+e^- \rightarrow$ hadrons with all inter-
         mediate quanta close to mass shell and (b) first subdomi-
         nant contribution with one quark far off shell.

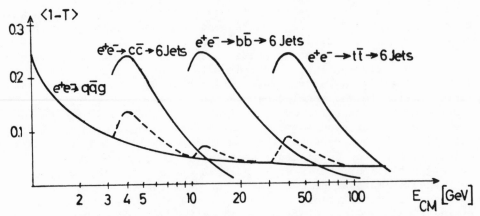

Fig. 18   <1-T> as a function[26] of energy for continuum (q$\bar{q}$g) and
          heavy quark thresholds assuming a flat fragmentation func-
          tion for Q → H(Q$\bar{q}$). The dashed line is the appropriately
          weighted superposition and hadronization has been neglected.

          Such an improvement program, if successful, would be parti-
cularly useful to the search for hard gluon bremsstrahlung effects
which are more elusive than the non-two-jettiness observed on resonance. The standard procedure is to make a cut in thrust or
spherocity, hopefully eliminating the dominant two jet background,
which should be increasingly effective as a cut at higher energies.
A caveat is the possibility that contributions to high (low) sphero-
citiy (thrust) from heavy quark production may drown the bremsstrahlung signal. This possibility is illustrated in Fig. 18, which
shows[26] <1 - T> as a function of energy for both gluon bremsstrahlung and heavy quark production, assuming a t quark mass of 15 GeV,
and neglecting hadronization effects. From this calculation it
would seem that gluon bremsstrahlung effects will be undetectable
except just below a new threshold. However, this depends on the
assumption made for the fragmentation function which governs the
fraction of heavy quark momentum transferred to the decaying heavy
hadron. The calculation of Fig. 18 assumed a flat distribution,
but it is generally anticipated that the distribution should become increasingly peaked towards z = 1 as the quark mass increases.
Assuming[14] a δ function at z = 1 gives thrust distributions which
fall off much more sharply above threshold.

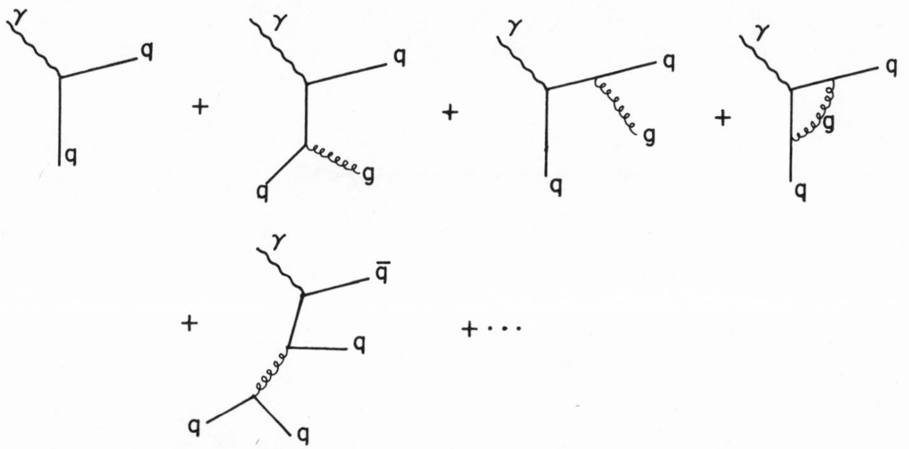

Fig. 19   Diagrams for deep inelastic lepton quark scattering.

5.   FACTORIZATION: NON-PERTURBATIVE EFFECTS

Factorization has been shown[1,2] to be valid in the leading
log approximation in perturbative QCD. For example, the probability
of finding a quark or gluon inside a quark $P_{q/q}$, $P_{g/q}$ is defined by
calculating the inclusive cross-section for deep inelastic lepton-
quark scattering via the diagrams of Fig. 19 and parametrizing the
result in the usual way in terms of parton distribution functions

$$\frac{d\sigma}{dx}\left("\gamma"+q \to X\right)= \sum_{i} P_{r_i/q}(x) \frac{d\hat{\sigma}}{dx}\left("\gamma"+p_i \to p_i'\right) \quad (5.1)$$

where the cross-sections on the right-hand side of (5.1) are for
"elementary" processes like the first one in Fig. 19. The distribution functions defined by Eq. (5.1) have infrared singularities
arising from soft or collinear gluon emission, but the important
point is that if one calculates "Drell-Yan" production of lepton
pairs by quarks

$$q+q \to \ell^+ + \ell^- + X \quad (5.2)$$

using the distribution functions so defined, one gets the same
result, in the leading log approximation, as from a direct perturbative calculation of diagrams like Fig. 20. In a similar way one
can define a quark fragmentation function D_{q/p_i}, i.e., the probability for a quark to be found among the fragments of a quanta p_i
by calculating the semi-inclusive cross-section for

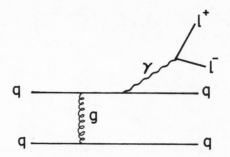

Fig. 20 Diagram contributing to production of Drell-Yan lepton
pairs by quarks.

$$e^+ e^- \to q + X \tag{5.3}$$

by the diagrams of Fig. 21, and parametrizing the result in terms
of fragmentation functions:

$$\frac{d\sigma}{dz}\left(e^+ e^- \to q(z) + x\right) = \sum_{p_i} \frac{d\sigma}{dz}\left(e^+ e^- \to p_i \bar{p}_i\right) D_{q/p_i}(z) \tag{5.4}$$

The resulting D's can then be combined, for example, with the p's
defined above to calculate the semi-inclusive cross-section for

$$"\gamma" + q \to q(z) + X \tag{5.5}$$

or with one another to calculate two-particle semi-inclusive cross-
sections

$$e^+ e^- \to q(z_1) + q(z_2) + X \tag{5.6}$$

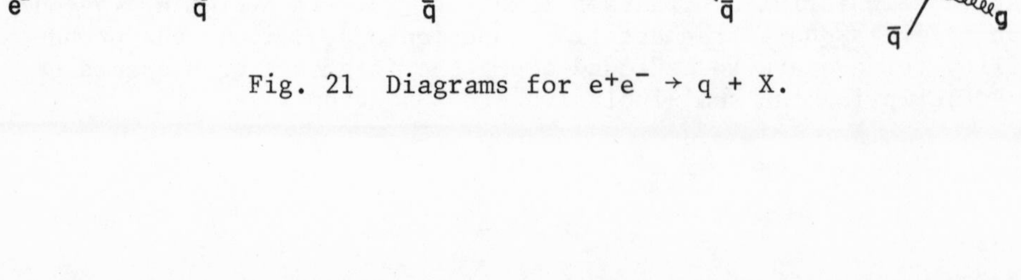

Fig. 21 Diagrams for $e^+ e^- \to q + X$.

The above procedure works because the calculated structure/ fragmentation functions always appear as universal factors multiplying point-like cross-sections. However, the results have been obtained for unbound quarks and gluons. In real life we have to deal with their bound states. The mechanism which confines quarks and gluons does not emerge from perturbation theory, and it has been hoped that whatever non-perturbative mechanism which is necessarily operating will not change this result, i.e., that its effects can also be absorbed into universal factors and will be essentially Q^2 independent. Preliminary studies, however, do not support this hope.

In classical QCD, non-perturbative phenomena have been found which induce vacuum fluctuations, and the physical vacuum is a superposition of the usual (zero field) one and of vacua with non-trivial field configurations. Attempts to investigate the influence of this phenomenon on observed processes use a semi-classical approximation in which quantum fluctuations are calculated in the presence of an external field. The simplest non-trivial solution[27] to the classical field equations is the one instanton configuration which can be represented as an external field of finite extension in space time. For example, vacuum polarization contributions to the photon propagator in the presence of an instanton,(Fig. 22a) are evaluated[28-30] using the known[31] form of the quark propagator in the presence of an instanton field, and then integrating over the instanton size ρ and position z, weighted by an appropriate density function[32] $d(\rho)$. This "dilute gas" approximation[33], in which instantons are assumed to be non-interacting and not very dense, may be valid for quantum fluctuations over short distances which are only sensitive to "small" instantons. What is relevant to e^+e^- annihilation is the absorptive part of the vacuum polarization, which is proportional to the ratio R in Eq. (3.2). It turns out[29,30] that for large Q^2 the result is indeed sensitive only to small instantons, $\rho \sim 1/Q^2$, and, to the extent that extrapolation of the semi-classical (and Euclidean space) calculation described above can be extended to the physical time-like region in Q^2, one finds[29]

$$\left(\frac{\Delta R}{R}\right)_{inst.} \simeq \left(\frac{Q}{1-2\,GeV}\right)^{-12} (\ln Q^2)^{3\ or\ 4} \tag{5.7}$$

The non-perturbative correction to R is thus O(1) in the resonance region and falls off rapidly above 2 GeV, which, if the calculation has any meaning, may provide a nice rationale for precocious scaling.

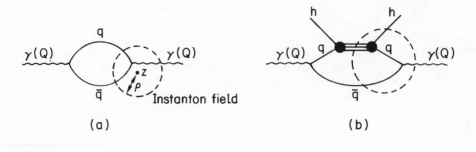

(a)   (b)

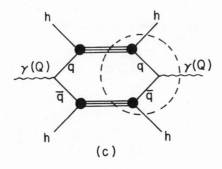

(c)

Fig. 22 Diagrams for evaluating instanton effects in a) $e^+e^- \to$
         $\to$ hadrons, b) $\ell + h \to \ell' + X$ or $e^+e^- \to h + X$, c) $h + h' \to$
         $\to \ell^+\ell^- + X$, $\ell + h \to \ell' + h' + X$ or $e^+e^- \to h + h' + X$.

         As the next step in complexity, consider the box diagram of
Fig. 22b, for which the absorptive part could correspond either to
the total cross-section in deep inelastic scattering, with h the
target nucleon, or to semi-inclusive e^+e^- annihilation, with h the
trigger hadron. The calculation has been performed[34] with h replaced by a not too massive virtual photon, and it is found that
a) corrections to the Q^2 dependence are again governed by small
instantons for large Q^2, but that b) the structure (fragmentation)
function for h receives a Q^2 independent contribution from large
instantons as well, so the dilute gas approximation breaks down.
This is not surprising since the bound state wave function is not
a short distance phenomenon, and it is not a problem since the
structure function is, in any case, not calculable from QCD perturbation theory alone.

Finally, we can consider a further degree of complexity, illustrated in Fig. 22c. In this case the absorptive part may represent Drell-Yan lepton pair production, semi-inclusive deep inelastic scattering, or two particle correlations in e^+e^- annihilation. One finds[35] again that the Q^2 dependence is unaffected at large Q^2 (i.e., effects are of order Q^{-12}) but that the process cannot be described in terms of the structure/fragmentation functions obtained by considering the diagram of Fig. 22b. The two bound states "feel" simultaneously the presence of large instantons. This simple analysis then suggests that while factorization may break down in predicting, say, the normalization of two body semi-inclusive cross-sections from that of one body ones in e^+e^- annihilation, it should still give a correct description of the Q^2 dependence. Then analysis in terms of double moments becomes relevant. If the nth moment of a fragmentation function is measured to be of the form

$$A_n(Q^2) = \int_0^1 z^n D_n(z, Q^2)\, dz = A_n^o \left[\ln\left(Q^2/\Lambda^2\right) \right]^{\gamma_n} \left[1 + o\left(\frac{1}{\ln Q^2}\right) \right] \quad (5.8)$$

where γ_n is calculable from an RGE analysis and the constant A_n^o is fixed by experiment, then a double moment for a two particle correlation should be of the form

$$A_{nm}(Q^2) = \int dz\, dz' \frac{d^2\sigma}{dz\, dz'} z^n z'^m \propto A_{nm}^o \left[\ln\left(Q^2/\Lambda^2\right) \right]^{\gamma_n + \gamma_m} \left[1 + o\left(\frac{1}{\ln Q^2}\right) \right] \quad (5.9)$$

with
$$A_{nm}^o \neq A_n^o A_m^o$$

However, it must be borne in mind that techniques for studying the phenomenology of non-perturbative phenomena are still at a very primitive stage, and the conclusions outlined above are based on the most naive assumptions possible.

6. WEAK INTERACTIONS

To conclude, I would like to recall some possible weak interaction effects which, if found, may contribute substantially to our understanding of the underlying theory.

One expectation is that the $B^0\bar{B}^0$ system may provide an observable source[36] of CP violating effects which could throw new light on this phenomenon which has so far been confined to the $K^0\bar{K}^0$ system. The idea is to look for "same sign" dileptons arising from $B^0\bar{B}^0$ production and decay with $B^0 \rlap{\,\not\;}{\rightleftarrows} \bar{B}^0$ mixing, e.g.,

$$e^+e^- \;\rightarrow\; B^0 + \overline{B}^0$$
$$\; \big\downarrow \;\rightarrow\; \bar{c}\,\ell^+ \nu_\ell$$
$$\; \big\downarrow \;\rightarrow\; \bar{B}^0 \rightarrow \bar{c}\,\ell^+ \nu_\ell \tag{6.1}$$

which is analogous to $K^0 \rlap{\,\not\;}{\rightleftarrows} \bar{K}^0$ mixing and is expected to be substantially more important than $D^0 \rlap{\,\not\;}{\rightleftarrows} \bar{D}^0$ mixing. The observations of a difference in rate for $\ell^+\ell^+$ and $\ell^-\ell^-$ events would be a signature for CP violation. In order for such effects to be observable, two phenomena must occur:

a) an appreciable B^0-$\bar{B}^0$ mixing: the number of "same sign" events is given by[37]

$$r_1 = \frac{\sqrt{N^{++}N^{--}}}{N^{++}} = \frac{\Delta}{1+\Delta} \quad,\quad \Delta = \frac{(\Delta\Gamma)^2/4 + (\Delta m)^2}{2\Gamma + (\Delta m)^2 - (\Delta\Gamma)^2/4} \tag{6.2}$$

where $\Delta\Gamma$ and Δm are the difference in decay rate and mass respectively, for the two decay eigenstates, and Γ is the average decay rate, and

b) a substantial CP violation which can be expressed as[37]

$$r_1 = \sqrt{\frac{N^{--}}{N^{++}}} \;=\; \left| \frac{\Delta\bar{m} - i\,\overline{\Delta\Gamma}/2}{\Delta\bar{m}^* - i\,\overline{\Delta\Gamma}^*/2} \right| \tag{6.3}$$

where here $\Delta\bar{m}$ and $\overline{\Delta\Gamma}$ are, respectively, the dispersive and absorptive parts of the (complex) $B^0 \rlap{\,\not\;}{\rightleftarrows} \bar{B}^0$ transition amplitude:

$$A(B^0 \rightarrow \bar{B}^0) \;\equiv\; \Delta\bar{m} - i\,\Delta\bar{\Gamma}/2 \tag{6.4}$$

The relevant quantities have been estimated[36] in the context of the Kobayashi-Maskawa six quark model[38] for CP violation as a function of the t quark mass. One finds, for example:

$$m_t = 8\,\text{GeV} : \quad r_1 \sim 10^{-2} \quad , \quad r_2 = (1+\sin\delta)/(1-\sin\delta)$$

$$m_t \gg 8\,\text{GeV} : \quad r_1 = O(1) \quad , \quad r_2 = 1 - 2\sin\delta\,(8\,\text{GeV}/m_t)^2 \quad (6.5)$$

where δ is a CP violating parameter which is <u>a priori</u> arbitrarily large. We see that a low t quark mass allows a large CP violation but little mixing, while a high mass generates large mixing but little CP violation. We can hope that this mass will fall in a region where the combined effect will not be negligible.

Same sign dileptons from the process (6.1) will have to compete with a large background from cascade dileptons, e.g.,

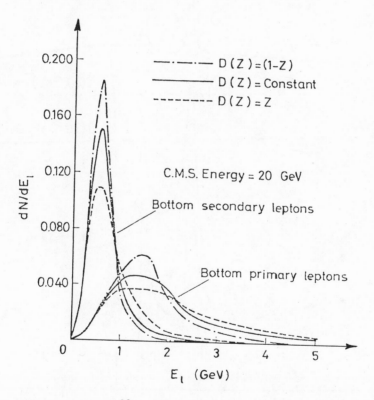

Fig. 23   Energy spectra[39] for primary and secondary leptons from bottom decay in the reaction $e^+e^- \to B^0\bar{B}^0 + X$ at $\sqrt{s}$ = 20 GeV with different assumptions on the b → B and c → D fragmentation functions.

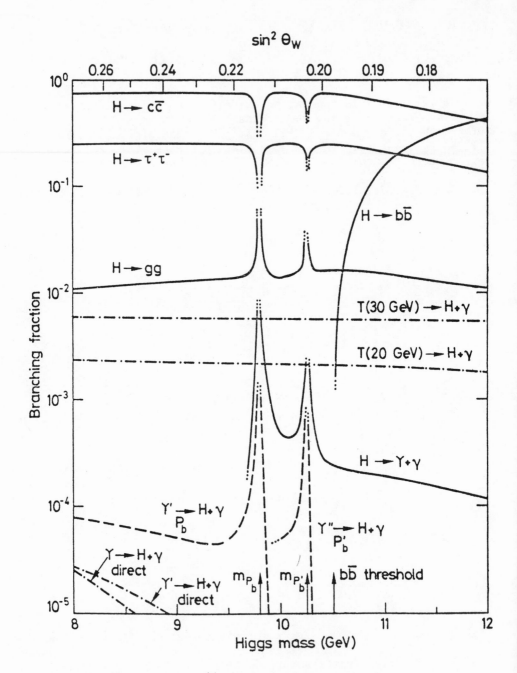

Fig. 24   Branching ratios[41] for decays of and into a Higgs boson as a function of its mass in the 10 GeV region.

$$e^+e^- \longrightarrow B^0 + \overline{B^0} + X$$
$$\quad\quad\quad\quad\quad \hookrightarrow \bar{c}\,\ell^+ \nu_\ell$$
$$\quad\quad\quad\quad\quad\quad\quad\quad c + \text{hadrons}$$
$$\quad\quad\quad\quad\quad\quad\quad\quad \hookrightarrow s + \ell^+ \nu_\ell$$

(6.6)

One way to reduce this background is by making a cut in the lepton energy. Leptons from the primary decay vertex will have a distinctly harder spectrum than those from the secondary c-decay vertex, as illustrated[39] in Fig. 23

Another weak interaction phenomenon about which little is understood is the Higgs particle. In particular, its mass is left undetermined by the theory. However, recent speculations[40] concerning the unification of strong, weak and electromagnetic forces suggest that its mass[41] may be in the 10 GeV region. If this is the case, it can have interesting decay and production properties because of the possibility of mixing with the $0^{++}(b\bar{b})$ bound states. Fig. 24 shows[41] branching ratios for Higgs decay and for its production via the radiative decays of the $1^{--}(b\bar{b})$ and $(t\bar{t})$ states as a function of its mass. Its production in conjunction with the Υ family could be appreciable if its mass were sufficiently close to a 0^{++} state accessible to 1^{--} radiative decay.

It is a pleasure to thank H. Meyer for many discussions and P. Binétruy for help with the manuscript.

REFERENCES

1. H.D. Politzer, Phys. Lett. 70B:403 (1977);
    I. G. Halliday, Nucl. Phys. B129:301 (1977);
    C.T. Sachrajda, Phys. Lett. 73B:185 (1978) and 76B:100 (1978);
    K.H. Craig and C.H. Llewellyn-Smith, Phys. Lett. 72B:349 (1978).
2. Yu. L. Dokshitzer, D.I. D'yakanov and S.I. Troyan, Materials
    for the XIIIth Leningrad Winter School (1978), p.3.
3. G. Sterman and S. Weinberg, Phys. Rev. Lett. 39:1436 (1977).
4. P. Binétruy and G. Girardi, Phys. Lett. 83B:339 (1979). See
    also, B.G. Weeks, Phys. Lett. 81B:377 (1979).
5. Ch. Berger et al., PLUTO Collaboration, Phys. Lett. 82B:449
    (1979).
6. I.I.Y. Bigi and T.F. Walsh, Phys. Lett. 82B:267 (1979).
7. J. Ellis, M.K. Gaillard and G.G. Ross, Nucl. Phys. B111:253
    (1976)
8. S. Brandt, Ch. Peyrou, R. Sosnowski and A. Wroblewski, Phys.
    Lett. 12:57 (1964);
    E. Farhi, Phys. Rev. Lett. 35:1609 (1975).
    The term thrust was adopted in Ref. 11.
9. H. Georgi and M. Machacek, Phys. Rev. Lett. 39:1237 (1977).

10. T.A. De Grand, Y.J. Ng and S.-H.H. Tye, Phys. Rev. D16:3257
    (1977).

11. A. de Rújula, J. Ellis, E.G. Floratos and M.K. Gaillard, Nucl.
    Phys. B138:387 (1978).

12. K. Shizuya and S.-H. Tye, Phys. Rev. Lett. 41:787, 1195(E) (1978);
    M.B. Einhorn and B. Weeks, SLAC-PUB-2164 (1978);
    See also, K. Konishi, A. Ukawa and G. Veneziano, Phys. Lett.
    80B:259 (1979).

13. H. Meyer, these proceedings.

14. H. Meyer, private communication.

15. S. Brandt and H.D. Dahmen, Zeitschrift für Phys. C1:61 (1979).

16. Ch. Berger et al., PLUTO Collaboration, Phys. Lett. 78B:176
    (1978).

17. K. Koller, H. Krasemann and T.F. Walsh, Zeitschrift für Phys.
    C1:71 (1979).

18. G. Fox and S. Wolfram, Phys. Lett. 82B:134 (1979), Phys. Rev.
    Lett. 41:1581 (1979) and Nucl. Phys. B149:413 (1979).

19. J. Ellis and I. Karliner, Nucl. Phys. B148:141 (1979).

20. S.-Y. Pi, R.L. Jaffe and F.E. Low, Phys. Rev. Lett. 41:142 (1978).

21. S.-Y. Pi, and R.L. Jaffe, MIT preprint CTP-735 (1978) to
    appear in Phys. Rev. D Comments and Addenda.

22. K. Koller and T.F. Walsh, Phys. Lett. 72B:227 (1977) and E
    73B:504 (1978);
    G. Parisi, Phys. Lett. 74B:65 (1978);
    C.L. Gasham, L.S. Brown, S.D. Ellis and S.T. Love, Phys. Rev.
    Lett. 41:1585 (1978), Phys. Rev. D17:2298 (1978), and Univ.
    of Washington preprint RLO-1388-759 (1978);
    G. Parisi and R. Petronzio, Phys. Lett. 82B:26 (1979).

23. M. Krammer and H. Krasemann, Phys. Lett. 73B:58 (1978);
    H. Krasemann, Zeitschrift für Physik 1:189 (1979).

24. C.H. Llewellyn Smith, Proc. XVII Internationale Universitäts-
    wochen für Kernphysik, Schladming (1978).

25. A. de Rújula, private remarks.

26. A. Ali, J.C. Körner, G. Kramer and J. Willrodt, Zeitschrift
    für Physik 1:203 (1979).

27. A.A. Belavin, A.M. Polyakov, A.S. Schwartz and Yu. S. Tyupkin,
    Phys. Lett. 59B:85 (1975).

28. N. Andrei and D.J. Gross, Phys. Rev. D18:468 (1978).

29. L. Baulieu, J. Ellis M.K. Gaillard and W.J. Zakrzewski, Phys.
    Lett. 77B:290 (1978).

30. T. Appelquist and R. Shankar, Phys. Lett. 78B:468 (1978).

31. L.S. Brown, R.D. Carlitz, D.B. Creamer and C. Lee, Phys. Lett.
    70B:180 (1977).

32. G. 't Hooft, Phys. Rev. Lett. 37:8 (1976) and Phys. Rev. D14:
    3432 (1976);
    C.G. Callan, R. Dashen and D.J. Gross, Phys. Lett. 63B:334 (1976)
    and 66B:375 (1977).

33. A.M. Polyakov, Nucl. Phys. B121:   (1977);
    C.G. Callan, R. Dashen and D.J. Gross, Phys. Rev. D17:2217 (1978).

34. L. Baulieu, J. Ellis, M.K. Gaillard and W.J. Zakrzewski, Phys.
     Lett. 81B:41 (1979);
     C.A. Flory, LBL preprint 8499 (1979).
35. J. Ellis, M.K. Gaillard and W.J. Zakrzewski, Phys. Lett. 81B:
     224 (1979).
36. J. Ellis, M.K. Gaillard, D.V. Nanopoulos and S. Rudaz, Nucl.
     Phys. B131:285 (1977); see also, M.K. Gaillard, Proc. SLAC
     Summer Institute on Particle Physics (1978) p. 397.
37. A. Pais and S.B. Treiman, Phys. Rev. D12:2744 (1975).
38. M. Kobayashi and K. Maskawa, Progr. Theor. Phys. 49:652 (1973).
39. A. Ali, Zeitschrift für Physik C1:25 (1979).
40. S. Weinberg, Phys. Lett. 82B:387 (1979).
41. J. Ellis, M.K. Gaillard, D.V. Nanopoulos and C.T. Sachrajda,
     Phys. Lett. 83B:382 (1979).

DEEP INELASTIC MUON AND ELECTRON SCATTERING

AS A PROBE FOR QUARKS AND GLUONS IN THE NUCLEON

H. L. Anderson

Los Alamos Scientific Laboratory
Los Alamos, New Mexico 87545
and
Enrico Fermi Institute
The University of Chicago
Chicago, Illinois 60637

The classical way to probe small distances is to carry out a scattering experiment. This was the method of Rutherford who used the scattering of α-particles from a thin gold foil to show that the gold atom had its positive charge concentrated in a small central core. Rutherford's experiments established the size of the gold nucleus to be of order 10^{-12} cm. Our present experiments with lepton probes are able to resolve distances in the 10^{-14} - 10^{-15} cm range, small enough compared to the radius of the nucleon, $\simeq 8 \times 10^{-14}$ cm, to be sensitive to whatever might be inside. Indeed, deep inelastic scattering measurements using electron and muon beams have provided persuasive evidence that the nucleons, and presumably all hadrons, are made of quarks and gluons. The measurements may be interpreted to give the momentum distribution, not only of the dominant flavors of quarks, but of the gluons as well.

These two talks are arranged to a) describe the muon data and how they were obtained and b) to show how they were combined with the electron data and interpreted from the point of view of quantum chromodynamics.

FERMILAB μP EXPERIMENT[1,2,3]

My connection with the deep inelastic muon scattering experiment was through my participation as a University of Chicago member of the CHIO (Chicago, Harvard, Illinois, Oxford) collaboration that carried out Fermilab experiments E98/398 over a 7-year period from

1971 to 1979. I show the members of the collaboration in Figure 1.
Included are some ten graduate students whose doctoral dissertations
were based on the work performed on these experiments.

Measurements were made on the hadrons produced as well as on the
scattered muons. However, in this report only the muon inclusive
cross-sections will be discussed.

INELASTIC SCATTERING

The deep inelastic scattering equipment is simple in principle.
As shown in Figure 2, incident muons of energy E are scattered by a
target through an angle θ and emerge with energy E'. Means are provided to measure these three quantities in the laboratory for each
scattered muon. This suffices to determine the two variables on
which the cross-section depends. A convenient choice is the two
Lorentz invariants, the square of the 4-momentum transfer

$$q^2 = 2m^2 - 2(EE' - \vec{p} \cdot \vec{p}') = -Q^2 \tag{1}$$

and the Bjorken variable

$$x = \frac{Q^2}{2M(E-E')} \tag{2}$$

which measures the elasticity of the collision.

Here, M is the nucleon mass, m the muon mass, E,E' are
the incident and scattered energy of the muon, $\vec{p}$, $\vec{p}'$ the corresponding momenta. The energy loss is frequently written $\nu = E-E'$.
For elastic scattering x = 1. Inelastic scatters have $0 < x < 1$.

CHICAGO	HARVARD	ILLINOIS	OXFORD
H. L. Anderson	B. A. Gordon	W. R. Francis	V. K. Bharadwaj
R. M. Fine	W. A. Loomis	R. G. Hicks	N. E. Booth
R. H. Heisterberg	F. M. Pipkin	T. B. W. Kirk	G. I. Kirkbride
W. W. Kinnison	S. H. Pordes		J. Proudfoot
H. S. Matis	A. L. Sessoms		T. W. Quirk
L. W. Mo	W. D. Shambroom		A. Skuja
L. C. Myrianthopoulos	C. Tao		M. A. Staton
S. C. Wright	L. J. Verhey		W. S. C. Williams
	Richard Wilson		

Fig. 1. Members of the Chicago, Harvard, Illinois, and Oxford collaboration that performed Fermilab Experiments E98/398.

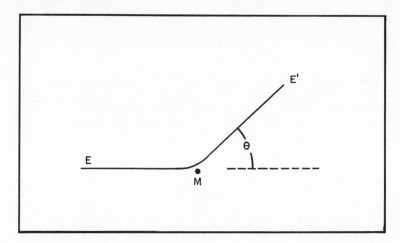

Fig. 2.   Schematic of mu-p scattering in the laboratory frame show-
          ing the quantities measured.

The Feynman diagram for the process is given in Figure 3.   The
muon interacts with the nucleon via one photon exchange.   The strength
of the interaction is governed by quantum electrodynamics.   The muon
(or electron) is taken to be a point charge.   The muon-photon vertex
has an interaction strength proportional to the square of the
electron charge e^2.   A second e^2 comes from the hadron-photon vertex.
The photon propagator contributes a factor Q^4 in the denominator of
the expression for the cross-section, which may be written

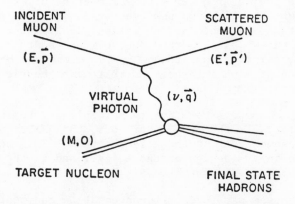

Fig. 3.   Feynman diagram for inelastic scattering by charged leptons
          in the 1 photon exchange approximation.

$$\frac{d^2\sigma}{dQ^2 dx} = \frac{2\pi e^4}{Q^4}\frac{1}{p^2}\frac{F_2(Q^2,x)}{x}\left\{(2EE'-Q^2/2) + \frac{(Q^2-2m^2)(1+Q^2/4M^2x^2)}{1+R(Q^2,x)}\right\}$$

$$(3)$$

The nucleon is not a point charge. Its behavior is described in terms of two structure functions $F_1(Q^2,x)$ and $F_2(Q^2,x)$ corresponding to the two independent degrees of polarization carried by the virtual photon. Since the small angle scattering is dominated by F_2, it has been the practice to introduce the quantity $R = \sigma_L/\sigma_T$ where σ_L and σ_T are the total photo-production cross-sections for longitudinal and transverse polarized photons, respectively. The quantity R is related to the structure functions F_2 and F_1 by the equation

$$R = \left(\frac{4M^2x^2}{Q^2} + 1\right)\frac{F_2(Q^2,x)}{xF_1(Q^2,x)} - 1$$

$$(4)$$

The object of the experiment is to measure the inclusive cross-section $d^2\sigma/dQ^2 dx$ for inelastic scatters over as large a range of Q^2 and x as possible and to use these to deduce the structure function F_2 and the ratio R as a function of these variables.

MUON BEAM

The muon beam was produced by the decay in flight of pions and kaons produced by the extracted proton beam striking an aluminum target. Figure 4 shows a schematic layout of the Fermilab muon beam. Following the target the secondaries are strong-focussed by a triplet of quadrupoles into a 500 meter evacuated decay pipe. The beam-line had four bending stations D1-D4, with a total bend at each station of 27 mr. These bends served to momentum-select the muons and to separate the muon beam line from the neutrino beam line. The beam coming out of the 500 meter decay pipe was bent at D1, re-focussed by quadrupole doublets Q2 and Q3 and bent again at D2. The gap in the dipole magnets at D3 contained approximately 23 meters of high density polyethylene. This absorbed the residual hadron contamination in the beam, mainly pions, to one in 10^7 muons and brought to 10^{-3}, the fraction of muon interactions that might have been faked by a pion.

Most of the halo muons were due to multiple scattering in the polyethylene filter. In the muon laboratory, the number of halo muons was of the same order as those in the beam. This required careful attention in the design of the trigger and in the reconstruction of genuine scattering events.

After D3 the beam was further focussed at Q4 and bent at D4 into the muon laboratory. The bending station at D4 was also used for momentum tagging the muons in the beam while Q4 focussed the beam onto the experimental target in the laboratory.

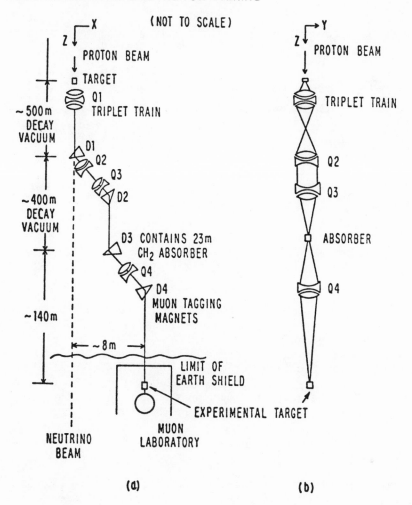

Fig. 4.   Fermilab muon beam. Pions produced by protons striking an
          aluminum target are momentum selected at the triplet train.
          A fraction decay into muons in the 500 meter drift space
          and are brought into the muon lab by a series of bending
          and focussing magnets in the beam line at Stations D1-D4.

BEAM TAGGING

        The trajectory of each incident muon was measured at four
points (as shown in Figure 5): one just after Q4, a second 70 m
downstream, just before D4, a third just after D4, and the last,
31 m further downstream, about 3 m before the target. The x coordinates were measured at all points but the y coordinates were
measured only after D4. The measurements were made with multiwire

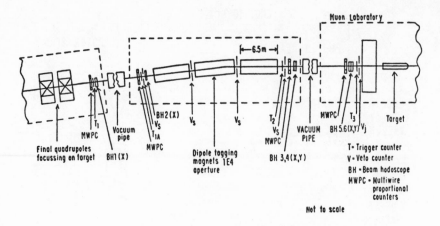

Fig. 5. Measurement of momentum and trajectory of the incident muons
         in Station D-4 using multiwire proportional chambers.

proportional counters, MWPC, combined with 8-element scintillation
counter hodoscopes (BH 1-6 in Figure 5) which covered the active area
of the MWPCs. The angle of bend in D4 was determined with a precision of 0.03 mr giving 0.1% error in the momentum at 150 GeV. The
system also gave the transverse position of the muon at the target
to a precision of 1/2 mm. In the reconstruction the beam muon could
be linked with the scattered muon and the vertex of the scattering
within the target determined in an unambiguous way.

MUON SPECTROMETER

     The arrangement of the spectrometer is shown in Figure 6. The
centerpiece of the spectrometer is the rebuilt Chicago cyclotron
magnet CCM, with a 432 cm diameter pole and a 129 cm high gap. The
magnet normally operated at 15,000 gauss and gave a transverse
momentum kick of 2.25 GeV/c.

     The scattered muon was detected in a cluster of 8-1m x 1m MWPC
planes between the target and the CCM. These chambers were arranged
with alternate vertical and horizontal wire planes giving 4 x and
4 y coordinates on the trajectory. They were used to measure the
tracks of charged particles emerging from the muon interaction point
and to determine the position of the event vertex in the target by
establishing a link with the beam muon.

     Downstream of the magnet were four groups of chambers, a total
of 20 planes, used to determine the trajectories of the particles
that had traversed the magnet. The wires were arranged either
vertically (x) or at an angle of $\pm \tan^{-1}(1/8)$ to the vertical (u,v).

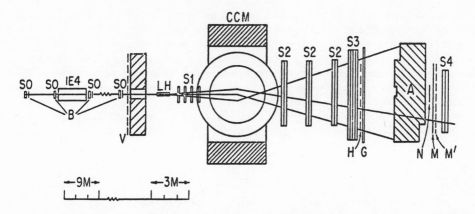

Fig 6. The Muon Spectrometer. The momentum of the scattered muons
        are measured by the amount they are deflected by the Chicago
        Cyclotron Magnet. The trigger depends on detecting the
        scattered muons after penetrating a 2-1/2 meter thick steel
        filter. Electron and neutral hadron identifiers not shown.

The first 3 groups of chambers were 4m x 2m with 4 planes arranged
as (uxxv), the fourth group was 6m x 2m with 8 planes arranged
(uxuxvuxv).

     Immediately downstream of the 6m x 2m spark chambers were two
large scintillation counter hodoscopes, G & H. The H hodoscope had
24 vertical elements covering a 4m x 2m area. The G hodoscope had
18 horizontal elements covering an area 6m x 2m. Both hodoscopes
were made insensitive in the region of the beam. Besides their use
in the trigger, these hodoscopes gave time information which was
used to identify event associated tracks on them.

     Behind the G and H hodoscopes were three systems of detectors,
a photon and electron identifier, a neutral hadron identifier, and
a muon identifier. The first consisted of 3 radiation lengths of
steel followed by a set of 4m x 2m magnetostrictive spark chambers.
Electrons or photons striking the steel wall had a good chance of
starting an electromagnetic shower which showed in the spark chambers
as a very large number of sparks. The system was used to help
identify μ-e elastic scatters at low Q^2.

     The second system consisted of a 40 cm thick lead wall followed
by a set of 4m x 2m magnetostrictive chambers. This thickness of
lead absorbed almost completely the electromagnetic showers but was
enough to start a nucleon cascade.

     The muon identifier consisted of a 250 cm thick steel wall
equivalent to 15 hadronic interaction lengths. Energetic muons in

the experiment penetrated this wall easily with very little contamination from hadrons. The M and M' hodoscopes behind it were used to identify muons in the trigger, the N hodoscope in the region of the beam defined the minimum scattering angle acceptance, and the K hodoscope in the beam served as a veto for unscattered muons. These hodoscopes were followed by a set of 4m x 2m magnetostrictive spark chambers to locate the scattered muon. This straightforward means for identifying the scattered muon is a major reason for preferring muons over electrons as a lepton probe.

TARGET

The target was liquid hydrogen or liquid deuterium in a 120 cm long, 18 cm diameter kapton flask. This was enclosed in an aluminum vacuum container with windows sufficiently far removed from the flask to make possible a clear separation of window and target scatters.

TRIGGER

The trigger was based on scintillation counters before the target and after the CCM, arranged to detect every incident muon within the beam acceptance and to trigger when one of these was scattered outside the beam region, but within the acceptance of the spectrometer.

The beam triggering and tagging system is shown in Figure 5. A B-signal was produced by a quadruple coincidence between the counters T1 and T1A upstream and T2 and T3 downstream of the final beam bending magnet. The counters (hodoscopes) downstream of the CCM were H and G upstream of the hadron shield and M, M' and N behind it as shown in Figure 6. To help assure that the muon detected downstream of the CCM was the same as the one that entered the target, a veto hodoscope K was placed in the beam region just behind the hadron shield. The absence of a signal from K signified that the entering muon was removed from the beam. The trigger logic used for the run at 219 GeV was $B\overline{K}$ (G or H)(M, M' or N). The H hodoscope was not included in the trigger logic in the running at 96 and 147 GeV.

The inclusion of a beam veto in the trigger was important because of the large number of halo muons outside the beam which would activate the trigger by being in accidental time coincidence with a beam muon.

To suppress the number of halo muons that could activate the trigger, veto counters were installed at various positions along the beam line (V_1-V_5 in Figure 5). A large veto counter hodoscope was installed covering the front face of a concrete block wall joint upstream of the target (Figures 5 and 6).

Under good operating conditions the beam delivered 10^6 muons in a one second spill. The number of halo muons tended to be somewhat larger than this. Without the K veto the beam halo could produce triggers at a rate of about 3 to 5 x 10^{-2} per incident muon. The inclusion of G or H suppressed events in which a beam muon scattered in the hadron absorber and struck M, M' or N rather than K. The beam contained positrons which could satisfy the beam and veto conditions at a rate of 2 x 10^{-4} per incident muon. In accidental coincidence with a halo muon these produced the bulk of the triggers which occurred at the rate of 6 to 10 x 10^{-6} per incident muon. The target associated trigger rate was visible at 1 x 10^{-6} if a full-empty sub-traction was done with adequate statistics.

Other undesired triggers were caused by muons interacting in the hadron absorber, firing M, M' or N and not K, in accidental coincidence with G (or H). Also, muons of low energy which scattered in or near the final bending magnet, satisfied the beam trigger requirements but were bent by the magnet into the acceptance region. Such events were sorted out from good events by the reconstruction. Out of 10 events normally collected per pulse, one was good.

Special precautions were taken because of the RF structure in the beam. Muons appeared in the apparatus in RF "buckets", 18.8 ns apart and 2 ns wide. As a guard against events with two muons per bucket no muon was accepted if more than one of the elements in each of the beam hodoscopes BH 2-6 was fired. This prevented the loss of a good event with one muon that was vetoed by the second. In addition no muon was counted or used which had a muon in the preceding RF bucket. This was done to avoid the effects of inefficiencies due to dead time in the K veto system. These could occur because muons emerging from the hadron wall were frequently accompanied by exten-sive electromagnetic showers which might disable the K system for more than 18 ns. The second muon could then satisfy the trigger condition by a chance coincidence with a halo muon.

TRACK RECONSTRUCTION

The reconstruction of events was greatly simplified because of the cylindrical symmetry of the magnetic field of the CCM. For such a field "Störmer's Theorem" applies and the impact parameter of the trajectory on entrance will be equal to that on leaving. Thus, it suffices to fit a straight line to the points measured along the trajectory in the low field region outside the magnet determining the perpendicular distance to the magnet center from an extension of this line. Upstream and downstream track segments were matched if they were found to have the same impact parameter.

The method is illustrated in Figure 7 which includes a plot of the difference of upstream and downstream impact parameters for good

events.   Matched impact parameters showed a standard deviation
σ = 1.9 mm.

We give two examples of reconstructed events, a μ–e elastic
scatter and a deep inelastic μ scatter producing hadrons in Figure 8.

The acceptance of the spectrometer from the locus of the scattered muon at the M hodoscope for various values Q^2 and ν is shown
in Figure 9.   The acceptance is limited by an inner hole in the beam
region and by the outer bounds of the hodoscope.   Contours of constant acceptance for the various runs are shown in Figure 10.

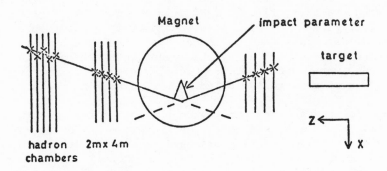

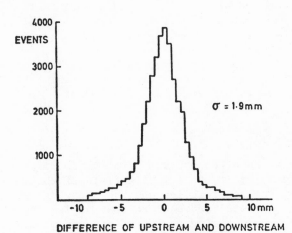

DIFFERENCE OF UPSTREAM AND DOWNSTREAM
TRACK IMPACT PARAMETERS

Fig. 7.   Method of reconstructing events using measurements outside
the magnet pole by application of Störmer's Theorem.   The
matching of impact parameters showed a standard deviation
of 1.9 mm.

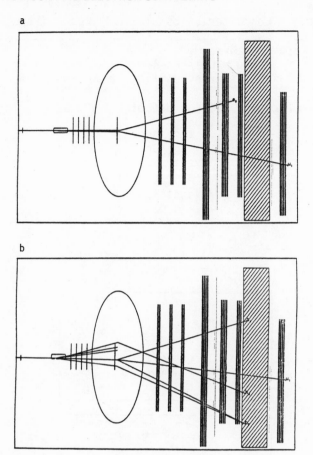

Fig. 8.   Example of a reconstructed event:
          a) an elastic μ-e scatter
          b) a deep inelastic μ-d scatter showing hadrons produced.

ENERGY CALIBRATION

     A check on the energy calibration and the resolution of the
spectrometer was obtained by examining the fairly large number of
μ-e elastic scatters that were collected. Except for radiative
effects, the sum of the muon and electron energies should equal the
energy of the beam muon. In Figure 11 the difference is plotted.
The curve centers on zero as it should and shows a standard deviation
of 2.3 GeV. A summary of the characteristics of the spectrometer is
given in Table 1.

Table 1. Characteristics of the Muon Spectrometer

1. Beam Chambers MWPC
       $\Delta p/p$ = 0.1% at 150 GeV/c
       Transverse Position at Target ± 0.5 mm

2. Chicago Cyclotron Magnet
       At 15 Kilogauss
           P_t = 2.25 GeV/c
           σ_x = 0.3-0.5 mm
           σ_p/p = 1.4 x 10^{-4} p   p in GeV/c
           σ_θ = 0.32 millirad
       From μe Events
           σ_{Q^2} = 0.014 $(GeV/c)^2$

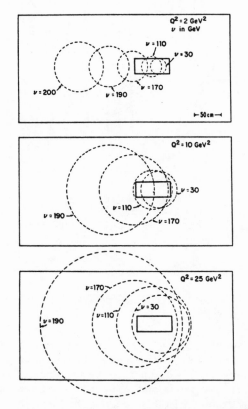

Fig. 9. The locus of the scattered muon at the M hodoscope for
         various values of Q^2 and ν, for 219 GeV incident muons,
         showing the fraction that come within the sensitive area.

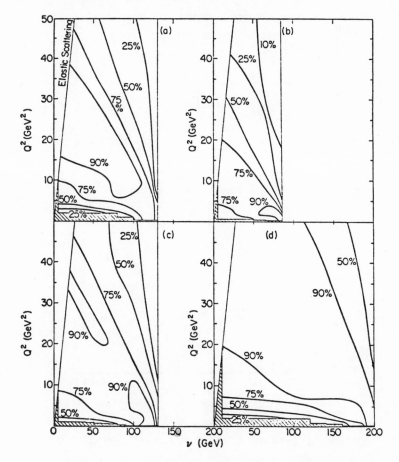

Fig. 10. Contours of constant acceptance for a) 1974, 147 GeV data,
          b) 96 GeV data, c) 1975, 147 GeV data, and d) 219 GeV data.

DETECTION EFFICIENCY

     We mention two of the problems that arose in evaluating the event
reconstruction efficiency. The first is due to the difficulty in
making a correct track identification in the region of the beam using
the downstream chambers. This effect was measured by taking data
from a real event and planting an extra track by a Monte Carlo method
which simulated known inefficiencies and spark spreads. The success
rate in finding the planted track measured the efficiency. Figure 12
shows this efficiency as a function of position across the 6m x 2m
chambers. The maximum inefficiency reaches 30%. By parametrizing
these results a correction could be made leaving a residual error of
up to 6% for low ν events, the worst affected.

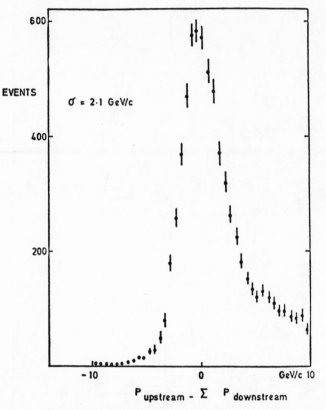

Fig. 11. Energy balance in μ-e events, an overall check on the
          energy calibrations and the resolution of the spectrometer.

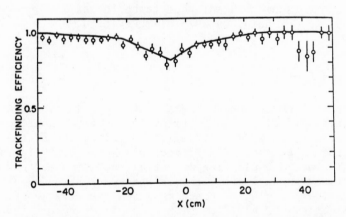

Fig. 12. Trackfinding efficiency across the 6 m chambers for the
          219 GeV data. Solid line is the parametrization used in
          correcting for this effect.

The second effect is the "suicide" correction. It arises when
a scattered muon inside the acceptance of the spectrometer produces
a δ-ray or an electromagnetic shower on emerging from the hadron
shield which strikes the K hodoscope and vetoes itself. The effect
was quantified by imbedded beam triggers. An imbedded beam trigger
is an event requiring only the B trigger but which is activated only
once per 10^6 beam muons. Its main purpose was to give an unbiased
sample of the phase space of the beam and the beam tagging efficiency. This effect could cause inefficiencies up to 10%, but when
corrected for the systematic error due to the uncertainty in this
correction was about 1%.

Other kinematic corrections were applied before extracting the
cross-sections from the number of events collected in each bin.
These were: 1) background subtraction obtained from target empty
running, 2) the geometric acceptance obtained by Monte Carlo simulation of the events, and 3) the radiative corrections made by
calculating the shift in the spectral distribution of scattered muons
by radiative processes either before or after the scattering.
A summary of the run statistics and data collected is given in Table 2.

MEASUREMENT OF R FOR THE PROTON

The extraction of the structure function $F_2(x, Q^2)$ from the
measured cross-section is done using Equation (3). This requires a
knowledge of R.

Table 2. Summary of the Data Collected

Incident Energy (year, target)	Total Incident Muons	Analyzed Events $Q^2 > 0.2$ GeV^2	Analyzed Events $Q^2 > 1$ GeV^2
147 GeV (1974, deuterium)	2.04×10^{10}	1.19×10^4	7.31×10^3
96 GeV (1975, hydrogen)	2.46×10^{10}	1.50×10^4	6.53×10^3
147 GeV (1974, hydrogen)	2.37×10^{10}	0.67×10^4	3.06×10^3
147 GeV (1975, hydrogen)	2.66×10^{10}	1.09×10^4	6.77×10^3
219 GeV (1976, hydrogen)	7.44×10^{10}	3.61×10^4	19.89×10^3

In the parton model, in the approximation that neglects the momentum transverse to the collision axis, the Callan-Gross relation[4] sets $1/x \ F_2(x,Q^2)/F(x,Q^2) = 1$ for spin 1/2 partons. If the transverse momentum of the parton P_T is taken into account a further contribution should be added.

$$R_T = 4 \ <P_T^2>/Q^2 , \qquad\qquad (5)$$

Other contributions coming from higher order QCD effects have been estimated. From the experimental point of view R can be extracted from measurements of the cross-section at the same values (Q^2,x) obtained at different incident muon energies. In the present experiments, where the bulk of the data was obtained with fairly small values of θ, the cross-section is largely determined by F_2 and is rather insensitive to R. Thus, while it is difficult to determine R well, the values of F_2 can be obtained with good accuracy.

To minimize the effect of systematic difference between runs at different energies, normalization factors were introduced for each energy which minimized the χ^2 in the cross-section comparison over those regions in (Q^2,x) which were most insensitive to R. When this was done for some of the data with intermediate values of x $(0.02 < x < 0.09)$, the normalization factors required of the 96 GeV and the 147 GeV data were 1.01 ± 0.055 and 1.01 ± 0.045 relative to the 219 GeV data. Thus the three data sets were multiplied by 1.01, 1.01, and 1.00 for the three energies, respectively.

In this experiment, the cross-section in the region x > 0.1 is quite insensitive to the value of R. In addition, muons having x > 0.1 were often close to the beam and the beam veto and were therefore more subject to systematic effects than the lower x data; these data were not used to evaluate R.

An example of the stability of the results in the face of changing systematics is shown in Figure 13, where the differing acceptances and radiative corrections at the three different energies are shown in conjunction with the extracted structure functions.

To determine R from the present data, the full overlap region was used and the deviations minimized by varying R using a χ^2 test. Various forms of R were used. Using R = constant the value R = $0.52^{+0.17}_{-0.15}$ was found with $\chi^2 = 168$ for 157 degrees of freedom. The errors indicated are statistical only. If the normalizations applied are allowed to change by their measured errors the resulting change in R is $^{+0.24}_{-0.20}$. Other uncertainties in the acceptance lead to a total estimated error, $\delta R = 0.35$.

If the data are split into three Q^2 ranges and the data available
in each range are averaged over x, the Q^2 variation shown in Figure 14
is obtained. It should be noted that the x range, while different for
each point, is confined to x < 0.1. The average x increases with Q^2.
Two fits are shown: a) $R = 1.20(1-x)/Q^2$ with Q^2 in GeV^2 [5], and b) $R =
1.18(1-x)/\ln(Q^2/\Lambda^2)$ with $\Lambda = 0.5$ GeV. These fits were no better than
$R = 0.52$, a constant.

The deviation of the data from the Callan-Gross relation is only
1-1/2 standard deviations. However, since the data does not clearly
distinguish any of the functional forms tried from R = constant and
since the muon data at x > 0.1 are quite insensitive to R, the structure function F_2 was extracted from the cross-sections using R = 0.52

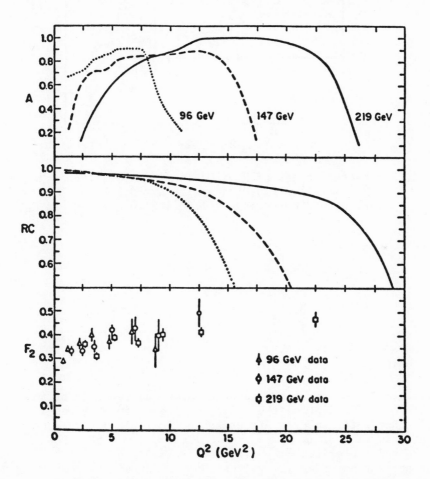

Fig. 13.   Values of the acceptance A, the inelastic radiative corrections RC, and the extracted structure function F_2 versus
Q^2 for $9 < \omega < 20$ for the 3 different incident energies.

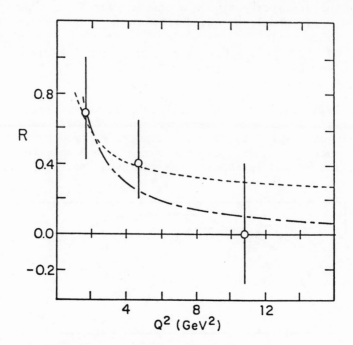

Fig. 14.   Extracted values of R vs Q^2 for x < 0.1. The dashed curve
           is $R = 1.18(1-x)/\ln(Q^2/\Lambda^2)$ with $\Lambda = 0.5$ GeV. The long-
           short dashed curve is $R = 1.20(1-x)/Q^2$ with Q^2 in GeV^2.

for all the muon data. There is no disagreement here with the
SLAC-MIT value[7] R = 0.21 ± 0.10 since this applies to the region
x > 0.2.

THE STRUCTURE FUNCTION $F_2(x,Q^2)$

     Values of the structure function $F_2(Q^2,x)$ for each of the three
muon energies incident on the hydrogen target, for the combined data,
and for the deuterium data are presented in Table 3. This data was
evaluated in bins of ($\omega = 1/x$ and Q^2) using R = 0.52 for both
hydrogen and deuterium. The values of $F_2(x,Q^2)$ are considered to
be the correct experimental values for a point at the center of the
bin. For ω large enough, the variation of F_2 over the bin is smooth
and linear enough that F_2 averaged over the bin, which is what is
measured, is almost the same as at the center. However, for $\omega < 5$,

Table 3. Values of $F_2(Q^2,x)$ Using R = 0.52

ω range	$\langle x \rangle$	$\langle Q^2 \rangle$ (GeV²)	F_2^D (147 GeV) 1974	F_2^P (96 GeV) 1975	F_2^P (147 GeV) 1974	F_2^P (147 GeV) 1975	F_2^P (219 GeV) 1976	F_2^P Combined

Table 3.   Values of $F_2(Q^2,x)$ Using R = 0.52

ω range	$\langle x \rangle$	$\langle Q^2 \rangle$ (GeV²)	F_2^D (147 GeV) 1974		F_2^P (96 GeV) 1975		F_2^P (147 GeV) 1974		F_2^P (147 GeV) 1975		F_2^P (219 GeV) 1976		F_2^P Combined	
	.16667	5.500	.628	.078	.380	.073	.309	.065	.274	.063	.362	.031	.336	.024
	.16667	6.500	.574	.083	.370	.088	.192	.047	.283	.073	.355	.035	.301	.028
	.16667	7.500	.506	.087	.346	.096	.187	.055	.338	.088	.322	.037	.288	.025
	.16667	9.000	.785	.069	.330	.080	.323	.057	.270	.075	.383	.033	.361	.025
	.16667	12.500	.056	.063	.432	.082	.306	.040	.273	.059	.357	.029	.339	.024
	.16667	22.500	.480	.060	.774	.450	2.165	.733	.293	.105	.491	.062	.508	.061
	.12500	1.375	.394	.113	.201	.060	.287	.295	.364	.063			.269	.052
	.12500	1.625	.262	.091	.349	.044	.408	.075	.306	.057			.354	.036
	.12500	2.125	.653	.160	.383	.054	.346	.107	.403	.073			.347	.039
	.12500	2.375	.641	.105	.445	.067	.480	.091	.368	.099			.453	.049
	.12500	2.625	.547	.092	.326	.065	.349	.075	.324	.097			.358	.047
	.12500	3.750	.583	.073	.379	.079	.298	.057	.363	.088			.241	.034
	.12500	4.500	.629	.087	.241	.070	.336	.072	.363	.069	.372	.091	.366	.040
	.12500	5.500	.728	.107	.318	.070	.364	.088	.405	.088	.334	.066	.389	.030
	.12500	7.500	.816	.129	.414	.090	.380	.075	.421	.083	.357	.058	.340	.033
	.12500	9.000	.618	.092	.356	.068	.298	.057	.405	.071	.392	.042	.335	.026
	.12500	12.500	.071	.092	.530	.084	.336	.072	.424	.086	.335	.045	.327	.032
	.12500	22.500	.628	.096	.417	.121	.364	.088	.448	.113	.336	.030	.393	.031
	.12500				.455	.114	.380	.075	.423	.120	.356	.034	.432	.033
	.12500				.303	.134	.543	.134	.440	.102	.263	.050	.427	.027
	.12500				(2.533 1.637)		(2.242 2.312)		.436	.103	.424	.036	.533	.100
	.10000	.950	.545	.151	.492	.247	.613	.173	.289	.062	.533	.100	.492	.247
	.10000	1.375	.509	.104	.258	.067	.276	.107	.440	.076			.258	.067
	.10000	1.625	.691	.178	.339	.047	.473	.133	.343	.068	.508	.086	.316	.034
	.10000	1.875	.645	.116	.328	.054	.246	.094	.306	.099	.265	.057	.318	.040
	.10000	2.125	.721	.101	.303	.072	.289	.110	.411	.098	.338	.065	.400	.042
	.10000	2.375	.601	.088	.408	.076	.550	.063	.438	.064	.425	.070	.294	.043
	.10000	2.625	.534	.069	.273	.115	.332	.110	.457	.086	.394	.047	.254	.038
	.10000	3.750	.651	.114	.193	.100	.315	.063	.319	.072	.380	.048	.398	.046
	.10000	4.500	.633	.102	.529	.078	.278	.079	.347	.062	.363	.036	.372	.034
	.10000	5.500	.705	.096	.293	.099	.271	.088	.360	.042	.191	.039	.367	.035
	.10000	9.000	1.316	.236	.352	.153	.427	.061	.514	.122	.396	.035	.385	.030
	.10000	12.500			.381	.108	.299	.088	.522	.126	.348	.037	.231	.032
	.10000	22.500			.946	.176	.752	.061	.522	.095	.427	.045	.401	.036
	.10000				.301	.262				.150		.043	.345	.030
	.10000				.471	(0.258 3.628)		.213					.250	.041
	.10000				.258	.262							.262	.063
	.08000	.650	.645	.153	.299	.063	.253	.100	.281	.055	.322	.058	.296	.040
	.08000	.760	.544	.141	.224	.040	.425	.134	.236	.050	.356	.059	.295	.036
	.08000	.950	.661	.140	.285	.052	.278	.101	.492	.074	.467	.066	.278	.032
	.08000	1.125			.302	.052	.188	.078	.358	.068			.361	.046
	.08000	1.375			.334	.063			.516	.071			.305	.037
	.08000	1.625			.374	.084			.293	.091			.319	.041
	.08000	1.875			.525	.094			.297	.079			.353	.040
	.04000				.403	.113								
	.04000				.443									
	.04000				.478									

Table 3. Values of $F_2(Q^2,x)$ Using R = 0.52

ω range	$\langle x \rangle$	$\langle Q^2 \rangle$ (GeV^2)	F_2^D (147 GeV) 1974	F_2^P (96 GeV) 1975	F_2^P (147 GeV) 1974	F_2^P (147 GeV) 1975	F_2^P (219 GeV) 1976	F_2^P Combined
1-1.4	.08000	3.250	.669 .103	.363 .076	.224 .063	.452 .079	.258 .034	.286 .026
1-1.4	.08000	3.750	.728 .112	.614 .112	.381 .083	.379 .083	.299 .038	.345 .031
1-1.4	.08000	4.500	.618 .078	.344 .072	.412 .063	.373 .069	.363 .035	.401 .026
1-1.4	.08000	6.500	.659 .104	.377 .092	.314 .063	.590 .107	.406 .044	.371 .029
1-1.4	.08000	7.500	.770 .129	.485 .126	.415 .080	.509 .114	.430 .052	.424 .037
1-1.4	.08000	12.500	.947 .101	.217 .099	.244 .072	.413 .120	.418 .043	.410 .033
1-1.4	.08000	22.500	.767 .096	.324 .111	.459 .089	.334 .094	.448 .049	.262 .049
1-1.4				1.710 .728	.449 .072	.342 .116 .617)	.519	.361 .038
2-2	.05882	5.750		.262 .032		.452 .079		.265 .036
2-2	.05882	6.750		.361 .038		.272 .045		.322 .041
2-2	.05882	7.750		.285 .036		.420 .056		.309 .025
2-2	.05882	1.125	.141	.326 .041		.367 .052		.371 .031
2-2	.05882	1.375	.135	.349 .037		.444 .071		.362 .032
2-2	.05882	1.625	.127	.364 .052		.204 .051		.305 .032
2-2	.05882	2.125	.089	.317 .055		.287 .077		.352 .030
2-2	.05882	2.675	.092	.420 .067		.253 .049		.387 .022
2-2	.05882	3.750	.094	.359 .065		.383 .070		.241 .021
2-2	.05882	4.500	.092	.340 .071		.434 .061		.386 .021
2-2	.05882	5.000	.114	.467 .082		.405 .069		.379 .033
2-2	.05882	7.500	.090	.508 .112		.436 .089		.379 .030
2-2	.05882	12.500	.156	.276 .112		.479 .111		.472 .031
2-2	.05882	22.500	1.003	.620 .277		.603 .086 2.736)		
2-3.5	.03036	.650	.368 .088	.238 .024	.429 .100	.272 .045	.376 .042	.238 .024
2-3.5	.03036	.650	.490 .093	.238 .025	.343 .102	.420 .056	.352 .039	.238 .025
2-3.5	.03036	.650	.514 .101	.293 .028	.301 .076	.367 .041	.398 .041	.293 .029
2-3.5	.03036	1.375	.757 .101	.279 .028	.375 .060	.469 .048	.383 .030	.377 .025
2-3.5	.03036	1.625	.767 .070	.358 .036	.333 .050	.500 .072	.382 .023	.342 .029
2-3.5	.03036	2.125	.808 .057	.344 .047	.387 .048	.357 .048	.478	.338 .028
2-3.5	.03036	2.675	.806 .060	.336 .053	.461 .040	.405 .056	.531 .036	.384 .028
2-3.5	.03036	3.750	.894 .066	.309 .048	.476 .062	.357	.389 .023	.377 .011
2-3.5	.03036	4.500	.750 .133	.401 .068	.521 .125)	.571 .088	.478 .036	.457 .025
2-3.5	.03036	5.000	.581 .105)	.413 .116	.662 .560)	.360 .094	.531 .036	.523 .029
2-3.5	.03036	7.500		.521 .271)	1.395	.716 .153)	.414 .031	.422 .031
2-3.5	.03036	9.000		.653 .714)		1.894 .847)	.529 .037	.414 .031
2-3.5	.03036	12.500					1.219 .271)	.529 .037
3.5-4.5	.02500	.650		.266 .038		.366		.266 .038
3.5-4.5	.02500	.650		.298 .042		.350 .064		.298 .045
3.5-4.5	.04500	.850		.181 .035				.181 .048
3.5-4.5	.04500	1.125		.322 .037				.329 .032

Table 3.   Values of $F_2(Q^2,x)$ Using R = 0.52

ω range	$\langle x \rangle$	$\langle Q^2 \rangle$ (GeV²)	F_2^D (147 GeV) 1974	F_2^P (96 GeV) 1975	F_2^P (147 GeV) 1974	F_2^P (147 GeV) 1975	F_2^P (219 GeV) 1976	F_2^P Combined
35–45	.02500	1.375	.744 .189					.367 .038
35–45	.02500	1.625	.480 .129	.405 .050				.425 .045
35–45	.02500	1.875	.605 .134	.334 .070				.398 .044
35–45	.02500	2.375	.436 .099	.382 .074				.369 .039
35–45	.02500	2.875	.958 .120	.345 .079	.300 .134	.317 .058	.377 .060	.348 .039
35–45	.02500	3.750	.958 .126	.368 .092	.293 .118	.316 .064	.363 .059	.362 .031
35–45	.02500	4.500	.958 .120	.232 .084	.458 .092	.504 .087	.356 .052	.362 .036
35–45	.02500	4.500		.370 .083	.368 .117	.455 .089	.405 .052	.475 .032
35–45	.02500	5.500		.261 .122	.348 .080	.360 .084	.509 .045	.449 .040
35–45	.02500	7.500	(1.023) .167	.714 .405	.448 .083	.287 .070	.467 .054	.450 .054
35–45	.02500	9.000	(.865) .277		.325 .165	.560 .089	.494 .054	.562 .063
30–40	.01905	.750		.123 .024	.572 .226	.491 .110	.456 .054	.123 .072
30–40	.01905	.850		.207 .033		.696 .293	.562 .072	.257 .033
30–40	.01905	.950		.257 .049				.318 .049
30–40	.01905	1.375	.824 .163	.328 .040		.373 .059	.254 .045	.337 .031
30–40	.01905	1.625	.622 .109	.316 .048	.280 .093	.331 .055	.412 .057	.388 .036
30–40	.01905	1.875	.816 .107	.342 .050	.466 .102	.323 .060	.326 .045	.363 .038
30–40	.01905	2.125	.636 .036	.302 .055	.484 .094	.358 .063	.426 .042	.326 .041
30–40	.01905	2.375	1.036 .011	.405 .074	.471 .088	.561 .078	.365 .037	.333 .037
30–40	.01905	2.625	1.068 .176	.400 .083	.371 .057	.407 .073	.407 .046	.410 .037
30–40	.01905	2.875		.355 .091	.189 .157	.324 .077	.513 .061	.301 .032
30–40	.01905	3.250		.420 .134	.366 .126	.577 .100	.591 .120	.419 .038
30–40	.01905	3.750		.480 .185		.638 .112		.402 .034
30–40	.01905	4.500				.945 .522		.477 .034
30–40	.01905	5.500						.513 .061
30–40	.01905	6.500						.516 .061
30–40	.01905	7.500						
40–60	.01429	.750	.259 .193	.256 .034	.611 .312	.296 .304	.333 .049	.256 .034
40–60	.01429	.850	.472 .102	.336 .043	.226 .054	.311 .050	.377 .050	.286 .044
40–60	.01429	.950	.031 .065	.280 .044	.298 .059	.323 .052	.326 .053	.262 .044
40–60	.01429	1.125	.509 .011	.265 .031	.408 .077	.419 .062	.402 .063	.271 .031
40–60	.01429	1.175	.486 .117	.283 .039	.438 .113	.319 .060	.416 .039	.297 .045
40–60	.01429	1.625	(1.376) .256	.339 .062	.728 .194	.565 .090	.454 .037	.372 .040
40–60	.01429	1.875		.554 .103		.436 .097	.487 .037	.396 .039
40–60	.01429	2.125		.453 .161		.539 .224	.611 .078	.318 .039
40–60	.01429	2.375		(1.033) 1.364		(1.733) 1.364		.341 .039
40–60	.01429	2.625				(1.343)		.392 .034
40–60	.01429	2.875						.487 .034
40–60	.01429	3.250						.454 .037
60–80	.01000	.650	.000 .120	.264 .031	.405 .098	.237 .124	.333 .049	.264 .031
60–80	.01000	.750	.726 .078	.318 .036	.341 .068	.371 .043	.377 .050	.318 .035
60–80	.01000	.950	.063 .009	.285 .034	.283 .056	.345 .044		.319 .035
60–80	.01000	1.125		.352 .042		.356 .044		.384 .027
60–80	.01000	1.625		.424 .075		.420 .052		.356 .037
60–80	.01000	1.875		.394 .165				.357 .036

Table 3. Values of $F_2(Q^2,x)$ Using R = 0.52

ω range	<x>	<Q²> (GeV²)	F_2^D (147 GeV) 1974	F_2^P (96 GeV) 1975	F_2^P (147 GeV) 1974	F_2^P (147 GeV) 1975	F_2^P (219 GeV) 1976	F_2^P Combined
80 - 120	.01000	2.125	.739 ± .083		.440 ± .086	.422 ± .063	.404 ± .039	.413 ± .031
80 - 120	.01000	2.375	.857 ± .124		.436 ± .125	.576 ± .101	.447 ± .041	.463 ± .036
80 - 120	.01000	2.625	1.026 (± .195)		.399 ± .232	.458 (± .129)	.461 ± .041	.461 ± .041
60 - 120	.01000	3.250	1.442 (± .321)		.667 (± .243)	.338 (± .205)	.432 ± .033	.436 ± .033
80 - 120	.01000	3.750					.486 ± .053	.488 ± .053
80 - 120	.01000	4.500					.690 ± .111	
120 - 160	.00714	.6500	.577 ± .073	.232 ± .031	.323 ± .054	.457 ± .181	.431 ± .055	.232 ± .031
120 - 160	.00714	.7500	.600 ± .000	.211 ± .033	.352 ± .041	.222 ± .000	.326 ± .036	.270 ± .032
120 - 160	.00714	.9500	.736 ± .084	.295 ± .043	.362 ± .082	.312 ± .043	.415 ± .041	.377 ± .035
120 - 160	.00714	1.125	.908 ± .106	.437 ± .053	.497 ± .290	.425 ± .047	.431 ± .036	.382 ± .048
120 - 160	.00714	1.625	1.147 (± .306)	.205 (± .299)	1.345 (± 1.423)	.304 ± .054	.444 ± .057	.358 ± .036
120 - 160	.00714	1.875				.634 ± .171	.466 ± .095	.415 ± .041
120 - 160	.00714	2.125						.444 ± .057
120 - 160	.00714	2.375						
120 - 160	.00714	2.625						
160 - 240	.00500	.4500	.522 ± .063	.240 ± .034	.694 ± .734	.347 ± .049	.371 ± .051	.240 ± .034
160 - 240	.00500	.5500	.511 ± .045	.282 ± .027	.418 ± .088	.287 ± .044	.401 ± .026	.280 ± .027
160 - 240	.00500	.6500	.738 (± .066)	.261 ± .039	.131 ± .087	.358 ± .041	.412 ± .039	.273 ± .037
160 - 240	.00500	.7500	.908 (± .140)	.382 ± .069	.302 ± .039	.336 ± .033	.490 ± .067	.369 ± .037
160 - 240	.00500	.8500		.414 (± .149)	.411 ± .118	.310 ± .077		.299 ± .023
160 - 240	.00500	.9500			1.642 (± 1.830)	.419		.401 ± .026
160 - 240	.00500	1.125					.496 ± .105	.490 ± .039
240 - 400	.00312	.3500	.384 ± .063	.213 ± .026	.138 ± .094	.274 ± .044	.474 ± .043	.213 ± .029
240 - 400	.00312	.4500	.465 ± .045	.224 ± .019	.181 ± .055	.283 ± .027	.322 ± .022	.263 ± .027
240 - 400	.00312	.5500	.538 ± .065	.290 ± .039	.192 ± .057	.301 ± .039	.411 ± .022	.281 ± .032
240 - 400	.00312	.6500	.738 (± .066)	.202 (± .110)	.203 ± .065	.322 ± .065	.502 ± .040	.326 ± .043
240 - 400	.00312	.7500	.908 (± .140)		.230 ± .183	.479 ± .113		.322 ± .022
240 - 400	.00312	1.125			2.162 (± 2.182)	.247 (± 1.424)		.411 ± .022
240 - 400	.00312	1.375						
400 - 600	.00200	.3500	.442 ± .035	.155 ± .013	.149 ± .049	.240 ± .018	.289 ± .025	.156 ± .017
400 - 600	.00200	.4500	.434 ± .030	.220 ± .027	.165 ± .035	.178 ± .021	.253 ± .021	.238 ± .018
400 - 600	.00200	.5500	.429 (± .068)		.269 ± .007	.264 ± .030	.208 ± .042	.289 ± .025
400 - 600	.00200	.6500			1.215 (± 1.005)	.168 (± .200)	.414 (± .312)	.253 ± .017
400 - 600	.00200	.7500				.814 (± 1.200)	.966	.268 ± .021
600 - 1000	.00125	.2500	.320 ± .022	.155 ± .015	.153 ± .021	.211 ± .011	.200 ± .015	.199 ± .015
600 - 1000	.00125	.3500	.421 ± .036	.220 ± .021	.043 ± .043	.220 ± .023	.276 ± .017	.200 ± .017
600 - 1000	.00125	.4500		.164 (± .021)	.582 (± .582)	.912 (± 1.458)	.314 ± .042	.276 ± .017
600 - 1000	.00125	.5500					.587 (± .217)	
1000 - 1400	.00063	.2500	.336 (± .047)	.164 (± .021)	.157 (± .049)	.229 (± .029)	.392 ± .018	.392
1000 - 1400	.00063	.3500					.401 ± .022	

F_2 varies considerably over a bin. Moreover, for the bins with ω near 1, part of the bin might be in an unphysical region and appropriate corrections were made to obtain bin centered values. For the largest ω bin for each Q^2, some parts of the bin were inaccessible, kinematically. The values are those obtained from averages over the portion of the bin where events were found. In Table 3, values obtained where the bin center is not kinematically accessible are shown in brackets. Care should be exercised in using such values. They were omitted in carrying out the moment analysis.

Figure 15 shows the combined hydrogen data as a function of Q^2 for various bands of ω. The horizontal bars shown at large ω and Q^2 indicate the effects of changing R by its standard deviation, to 0.69 and 0.37. Some of the SLAC-MIT ep data are also shown in the figures

The data show a pattern of Q^2 dependence that has now become familiar as characteristic of QCD. For $3 < \omega < 5$ F_2 shows a flat Q^2

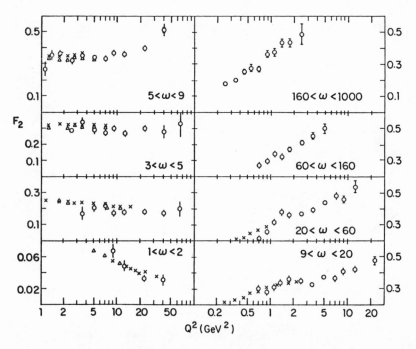

Fig. 15. F_2 for hydrogen as a function of Q^2 for various ω bins, using R = 0.52, showing the combined Fermilab muon data (circles), the MIT-SLAC electron data (triangles)(Ref. 8) and the SLAC electron data (crosses)(Ref. 8). The errors shown are statistical only. Horizontal bars indicate the effect of changing R to 0.69 and 0.37. Note the difference of scale and suppressed zeros on the ordinates.

dependence. This flat behavior, found originally by the MIT-SLAC
experimenters, was the primary evidence for scaling, and the precipi-
tating cause of the parton model. Here we see that the new μp data
extends the range over which F_2 is seen to be Q^2 independent from
8 to 40 GeV^2. However, for lower values of ω, e.g., for $2 < \omega < 3$
the trend of F_2 with Q^2 is decreasing, while for the large values of
ω the trend is increasing, the increase becoming more pronounced the
larger ω becomes. In judging these trends from the point of view of
QCD, the data with the largest Q^2 carry the most weight. Only in
perturbative QCD can the analysis be made quantitative. The restriction here is that $Q^2 \gg M^2$. In the moment analysis given below,
only the data with $Q^2 \geq 3.0$ GeV^2 is used.

The same pattern has been found for F_2 from the neutrino experiments. This is seen in Figure 16 which gives the data published by
the ABCLOS group[9] and in Figure 17 for the CDHS group[10].

Moments of F_2

To see to what extent the deviations from scaling, made evident
in these experiments, correspond to the predictions of QCD, it is
useful to consider the moments,

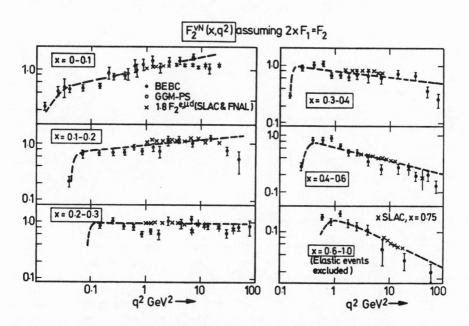

Fig. 16. Values of F_2 determined from the neutrino measurements in
the bubble chamber by the ABCLOS group (Ref. 9).

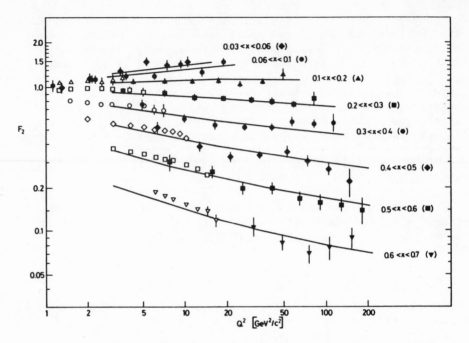

Fig. 17.   Values of F_2 determined from the neutrino measurements
           using the drift chamber spectrometer of the CDHS group.

$$I_i(n,Q^2) = \int_0^1 x^{n-2} F_i(x,Q^2) dx \qquad (5)$$

rather than the structure functions themselves. This is because in
QCD the moments are given by a sum of terms which vary as negative
powers of $\ln(Q^2/\Lambda^2)$ as Q^2 becomes large. Here Λ is a scale parameter
which must be determined from the data. These powers are the anoma-
lous dimensions given explicitly by the theory.

It is important to emphasize that the theory we use is better
referred to as perturbative QCD and that the predictions can be made
cleanly in leading order in the theory. To the extent that higher
order corrections are important, the theory is less explicit and
ambiguities appear. The restriction to Q^2 large is intended to limit
the effect of the higher order corrections. In practice, the present
data are mainly below 50 GeV^2. In the analysis presented here the
data in the range $3 \geq Q^2 \geq 50$ GeV^2 was chosen to keep Q^2 as high as
possible while retaining a statistically ample data sample over a
wide enough Q^2 range. With this choice the higher order corrections
are present but in manageable proportions, as will be discussed
below.

In this range of Q^2 one class of effects can be corrected for by using the Nachtmann moments[11] in place of Equation (5). The Nachtmann form for the moments was obtained by projecting out of the Wilson operator product expansion all the contributions to a given spin n. These are the so-called target mass effects. Thus, the leading order expressions in QCD become more correct if they are used with the Nachtmann moments given by

$$M_2(n,Q^2) = \int_o^1 x^{n-2} N_n(x,Q^2) F_2(x,Q^2) dx$$

$$N_n = [2/(1+\sqrt{1+Q^2/\nu^2})]^{n-1}(1-M^4\xi^4/Q^4)(1+3\eta_n)(1+Q^2/\nu^2)^{1/2}$$

$$\xi = 2x/(1+\sqrt{1+Q^2/\nu^2}) \tag{6}$$

$$\eta_n = [(n+1)M\nu\xi - (n+2)Q^2]/[(n+2)(n+3)(\nu^2+Q^2)]$$

To obtain the moments it is necessary to integrate the structure function over the full range of x, from 0 to 1. Since the QCD theorems we use are obtained[12] from the imaginary part of the Compton cross-section, the moment integrals are derived from a total cross-section and should include the elastic scattering contribution at x = 1.

The data for inelastic scattering over the full range of x may be obtained by combining the new Fermilab μp (and μd) data with the electron data obtained from the SLAC and MIT-SLAC experiments[8]. In Figure 18 we show the data and indicate the experiments from which they came. The data are plotted for given Q^2 corresponding to the bin centered values used for the muon results. For the electron results we converted each F_2 to the bin centered value Q_c^2 by assuming a Q^2 variation given by

$$F_2(Q^2,x) = F_2(Q_c^2,x) \left(\frac{Q^2}{Q_c^2}\right)^{b(x)}$$

with b(x) = 0.25 - x as given by Perkins[13]. Where the data points overlapped within the plotting resolution in x, an average was taken.

The plots show that the measurements of F_2 cover rather completely the full range of x. There are gaps here and there and bumps here and there and for Q^2 = 40 GeV^2, the data is quite sparse. Nevertheless it was a straightforward matter to obtain the moments by numerical integration. The contribution of the unmeasured portion from x_{min} to x = 0 and from x_{max} to the inelastic limit were

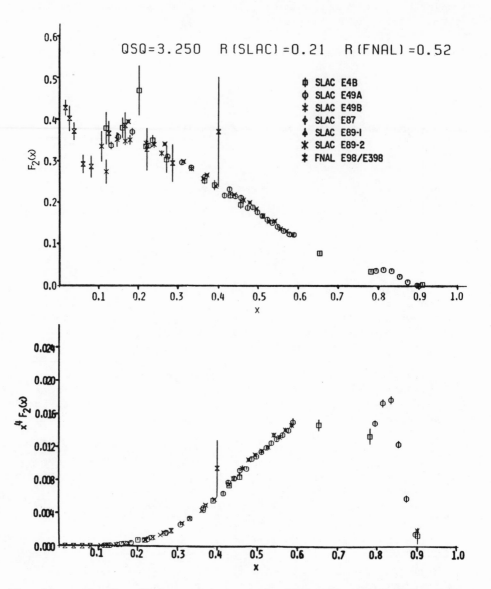

Fig. 18a.   The Fermilab μp data and the MIT-SLAC and SLAC ep data
             used to calculate the n = 2 and n = 6 moments of F_2 at
             Q^2 = 3.25 GeV^2. Upper curve $F_2(x)$ used to calculate the
             2nd moment, lower curve $x^4 F_2(x)$ used to calculate the
             6th moment.

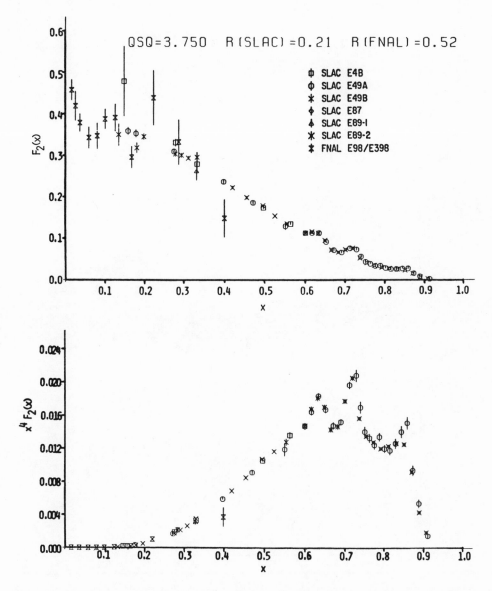

Fig. 18b. The Fermilab μp data and the MIT-SLAC and SLAC ep data used to calculate the n = 2 and n = 6 moments of F_2 at Q^2 = 3.75 GeV^2. Upper curve $F_2(x)$ used to calculate the 2nd moment, lower curve $x^4 F_2(x)$ used to calculate the 6th moment.

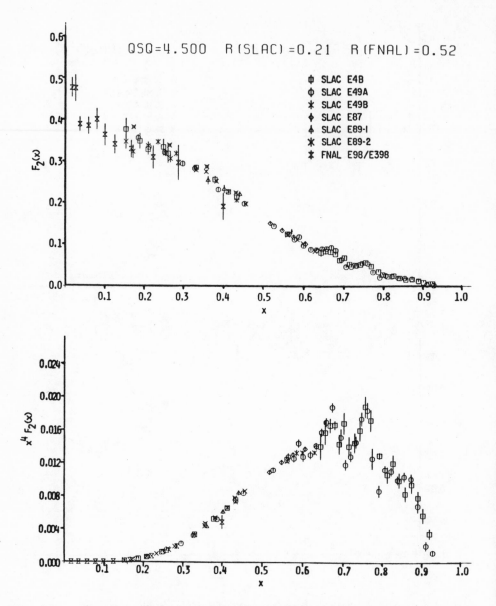

Fig. 18c.   The Fermilab μp data and the MIT-SLAC and SLAC ep data
used to calculate the n = 2 and n = 6 moments of F_2 at
Q^2 = 4.50 GeV^2. Upper curve $F_2(x)$ used to calculate the
2nd moment, lower curve $x^4 F_2(x)$ used to calculate the
6th moment.

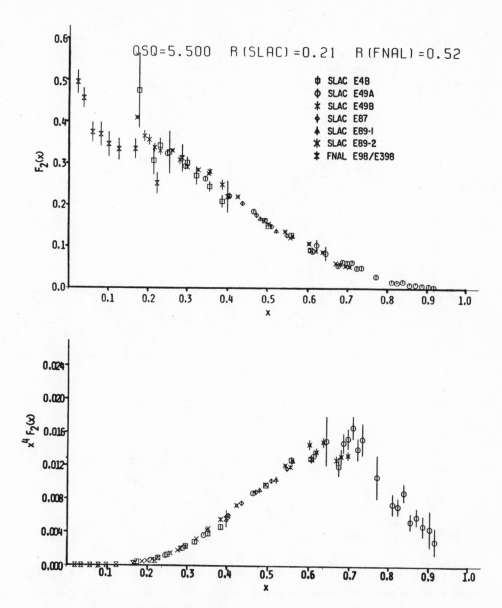

Fig. 18d.   The Fermilab μp data and the MIT-SLAC and SLAC ep data
            used to calculate the n = 2 and n = 6 moments of F_2 at
            Q^2 = 5.50 GeV^2. Upper curve $F_2(x)$ used to calculate the
            2nd moment, lower curve $x^4 F_2(x)$ used to calculate the
            6th moment.

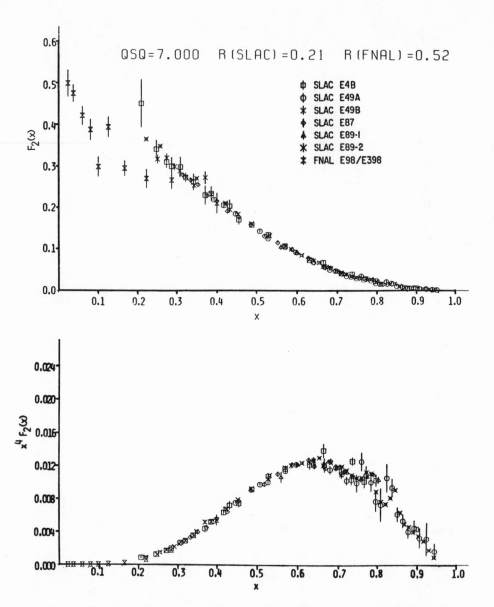

Fig. 18e. The Fermilab μp data and the MIT-SLAC and SLAC ep data
used to calculate the n = 2 and n = 6 moments of F_2 at
Q^2 = 7.00 GeV^2. Upper curve $F_2(x)$ used to calculate the
2nd moment, lower curve $x^4 F_2(x)$ used to calculate the
6th moment.

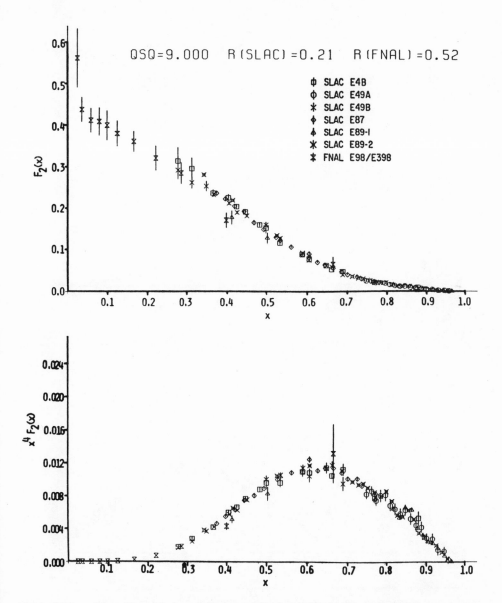

Fig. 18f.   The Fermilab µp data and the MIT-SLAC and SLAC ep data
used to calculate the n = 2 and n = 6 moments of F_2 at
Q^2 = 9.00 GeV^2. Upper curve $F_2(x)$ used to calculate the
2nd moment, lower curve $x^4F_2(x)$ used to calculate the
6th moment.

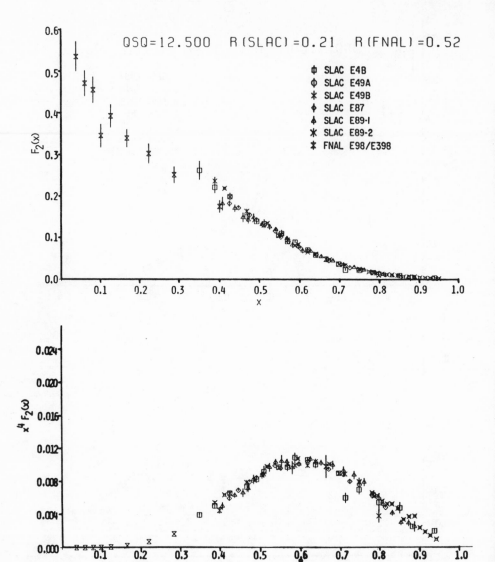

Fig. 18g.   The Fermilab μp data and the MIT-SLAC and SLAC ep data
            used to calculate the n = 2 and n = 6 moments of F_2 at
            Q^2 = 12.50 GeV^2. Upper curve $F_2(x)$ used to calculate
            the 2nd moment, lower curve $x^4 F_2(x)$ used to calculate
            the 6th moment.

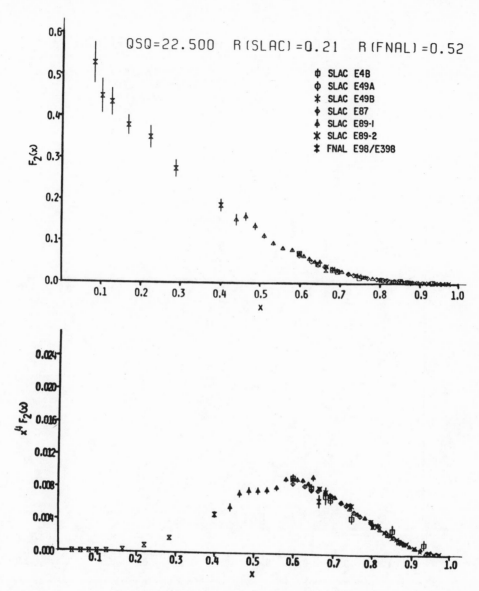

Fig. 18h. The Fermilab µp data and the MIT-SLAC and SLAC ep data
used to calculate the n = 2 and n = 6 moments of F_2 at
Q^2 = 22.50 GeV^2. Upper curve $F_2(x)$ used to calculate
the 2nd moment, lower curve $x^4 F_2(x)$ used to calculate
the 6th moment.

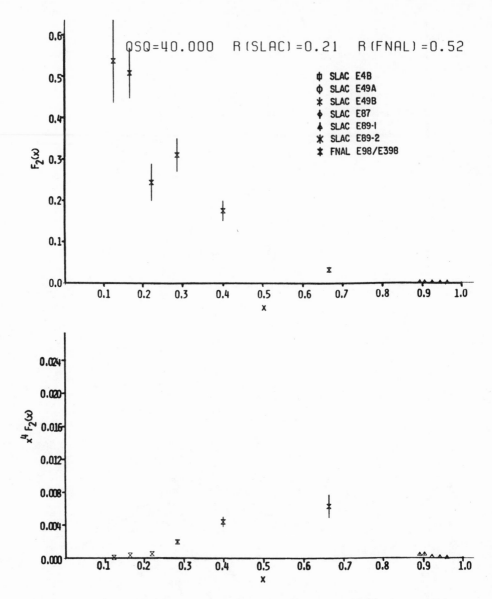

Fig. 18i. The Fermilab μp data and the MIT-SLAC and SLAC ep data used to calculate the n = 2 and n = 6 moments of F_2 at Q^2 = 40.00 GeV². Upper curve $F_2(x)$ used to calculate the 2nd moment, lower curve $x^4 F_2(x)$ used to calculate the 6th moment.

estimated by taking $dF_2/dx = 0$ at $x = 0$ and $F_2 \sim (1-x)$ near $x = 1$. These contributions were small compared with other errors.

In Figure 18 we also give plots of the function $x^4 F_2(x,Q^2)$ to show how the weight of the data shifts to higher x as n increases. In these plots the main contribution to the integral comes from the electron data. The bumps in the data, due to the excitation of the Δ and N^* resonances, are more prominent in the higher moments, especially in the low Q^2 plots.

The numerical integration procedure includes the effects of the nucleon resonances since there is electron data right through the resonances. However, it is not clear whether the procedure gives the resonances their correct weight. This question becomes more serious at high n and low Q^2. The contribution to the integral from elastic scattering was included using the dipole formula calculated assuming that[14]

$$G_M^P(Q^2)/\mu_P = G_E^P(Q^2) = G_M^N(Q^2)/\mu_N = (1+Q^2/0.71)^{-2}$$

$$G_E^N(Q^2) = 0$$

(7)

These contributions are tabulated in Table 4.

Table 4. Elastic Scattering Contribution to the Nachtmann Moments

ELASTIC CONTRIBUTION FOR HYDROGEN

Q**2 GEV**2	2	3	4	5	6	7	8	9	10
3.25	.003945	.003400	.002877	.002411	.002008	.001666	.001378	.001138	.000938
3.75	.002625	.002299	.001981	.001692	.001437	.001216	.001027	.000865	.000728
4.50	.001527	.001363	.001199	.001047	.000909	.000787	.000680	.000587	.000505
5.50	.000820	.000745	.000668	.000596	.000529	.000468	.000414	.000365	.000322
7.00	.000375	.000347	.000318	.000290	.000263	.000238	.000215	.000195	.000176
9.00	.000161	.000151	.000141	.000131	.000121	.000112	.000103	.000095	.000088
12.50	.000051	.000049	.000046	.000044	.000041	.000039	.000037	.000034	.000032
22.50	.000006	.000006	.000006	.000005	.000005	.000005	.000005	.000005	.000005
40.00	.000001	.000001	.000001	.000001	.000001	.000001	.000001	.000001	.000001

ELASTIC CONTRIBUTION FOR DEUTERIUM

Q**2 GEV**2	2	3	4	5	6	7	8	9	10
3.25	.005563	.004792	.004054	.003396	.002827	.002344	.001939	.001600	.001319
3.75	.003719	.003256	.002805	.002395	.002033	.001720	.001452	.001223	.001028
4.50	.002176	.001941	.001708	.001490	.001294	.001120	.000967	.000834	.000718
5.50	.001174	.001066	.000957	.000853	.000757	.000670	.000592	.000522	.000460
7.00	.000540	.000500	.000458	.000417	.000378	.000343	.000310	.000280	.000252
9.00	.000233	.000219	.000204	.000189	.000175	.000162	.000149	.000137	.000126
12.50	.000074	.000071	.000067	.000063	.000060	.000056	.000053	.000050	.000047
22.50	.000009	.000008	.000008	.000008	.000008	.000007	.000007	.000007	.000007
40.00	.000001	.000001	.000001	.000001	.000001	.000001	.000001	.000001	.000001

A tabulation of the Nachtmann moments for hydrogen and deuterium
is given in Table 5. The values of F_2 used were obtained from the
cross-sections using R = 0.21 for the electron data and 0.52 for the
muon data. The deuteron moments are given per deuteron (not per
nucleon) after correction for the Fermi motion. The elastic correction is included, obtained by summing the contribution of the neutron
and the proton, taken as free. The correction for Fermi motion in
deuterium was made[15] by dividing the inelastic part of the deuteron
moments by 0.985, 0.989, 1.000, 1.018, 1.043, 1.078, 1.119, 1.177,
1.252, for n = 2 through 10, respectively. These values were obtained
by using the formalism of Atwood and West[16] with the Reid hard core

Table 5.   The Nachtmann Moments of F_2 for Hydrogen and Deuterium Using
           R = 0.21 for the Electron Data, R = 0.52 for the Muon Data

NACHTMANN MOMENTS FOR HYDROGEN

Q**2 GEV**2	N 2	3	4	5	6	7	8	9	10
3.25	.186262	.053662	.022663	.011624	.006764	.004302	.002917	.002073	.001524
+-	.005102	.001477	.000667	.000391	.000269	.000201	.000158	.000126	.000102
3.75	.186974	.052362	.021904	.011167	.006446	.004056	.002718	.001910	.001393
+-	.005665	.001486	.000610	.000333	.000218	.000159	.000123	.000099	.000081
4.50	.184071	.051032	.021229	.010738	.006141	.003829	.002546	.001779	.001291
+-	.004883	.001315	.000556	.000296	.000183	.000127	.000095	.000075	.000061
5.50	.182328	.049092	.019867	.009769	.005432	.003297	.002141	.001465	.001047
+-	.005074	.001314	.000558	.000298	.000183	.000123	.000088	.000066	.000051
7.00	.179882	.046907	.018553	.008960	.004907	.002940	.001885	.001276	.000901
+-	.005276	.001261	.000481	.000231	.000128	.000079	.000053	.000039	.000030
9.00	.178386	.045097	.017478	.008298	.004461	.002619	.001644	.001088	.000753
+-	.005772	.001271	.000466	.000218	.000118	.000070	.000045	.000030	.000022
12.50	.172521	.042447	.016222	.007652	.004087	.002383	.001486	.000977	.000672
+-	.005498	.001270	.000456	.000205	.000107	.000062	.000039	.000026	.000019
22.50	.176594	.041830	.014749	.006634	.003438	.001953	.001187	.000760	.000508
+-	.007050	.001395	.000462	.000197	.000097	.000054	.000032	.000021	.000014
40.00	.184841	.043107	.013499	.005432	.002613	.001420	.000838	.000523	.000339
+-	.016888	.003251	.001168	.000605	.000363	.000231	.000150	.000099	.000066

NACHTMANN MOMENTS FOR DEUTERIUM

Q**2 GEV**2	N 2	3	4	5	6	7	8	9	10
3.25	.323238	.087463	.035764	.017913	.010219	.006376	.004252	.002973	.002153
+-	.009481	.002342	.001020	.000584	.000395	.000293	.000228	.000181	.000146
3.75	.315272	.082959	.033717	.016770	.009465	.005826	.003829	.002642	.001893
+-	.010124	.002321	.000928	.000495	.000318	.000229	.000177	.000141	.000115
4.50	.314586	.080999	.032038	.015537	.008572	.005170	.003337	.002266	.001603
+-	.009624	.002103	.000835	.000425	.000256	.000175	.000131	.000103	.000084
5.50	.317479	.079701	.030965	.014786	.008020	.004746	.003003	.001997	.001385
+-	.010726	.002144	.000843	.000431	.000256	.000169	.000119	.000089	.000069
7.00	.302205	.074342	.028288	.013293	.007089	.004118	.002556	.001668	.001134
+-	.010264	.002105	.000743	.000344	.000187	.000113	.000075	.000053	.000041
9.00	.293383	.073343	.027048	.012370	.006449	.003669	.002233	.001429	.000951
+-	.012515	.002378	.000752	.000326	.000169	.000097	.000061	.000041	.000029
12.50	.304917	.070884	.025455	.011509	.005937	.003333	.001994	.001248	.000811
+-	.015662	.002657	.000809	.000327	.000160	.000089	.000053	.000034	.000023
22.50	.333850	.072937	.024072	.010456	.005284	.002914	.001712	.001054	.000671
+-	.029112	.003915	.000960	.000363	.000166	.000085	.000048	.000030	.000020
40.00	.286310	.079770	.025735	.010062	.004761	.002586	.001529	.000946	.000598
+-	.063503	.015225	.004480	.001994	.001172	.000755	.000493	.000325	.000215

model for the deuteron[17]. In this work we have neglected the Q^2 dependence of this correction since this depends on the details of the shape of the structure function near x = 1, not yet in hand. Estimates show that the Q^2 dependence of the correction can be appreciable for large values of n at low values of Q^2. For this reason, in addition to others that have been mentioned, the moments with high n and low Q^2 should be applied with caution and reservation.

The errors stated include, besides the statistical errors, errors of extrapolation, estimated as 25% of the amount of extrapolation, an uncertainty of 10% of the amount of the elastic scattering contribution, and a 2.5% systematic error added in quadrature to take into account uncertainties in normalization. A further uncertainty, not included in Table 5 is due to the lack of knowledge of R. We studied this effect by calculating moments using values of R greater or smaller by one standard deviation, namely, R = 0.52 ± 0.35 for the muon data and R = 0.21 ± 0.10 for the electron data. Such changes in R resulted in a change of 7% for the moments with n = 2, Q^2 = 40 GeV^2. It was less for smaller Q^2 and higher n.

LECTURE 2

The predictions of leading order QCD are simplest for the flavor non-singlet structure functions such as $F_2^{ep} - F_2^{eN}$ in deep inelastic electron (and muon) scattering, or in the vector-axial vector interference structure function xF_3 in charged current νN and $\bar{\nu} N$ scattering. In these cases the Q^2 behavior of each moment is given by a single term,[12,18]

$$M_{NS}(n,Q^2) = M_{NS}(n,Q_o^2)e^{-\lambda_{NS}^{(n)}s} \tag{8}$$

Here $M_{NS}(n,Q_o^2)$ is the value of the moment at some reference value Q_o^2 and $s = \ln[\ln(Q^2/\Lambda^2)\ln(Q_o^2/\Lambda^2)]$. The quantities $\lambda_{NS}^{(n)}$ are the "anomalous dimensions" of the theory, given by

$$\lambda_{NS}^{(n)} = \frac{4}{33-2f}\left[1 - \frac{2}{n(n+1)} + 4\sum_{j=2}^{n}\frac{1}{j}\right] \tag{9}$$

where f is the number of flavors.

The moments of the structure function F_2 are complicated by the presence of two additional singlet terms M_+ and M_-. In this case leading order QCD gives[18,19]

$$M_2^N(n,Q^2) = M_{NS}^N(n,Q_0^2)e^{-\lambda_{NS}^{(n)}s} + M_+(n,Q_0^2)e^{-\lambda_+^{(n)}s} + M_-(n,Q_0^2)e^{-\lambda_-^{(n)}s}$$

(10)

where N is either P for proton or N for neutron. QCD specifies the values of $\lambda_i^{(n)}$, we give them in the Appendix. The coefficients $M_i(n,Q_0^2$ are not calculated in the theory but have to be determined from the data. They may have negative as well as positive values. However, the observed moments M_2 (n,Q^2) are necessarily positive.

PARTON THEORY

In parton theory the structure function F_2 is given by[20]

$$F_2(x) = \sum_i e_i^2 x[q_i(x) + \overline{q}_i(x)]\delta\left(x - \frac{Q^2}{2M\nu}\right)$$

(11)

This is simply the contribution to F_2 of a collection of points, quarks and anti-quarks, with charges e_i and spin 1/2 and with momentum fraction x of the nucleon mass M. Momentum transverse to the collision axis is neglected. The δ function prescribes that the collision be elastic. For each quark [anti-quark] flavor $q_i(x)[\overline{q}_i(x)]$ is the number that have the momentum fraction x per unit interval in x.

An important point is that x, which is the momentum fraction carried by the quark, is directly measured in the experiment, insofar as the collision is elastic, since then $x = Q^2/2M\nu$.

ENERGY-MOMENTUM SUM RULE

From Equation (11) it follows that the integral

$$I_2(2,Q^2) = \int_0^1 F_2(Q^2,x)dx,$$

(12)

or more accurately, the Nachtmann n = 2 moment of F_2, measures the fraction of energy-momentum of the proton carried by the quarks, weighted by their charge squared. This is the energy-momentum sum rule[12,20,21].

In the naive parton model, the expected value for the proton, made of two "up" quarks and one "down" quark, is 0.333. For the neutron, two "down" and one "up", the value would be 0.222. Instead,

as Table 5 shows, the n = 2 Natchmann moments are close to 0.18 for
the proton and for the neutron, 0.13, which is obtained as the
difference of deuteron minus proton. Thus, in both cases, about
45% of the energy-momentum is missing. In the light of QCD, the
missing energy-momentum is carried by the gluons.

        In QCD the energy-momentum sum rule at large Q^2 takes a simple
form,[12] an equipartition principle makes it easy to calculate the
expected value of the 2nd moment. In a 3 color-4 quark model there
are 12 distinct quarks. There are also 2 x 8 equivalent gluons; the
factor 2 coming from the group properties of the theory. At large
Q^2 all share equally the energy-momentum and the quark share is
12/(12+16) = 3/7. Multiplying this by the average squared charge
(= 5/18) gives 0.119, the expected value of the sum rule in the
large Q^2 limit, the same for the neutron as for the proton. Since
the values are not the same, we conclude that at our relatively low
values of Q^2 the quarks in evidence in the proton and the neutron
are mainly the valence quarks, carrying 54% and 59% of the momentum,
respectively. The gluons carry the remainder, 46% and 41% for the
proton and neutron, respectively.

QUARK AND GLUON MOMENTS

        To pursue the analysis in more detail we introduce the quark
and gluon moments[19,22,23] at $Q^2 = Q_o^2$

        For up quarks

$$<u>_n = \int_1 x^{n-1}[u(Q_o^2,x) + \bar{u}(Q_o^2,x)]dx, \qquad (13)$$

where $u(Q_o^2,x)$ and $\bar{u}(Q_o^2,x)$ are the number of up quarks per unit momentum interval, and similarly for the d and s quarks (we neglect the
c quarks). For gluons

$$<G>_n = \int_o^1 x^{n-1} G(Q_o^2,x)dx \qquad (14)$$

where $G(Q_o^2,x)$ is the number of gluons per unit momentum interval.

        We now choose the non-singlet and singlet combinations of the
quarks and gluons and write

$$M_{NS}^P (n,Q_o^2) = \frac{1}{6}[<u>_n - <d>_n - <s>_n] \qquad (15)$$

$$M_{NS}^{N}(n,Q_o^2) = \frac{1}{6}[-<u>_n + <d>_n - <s>_n] \tag{16}$$

$$M_+(n,Q_o^2) = \left[\frac{5}{18}\, a_n^-(<u>_n + <d>_n + <s>_n) - <G>_n\right]/(a_n^- - a_n^+) \tag{17}$$

$$M_-(n,Q_o^2) = \left[\frac{5}{18}\, a_n^+(<u>_n + <d>_n + <s>_n) - <G>_n\right]/(a_n^+ - a_n^-) \tag{18}$$

The constants a_n^+ and a_n^- are given by the theory. The formulas are given in the Appendix. For n = 2, $a_2^+ = -18/5$, $a_2^- = 24/5$.

A fit to the data is found by choosing a value of Λ and then finding the values of the $<q_i>$ and $<G>$ by least squares. The value of Λ is then varied until the best overall fit is found. The moments $<q_i>_2$ give directly the partition of energy-momentum among the quarks. Energy-momentum conservation gives the relation,

$$\sum_i <q_i>_2 + <G>_2 = 1. \tag{19}$$

The quark and gluon moments are subject to positivity constraints[11,19]:

$$<q_i>_n \geq <q_i>_{n+1} \geq 0, \quad <G>_n \geq <G>_{n+1} \geq 0$$

and   (20)

$$<q_i>_{n+2m}<q_i>_n - <q_i>_{n+m}^2 \geq 0$$

$$<G>_{n+2m}<G>_n - <G>_{n+m}^2 \geq 0$$

Using the n = 2 moments listed in Table 5 and setting $<s>_2 = 0$, we found $<u>_2 = 0.350 \pm 0.006$, $<d>_2 = 0.199 \pm 0.010$, $<G>_2 = 0.451 \pm 0.065$. The solid curves of Figure 19 show that a good fit is obtained. The fit is rather insensitive to the value of Λ which turned out to be $\Lambda = 0.24 \pm 0.24$ GeV.

A simultaneous fit of the n = 2, 4, 6 moments with common Λ is more restrictive and gave $\Lambda = 0.65 \pm 0.21$ GeV. The moments were essentially unchanged with $<u>_2 = 0.344 \pm 0.006$, $<d>_2 = 0.195 \pm 0.010$, $<G>_2 = 0.461 \pm 0.060$. The moment $<s>_2$ was set equal to zero.

In view of the current interest in the 6 quark model we tried a fit with 6 flavors but setting the moments of the new quarks equal to zero. In this case, the QCD prediction of the energy-momentum

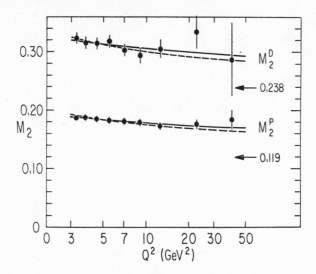

Fig. 19.   Energy-momentum sum rule. The experimental Nachtmann
           moments $M_2(n=2,Q^2)$ for hydrogen and deuterium. Leading
           order QCD fits are shown. The solid curve is for Λ =
           0.24 GeV obtained by fitting the n=2 moments alone. The
           dashed curve is for Λ = 0.65 GeV obtained by a simultane-
           ous fit of the n=2,4,6 moments with common value of Λ.
           The values 0.238 and 0.119 indicated by the arrows are
           the asymptotic $Q^2 \to \infty$ limits expected in the 4 flavor, 3
           color quark-gluon model in QCD for deuterium and hydrogen,
           respectively. For 6 flavors these values are 0.294 and
           0.147, respectively.

sum rule at large Q^2 is 0.147. This lies between the values M_2^P =
0.18 and M_2^N = 0.13 found at low Q^2 and in better accord with the
almost flat behavior of these moments. The simultaneous fit of the
n = 2, 4, 6 moments with common Λ gave $<u>_2$ = 0.333 ± 0.021, $<d>_2$ =
0.189 ± 0.026, $<s>_2$ = 0.071 ± 0.094, and $<G>_2$ = 0.406 ± 0.050, with
Λ = 0.50 ± 0.29 GeV.

     To test leading order QCD we look to see whether the same value
of Λ is obtained when the data is analyzed for each n separately.
The effect of higher order corrections can be absorbed into Λ and
will give values of Λ which are different for each n [24,25]. We denote
these values by Λ_n. Such values of Λ may give a good fit to the data
but will no longer be related to the coupling strength in a simple
way.

The results of such fits are shown in Table 6. In the analysis we set $<s>_n = 0$ to reduce the number of free parameters in the fit, since this moment is known to be small in any case. The values of Λ_n show a regular procession, increasing with n, as shown in Figure 20 Good fits to the data were obtained as indicated by the small χ^2 values obtained for each n compared to the number of degrees of freedom, 14. The fits are shown in Figure 21 for hydrogen, and in Figure 22 for deuterium. We show only the even moments. There is a high degree of correlation between one Λ_n and the next because for each n the same data set is used with emphasis shifting from low x to high x as n increases. Thus, the error bars indicate the uncertainty of a given Λ_n taken alone, but the uncertainty in comparing the trend in Λ_n with n is much less. The increase in Λ_n from n = 4 to n = 10 amounts to 63%. The sensitivity of the fit to Λ_n is poor, however, and an acceptable fit with constant Λ can also be obtained.

The evidence presented here, though not conclusive, is indicative of the presence of higher order QCD effects. This is in contrast to the neutrino data [9,10] for which such evidence is lacking. In Bosetti et al.,[9] a quantitative verification of QCD is given based on the leading order expression:

Table 6. Quark and Gluon Moments and Λ_n. Leading Order Fit at $Q_0^2 = 10$ GeV^2. The error in the least significant digits is given in brackets.

(a) Each moment analyzed separately.

n	$<u>_n$	$<d>_n$	$<G>_n$	Λ_n(GeV)	χ^2
2	.3499(60)	.199(10)	.4507(65)	.24(24)	5.2
3	.0915(14)	.0402(24)	.085(32)	.38(26)	10.1
4	.03553(54)	.0126(8)	.025(11)	.53(21)	4.9
5	.01701(26)	.00500(39)	.0099(55)	.60(18)	5.2
6	.00926(14)	.00223(22)	.0039(33)	.65(14)	7.7
7	.00551(9)	.00104(13)	.0016(16)	.686(71)	10.7
8	.00350(6)	.00049(8)	.00062(62)	.748(53)	11.7
9	.00234(4)	.00023(6)	.00024(24)	.807(43)	13.0
10	.00164(3)	.00011(4)	.00010(10)	.861(36)	20.5

(b) Common Λ for n = 2, 4, 6

n	$<u>_n$	$<d>_n$	$<G>_n$	Λ_n(GeV)	χ^2
2	.3441(58)	.195(10)	.4612(60)	.65(21)	8.5
4	.03537(3)	.0129(8)	.0495(92)	.65(21)	5.3
6	.00926(14)	.00224(21)	.0053(33)	.65(21)	8.0

$$\frac{d \log M_{NS}(n,Q^2)}{d \log M_{NS}(m,Q^2)} = \frac{\lambda_{NS}^{(n)}}{\lambda_{NS}^{(m)}} \qquad (21)$$

The experimental values of the lefthand side are shown to agree
remarkably well with the QCD value of the ratio $\lambda_{NS}^{(n)}/\lambda_{NS}^{(m)}$. A similar
agreement has been obtained by de Groot et al.[10] Such an agreement
implies $\Lambda_n = \Lambda_m$. However, if $\Lambda_n > \Lambda_m$, as our results indicate, the
ratio of the slopes would increase, with $\Lambda_4 = 0.53$ and $\Lambda_6 = 0.65$ as
in Table 6, the expected value of the slope would be raised to 1.56
for $Q^2 = 3$ GeV^2.

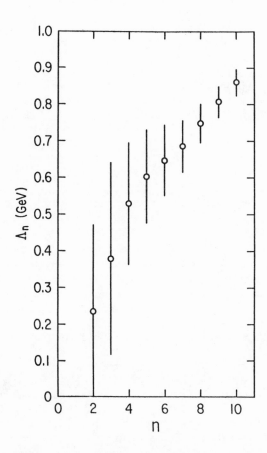

Fig. 20. Variation of Λ_n with n from the leading order QCD fits.
Every point represents the same data set and carries its
full statistical weight. The point to point errors are
highly correlated and the relative errors are much smaller.

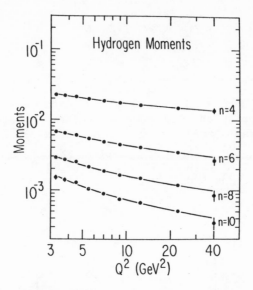

Fig. 21.   The experimental Nachtmann moments for n=4 to n=10 for
hydrogen and the fit in leading order QCD.   Each n was
was analyzed separately using Q_0^2 = 10 GeV .

        In general, the moments of the structure function F_2 as given
in Figures 21 and 22 include a large singlet contribution.   Contributions come from three terms with different exponential coefficients.
However, the singlet contribution M_+ is relatively small while the
anomalous dimensions $\lambda_{NS}^{(n)}$ and $\lambda_-^{(n)}$ are close to one another and
become closer as n increases for $n \geq 4$.   This can be seen in Table 7
where the anomalous dimensions for the non-singlet and singlet terms
are given for the 4 flavor, 3 color quark-gluon model.   Thus, the
moments of F_2 for $n \geq 4$ can be expected to exhibit a simple exponential behavior with a slope approaching that of the NS term alone.

        In Figure 23 we display log M_6 versus log M_4. **The least squares**
fit to the points is d log M_6/d log M_4 = 1.58 ± 0.14, a value
appreciably larger than the value 1.31 expected from leading order
QCD if the term M_+ were completely negligible.   Since M_+ is given in
the analysis we can calculate that its contribution increases the
ratio to 1.37.   The effect of the change in Λ_n accounts for the
remaining discrepancy.   We also display d log M_{10} versus d log M_6.
Here the ratio of the slopes is found to be 1.61 ± 0.10.   The value
$\lambda_-^{(10)}/\lambda_-^{(6)}$ = 1.29. The effect of M_+ raises this to 1.33.   The effect
of $\Lambda_{10} > \Lambda_6$ brings the value to 1.66, close to that observed.

Table 7.   The anomalous dimensions in QCD for the 4 flavor, 3 color
           quark-gluon model.

n	$\lambda_{NS}^{(n)}$	$\lambda_-^{(n)}$	$\lambda_+^{(n)}$	a_n^+	a_n^-
2	.42666	.00000	.74666	-3.59999	4.80000
3	.66666	.60854	1.38611	-11.56261	.93404
4	.83733	.81698	1.85234	-20.76160	.41615
5	.97066	.96044	2.19193	-30.05466	.25154
6	1.08038	1.07427	2.46039	-39.51854	.17490
7	1.17371	1.16967	2.68347	-49.19723	.13171
8	1.25498	1.25212	2.87486	-59.10364	.10441
9	1.32698	1.32485	3.04274	-69.23672	.08579
10	1.39163	1.38999	3.19244	-79.58946	.07237

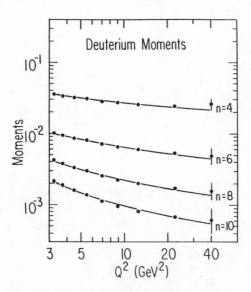

Fig. 22.   The experimental Nachtmann moments for n=4 to n=10 for
           deuterium and the fit to leading order QCD. Each n was
           analyzed separately using $Q_0^2 = 10$ GeV^2.

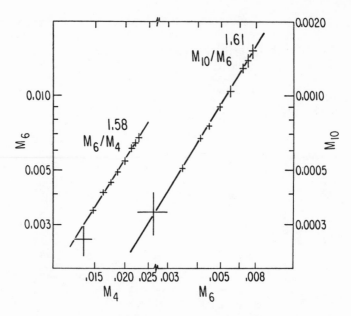

Fig. 23. Plot of log $M_6(Q^2)$ versus log $M_4(Q^2)$ and log $M_{10}(Q^2)$
versus log $M_6(Q^2)$ for the moments of F_2.

SECOND ORDER CORRECTIONS

Recently, Floratos, Ross and Sachrajda[26] have calculated the
2nd order corrections in QCD. They included both the singlet and
the non-singlet terms, making the corrections applicable to the
structure function F_2. There were some errors in the numbers given
in Reference 26, but these were corrected for us by W. A. Bardeen,
to whom we are grateful. Using the corrected coefficients we
carried out an analysis in the 2nd order, setting $<s>_n = 0$, as
before.

In Table 8 we give the best fit values of the quark and gluon
moments obtained by allowing Λ to vary for each n separately. The
fits obtained were indistinguishable from those obtained in the
leading order analysis. The quark moments (Figures 24 and 25) are
essentially unchanged; the gluon moments (Figure 26) are shifted
slightly. The main effect is to reduce the value of Λ_n and its
variation with n as shown in Figure 27. The effect of changing R
by one standard deviation is also shown.

Although a rising trend is still evident, it is less marked
than in the analysis done in leading order. This may imply that
additional corrections, e.g., twist four,[6] need to be included.

Table 8. Quark and gluon moments and Λ. QCD fit including 2nd order correction at $Q_0^2 = 10 \ GeV^2$. The error in the least significant digits is given in brackets.

a. Each moment analyzed separately.

n	$<u>_n$	$<d>_n$	$<G>_n$	Λ_n(Gev)	χ^2
2	.3501(60)	.199(10)	.4507(65)	.14(13)	5.2
4	.03547(54)	.0128(8)	.0295(79)	.33(13)	4.9
6	.00926(14)	.00226(21)	.0054(21)	.370(80)	8.2
8	.00351(6)	.00049(8)	.0010(10)	.397(42)	14.3
10	.00165(3)	.00011(4)	.00018(18)	.436(24)	25.1

b. Common Λ for n = 2, 4, 6

n	$<u>_n$	$<d>_n$	$<G>_n$	Λ_n(Gev)	χ^2
2	.3452(59)	.195(10)	.4600(61)	.34(13)	7.8
4	.03543(54)	.0128(8)	.0344(75)	.34(13)	4.9
6	.00928(14)	.00223(22)	.0026(23)	.34(13)	8.4

On the other hand, these will mainly affect the higher moments and we have already commented on the questionable validity of the higher moments at low Q^2. There are also uncertainties in the values used for R. When the smaller values of R are used the variation in Λ_n with n is reduced.

We can make a global fit to the data for n = 2, 4, 6 and obtain a good fit with a common value $\Lambda = 0.34 \pm 0.13$ GeV. The results of the analysis are given in Table 8. If we take this value of Λ as representative of our data and insert it into the relation between Λ and the strong coupling strengths to 2nd order in QCD as given by Floratos et al.,[26] after correcting an obvious error, namely

$$\frac{\alpha_s(Q^2)}{\pi} = \frac{4}{\beta_o \ln(Q^2/\Lambda^2)} \left(1 - \frac{\beta_1}{\beta_o^2} \frac{\ln\ln(Q^2/\Lambda^2)}{\ln(Q^2/\Lambda^2)} \right) \tag{22}$$

where $\beta_o = 11 - \frac{2}{3} f$

$\beta_1 = 102 - \frac{38}{3} f$

and f is the number of flavors, we obtain $\alpha_s/\pi = 0.108 \pm 0.025$ at $Q^2 = 3$ GeV^2.

In view of the ambiguities in the meaning of Λ whenever higher order corrections are significant, only an effective value of Λ adjusted to fit the data can be given. However, the quark and gluon moments are well determined, at least for values of $n \leq 6$. These provide an interesting view of what the nucleon is made of.

ACKNOWLEDGEMENT

The author is especially indebted to Drs. R.J. McKee and L.C. Myrianthopoulos for their help in carrying out the analyses reported here. Much of the material presented here is from a paper to be published in the Physical Review by the members of the CHIO collaboration who performed Fermilab experiments E98/398.

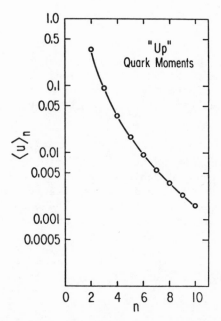

Fig. 24.   Up quark moments evaluated at $Q_0 = 10$ GeV . The values obtained including the 2nd order corrections are indistinguishable from the leading order values.

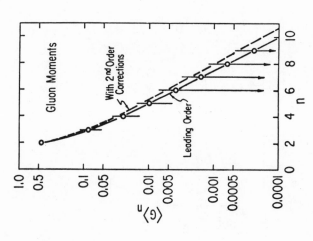

Fig. 26.   Gluon moments evaluated at Q_0^2 = 10 GeV^2. The values obtained including the 2nd order corrections are shifted higher with respect to the leading order values.

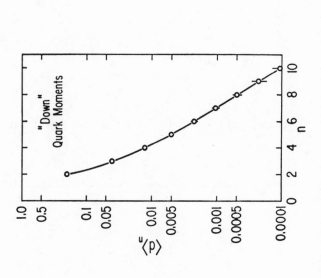

Fig. 25.   Down quark moments evaluated at Q_0^2 = 10 GeV^2. The values obtained including the 2nd order corrections are unchanged from the leading order values.

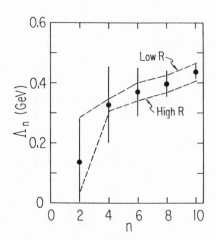

Fig. 27. Variation of Λ_n with n in 2nd order QCD analysis. Low R
means R = 0.11 for ep, 0.17 for μp. High R means R = 0.31
for ep, 0.87 for μp. Every point represents the same data
set and carries the full statistical weight giving the
error shown. The weight of the data is shifted to larger
x as n increases so that there is a strong correlation and
small relative error from one point to the next.

APPENDIX

    The constants a_n^+ and a_n^- are given by   (We use Nachtmann's
formulation)

$$a_n^+ = \frac{36}{5}\, \frac{d_{G\psi}^n}{d_{\psi\psi}^n - \lambda_-^n}$$

$$a_n^- = \frac{36}{5}\, \frac{d_{G\psi}^n}{d_{\psi\psi}^n - \lambda_+^n}$$

where

$$d_{\psi\psi}^n = \frac{4}{33 - 2f}\left[1 - \frac{2}{n(n+1)} + 4\sum_{j=2}^{n}\frac{1}{j}\right] = \lambda_{NS}^n$$

$$d_{G\psi}^n = -\frac{4}{33 - 2f}\frac{n^2 + n + 2}{n(n^2 - 1)}$$

$$d_{\psi G}^n = -\frac{12f}{33 - 2f}\frac{n^2 + n + 2}{n(n+1)(n+2)}$$

$$d_{GG}^n = \frac{9}{33 - 2f}\left[\frac{1}{3} + \frac{2f}{9} - \frac{4}{n(n-1)} - \frac{4}{(n+1)(n+2)} + 4\sum_{j=2}^{n}\frac{1}{j}\right]$$

$$\lambda_{\pm}^n = \frac{1}{2}(d_{\psi\psi}^n + d_{GG}^n) \pm \frac{1}{2}\left((d_{\psi\psi}^n - d_{GG}^n)^2 + 4d_{G\psi}^n d_{\psi G}^n\right)^{1/2}$$

REFERENCES

1. H. L. Anderson, V. K. Bharadwaj, N. E. Booth, R. M. Fine, W. R. Francis, B. A. Gordon, R. H. Heisterberg, R. G. Hicks, T. B. W. Kirk, G. I. Kirkbride, W. A. Loomis, H. S. Matis, L. W. Mo, L. C. Myrianthopoulos, F. M. Pipkin, S. H. Pordes, T. W. Quirk, W. D. Shambroom, A. Skuja, M. A. Staton, W. S. C. Williams, L. J. Verhey, Richard Wilson, and S. C. Wright, Phys. Rev. Lett. 37, 4 (1976) and Phys. Rev. Lett. 38, 1450 (1977).
2. B. A. Gordon, T. W. Quirk, H. L. Anderson, N. E. Booth, W. R. Francis, R. G. Hicks, W. W. Kinnison, T. B. W. Kirk, W. A. Loomis, H. S. Matis, L. W. Mo, L. C. Myrianthopoulos, F. M. Pipkin, J. Proudfoot, A. L. Sessoms, W. D. Shambroom, A. Skuja, M. A. Staton, C. Tao, W. S. C. Williams, Richard Wilson, and S. C. Wright, Phys. Rev. Lett. 41, 615 (1978).
3. Many details about the apparatus and the analysis can be found in the following PhD theses which dealt with the muon inclusive aspect of the experiments.
      S. H. Pordes, PhD Thesis, Harvard University (1976).
      R. H. Heisterberg, PhD Thesis, University of Chicago (1976).
      G. I. Kirkbride, PhD Thesis, Oxford University (1976).
      V. K. Bharadwaj, PhD Thesis, Oxford University (1977).
      G. A. Gordon, PhD Thesis, Harvard University (1978).
4. C. Callan and D. Gross, Phys. Rev. Lett. 22, 156 (1969).
5. R. P. Feynman, private communication.

6. A. de Rujula, H. Georgi, and H. D. Politzer, Ann. Phys. <u>103</u>, 315 (1977).

7. R. E. Taylor, 19th International Conference on High Energy Physics, Tokyo, Japan (1978). M. D. Mestayer, "A Measurement of the Proton Structure Functions Using Inelastic Electron Scattering", SLAC-214 Thesis (1978).

8. J. I. Friedman and H. W. Kendall generously supplied tables of the MIT-SLAC electron scattering results. We used tapes kindly made available by R. E. Taylor, W. B. Atwood, and M. D. Mestayer containing a complete set of the electron-scattering results from the SLAC and MIT-SLAC experiments: SLAC E4B, G. Miller, "Inelastic Electron Scattering at Large Angles", SLAC-129 Thesis (1971); SLAC E49A, J. S. Poucher, et. al., Phys. Rev. Lett. <u>32</u>, 118 (1974) and "High Energy Single Arm Inelastic e-p and e-d Scattering at 6° and 10°", SLAC-1309 (1973); SLAC E49B, A. Bodek, "Inelastic Electron-Deuteron Scattering and the Structure of the Neutron", Technical Report 93, MIT Thesis (1972); SLAC E87, E. M. Riordan, "Extraction of Structure Functions and R = σ_L/σ_T from Deep Inelastic e-p and e-d Cross Sections", SLAC-PUB-1634 (1975); SLAC E89-1, W. B. Atwood, et. al., Phys. Lett. <u>64B</u>, 479 (1976); W. B. Atwood, "Electron Scattering of Hydrogen and Deuterium at 50° and 60°", SLAC-185 Thesis (1978); SLAC E89-2, M. D. Mestayer, "A Measurement of the Proton Structure Functions Using Inelastic Electron Scattering", SLAC-214 Thesis (1978); A. Bodek, et. al., "Experimental Studies of the Neutron and Proton Structure Functions", SLAC-PUB-2242 (1979).

9. P. C. Bosetti, et. al., Nucl. Phys. <u>B142</u>, 1 (1978).

10. J. de Groot, et. al., Phys. Lett. <u>82B</u>, 456 (1979).

11. O. Nachtmann, Nucl. Phys. <u>B63</u>, 237 (1973); <u>B78</u>, 455 (1974).

12. H. D. Politzer, Phys. Reports <u>14C</u>, 130 (1974). Also see D. Gross and F. Wilczek, Phys. Rev. <u>D8</u>, 3633 (1973) and Phys. Rev. <u>D9</u>, 980 (1974).

13. D. H. Perkins, P. Schreiner, and W. G. Scott, Phys. Lett. <u>67B</u>, 347 (1977).

14. D. H. Perkins, "Introduction to High Energy Physics", p. 204, Addison Wesley (1972).

15. R. J. McKee, private communication.

16. W. B. Atwood and G. B. West, Phys. Rev. <u>D7</u>, 733 (1973).

17. R. V. Reid, Jr., Ann. Phys. (N.Y.) <u>50</u>, 411 (1968).

18. M. Glück and E. Reya, Phys. Rev. <u>D14</u>, 3034 (1976).

19. H. L. Anderson, H. S. Matis, and L. C. Myrianthopoulos, Phys. Rev. Lett. <u>40</u>, 1061 (1978).

20. R. P. Feynman, "Photon-Hadron Interactions", W. A. Benjamin (1972).

21. C. G. Callan and D. J. Gross, Phys. Rev. Lett. <u>21</u>, 311 (1968).

22. O. Nachtmann, private communication, see Ref. 19.

23. Wu-ki Tung, Phys. Rev. <u>D16</u>, 2769 (1977).

24. E. G. Floratos, D. A. Ross, and C. T. Sachrajda, Nucl. Phys.
     B129, 66 (1977).
25. W. A. Bardeen, A. J. Buras, D. W. Duke, and T. Muta, Phys. Rev.
     D18, 3998 (1978).
26. E. G. Floratos, D. A. Ross, and C. T. Sachrajda, Phys. Lett.
     80B, 269 (1979).

RECOIL HADRON MEASUREMENTS IN MUON DEEPLY INELASTIC SCATTERING

Thomas B. W. Kirk

Neutrino Department Head
Fermilab
P. O. Box 500
Batavia, IL   60510

INTRODUCTION

Most of the results to be discussed in this paper were obtained by a group of physicists from the University of Chicago, Harvard University, the University of Illinois, and the University of Oxford who performed a series of muon scattering experiments at Fermilab in the period 1973-1976. The principal reference for this paper and for the work on recoil hadrons done by this group is a paper in Physical Review listed here as Reference 1. For that paper only, all the authors names are listed to illustrate the number of scientists whose efforts were required over a six year period to produce the results to be discussed. I was privileged to work with this group the entire time and want to acknowledge here my debt to their hard work and cooperation; it made this talk possible.

In order to make the results more comprehensible and to place them properly in a larger physics context, I have attempted to use the related data from several other groups which have published recoil inclusive hadron data on electron, muon, neutrino and even hadron scattering experiments[2-8]. Of equal interest are the recoil hadrons from e^+e^- annihilation reactions and I have included these data at appropriate places. I am particularly indebted to Dr. G. Wolf who kindly allowed me access to his very recent and unpublished inclusive hadron data from the TASSO experiment.

NOTATION AND PHENOMENOLOGY

The discussion of inclusive recoil hadron properties of deeply inelastic collisions requires a set of kinematically defined variables which are typically constructed to have useful properties under Lorentz transformation and which reflect the limited transverse momentum property that is nearly universally observed in particle dynamics and which gives rise to the concept of hadron "jets". In this paper, and for its knowledgeable readers, I will assume familiarity with the idea of the single γ (W, Z^0) exchange picture of deeply inelastic electromagnetic (weak) scattering and the basic quark/gluon parton model of the nucleon. I now give a list of the variables that will be used in this paper to define and measure the recoil properties of deeply inelastic collisions and e^+e^- annihilations:

1) $\nu \equiv (E_\ell - E_\ell')$

   $\vec{q} \equiv \vec{P}_\ell - \vec{P}_\ell'$;   energy and momentum lost by the scattered lepton

2) $Q^2 \equiv P_\ell \cdot P_\ell \simeq 2M\nu$;   invariant 4 momentum transfer to the nucleon, also the "mass" of the virtual photon, W or Z.

3) $\dfrac{E^*}{\sigma_{Tot}} \dfrac{d^3\sigma}{dP^{*3}} = \dfrac{E^*}{\pi\sigma_{Tot}} \dfrac{d^3\sigma}{dP_\perp^2 dP_{||} d\phi}$;   single hadron phase space density normalized to the total cross section (a Lorentz invariant)

4) $\left\langle \dfrac{E^*}{\sigma_{Tot}} \dfrac{d^3\sigma}{dP^{*3}} \right\rangle = \dfrac{E^*}{\pi\sigma_{Tot}} \dfrac{d^2\sigma}{dP_\perp^2 dP_{||}}$;   azimuthally averaged hadron phase space density

5) $x' \equiv P_{||}^* \left/ \sqrt{P_{max}^2 - P_\perp^2} \right.$;   ratio of hadron longitudinal momentum to its maximum possible value, "Feynman x-prime"

6) $y \equiv 1/2 \ln \left[\dfrac{E^* + P_{||}^*}{E^* - P_{||}^*} \right]$;   C.M. rapidity

7) $z \equiv \dfrac{E}{\nu}$;   fraction of the hadron laboratory energy to its maximum possible value (a Lorentz invariant)

8) $F(x') = \dfrac{E^*}{\pi\sigma_{Tot}} \displaystyle\int_{0}^{P^2_{max}} \dfrac{d^2\sigma}{dP^2_\perp dP_\parallel} \, dP^2_\perp$;   hadron longitudinal structure function

9) $\dfrac{1}{\pi}\dfrac{dN}{dy}$;   rapidity distribution of hadrons

In the interests of simplicity, I have suppressed indices for particle type in all the recoil particle definitions. In most cases, the meaning will be clear in context. For the few cases where confusion is possible, appropriate particle type indices will be added.

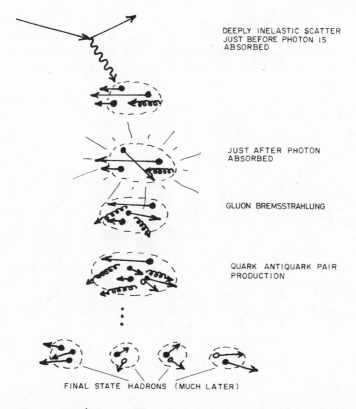

DEEPLY INELASTIC SCATTER JUST BEFORE PHOTON IS ABSORBED

JUST AFTER PHOTON ABSORBED

GLUON BREMSSTRAHLUNG

QUARK ANTIQUARK PAIR PRODUCTION

FINAL STATE HADRONS (MUCH LATER)

Fig. 1. The quark/gluon parton model of a nucleon struck by a virtual photon. A single quark absorbs the entire four momentum of the photon and then "fragments" into a number of quarks, antiquarks and gluons by the process of gluon bremsstrahlung. Eventually, the partons combine in stable configurations we know as mesons and baryons.

As noted, we will discuss the recoil hadron distributions in the context of the parton picture. In this picture, as illustrated in Fig. 1, a single quark in the initial nucleon is assumed to absorb the entire four momentum of the virtual photon. As it tries to move apart from its sister quarks in configuration space, it emits gluon bremsstrahlung quanta which in turn materialize as quark-antiquark pairs which themselves emit gluons, etc., until a contact in momentum phase space is achieved between the initial struck quark and the two spectator quarks. This process is called "quark fragmentation".

Since the individual quarks and gluons cannot exist alone in the final state due to color confinement, quark-antiquark pairs in nearby regions of momentum space combine to form colorless primordial mesons. These mesons are often highly unstable, perhaps without definite mass and quantum numbers (clusters) and they rapidly decay into the stable hadrons π, K, ρ, ϕ, etc., that we are familiar with in ordinary strong interactions. These stable hadrons are the particles that we measure in deeply inelastic scattering. We hope the measurements will teach us something about the underlying quark fragmentation process.

CHIO SPECTROMETER

The muon experiment apparatus, from which the bulk of the hadron spectra reported here come, is shown in Fig. 2. Not all the equipment shown was in use at all times, but the basic multi-particle magnetic spectrometer used as a source of charged particle momentum distributions remained intact the entire time and forms the data base upon which Ref. 1 and this talk are based.

The spectrometer was of the "open geometry" dipole type and was capable of taking muon scattering events simultaneously over a wide kinematic spectrum as well as recording essentially all accompanying charged hadron trajectories with Feynman x greater than about 0.1. The scattered muon acceptance is shown in Fig. 3 and the kinematic regions by which the recoil hadron data was binned in Fig. 4. The large acceptance in momentum space for charged recoil hadrons is shown in Fig. 5. Regions of zero hadron acceptance are shown shaded.

No more will be said about the apparatus itself except to note that it served gallantly and successfully for our muon experiments as well as for several experiments performed by other groups of experimenters over a period of six years; the detectors have now been retired with honor in favor of new, updated instrumentation that will increase the data rate capabilities of the spectrometer for upcoming hadron experiments.

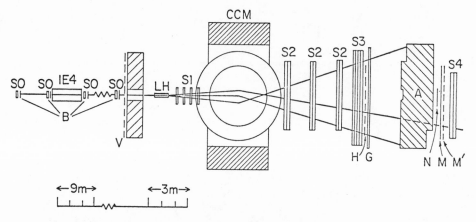

Fig. 2. The Chicago Cyclotron Spectrometer as it was configured
         for most of the data reported here. S0–S4 represent
         planes of wire chambers for tracking charged particles.

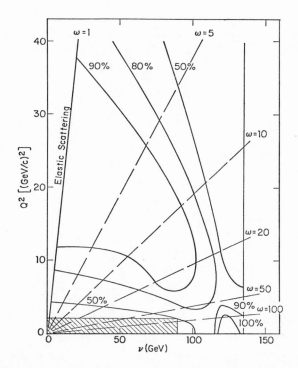

Fig. 3. Kinematic acceptance curves for the scattered muon in
         ν and Q^2.

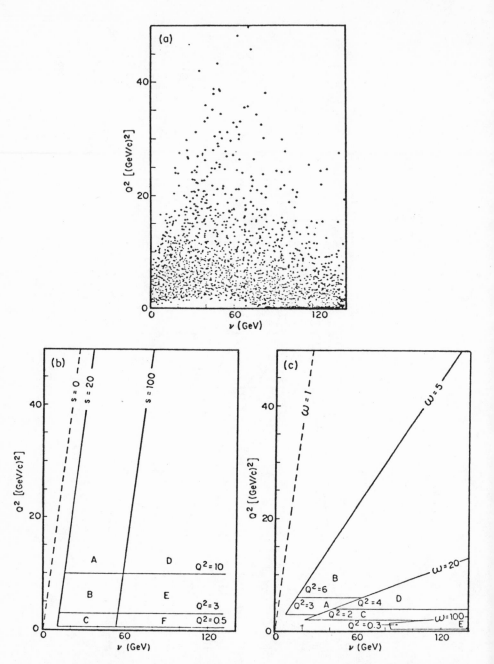

Fig. 4. Kinematic regions chosen for classifying the recoil hadron
         data. Since the dependence of hadron distributions on
         Q^2 and Feynman x is small, large bins are adequate.

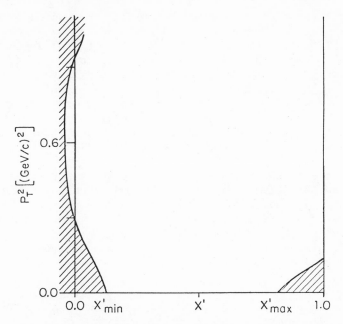

Fig. 5. Acceptance in transverse momentum and Feynman x for
         hadrons from the CHIO experiment. The results cover
         charged hadrons from the projectile hemisphere in the
         center of momentum.

RECOIL HADRON STRUCTURE FUNCTION

        The hadron spectra to be shown have all been averaged in
azimuth about the virtual photon direction. The kinematics of
single photon exchange allow a distribution in this angle of the
form:

$$10) \quad \frac{d^3\sigma}{dP^3} = \frac{d^3\sigma_t}{dP^3} + \frac{\varepsilon d^3\sigma_s}{dP^3} + \frac{\varepsilon d^3\sigma_{tt}}{dP^3} \quad \cos 2\phi$$

$$+ \frac{\varepsilon(\varepsilon + 1)}{2}^{1/2} \frac{d^3\sigma_{st}}{dP^3} \quad \cos \phi,$$

where the s and t subscripts denote contributions from the virtual
photon field corresponding to scalar (or longitudinal) and transverse states of polarization. In fact, the data is consistent
only with small contributions from the ϕ dependent interference
terms as shown in Fig. 6, and from here on, we shall average all
distributions over ϕ and ignore a possible ϕ dependence.

We will first consider the behavior of the hadron structure
function of Eq. 3. Figures 7 and 8 show the inclusive hadron data
of Ref. 1 binned to exhibit the (Q^2, ω) and (Q^2, s) dependences,
respectively. Note that there is a striking constancy about these
distributions. The constancy is the basic manifestation of the
principle of "universal quark fragmentation", by which phrase we
mean that the average properties of a struck quark evolving into
hadrons are independent of the Q^2 (violence) of the collision.
This seems to be true above some threshold value around Q^2 ~
~ $1(GeV/c)^2$, and is probably broken at the level of a few per cent by
the QCD process of gluon bremsstrahlung (a process which should be
weakly Q^2 dependent). A wider range of testing for the s dependence of the inclusive hadron distribution is given in Fig. 9,
where several virtual photon experiments are compared. Note that
the low energy experiments show a large electric charge asymmetry.
This charge asymmetry is probably connected with the small range
of rapidity available in the low s region and disappears in the
CHIO data at high s.

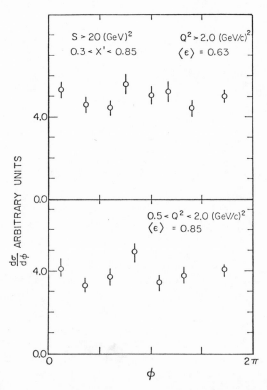

Fig. 6. Azimuthal distributions of hadron momentum measured about
         the virtual photon direction.

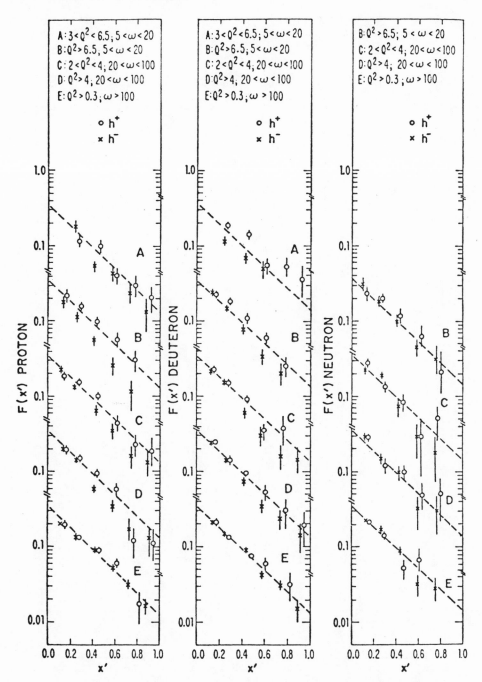

Fig. 7. CHIO data for the inclusive hadron structure function binned to exhibit the Q^2, ω behavior for proton and deuteron targets.

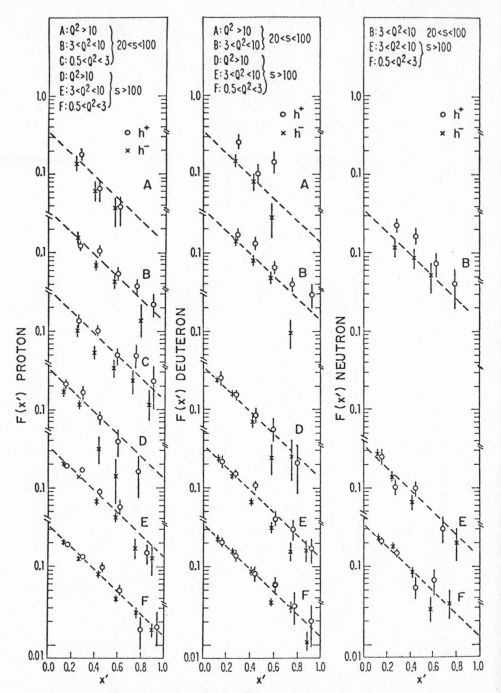

Fig. 8. CHIO data for the inclusive hadron structure function
binned to exhibit the Q^2, s behavior for proton and
deuteron targets.

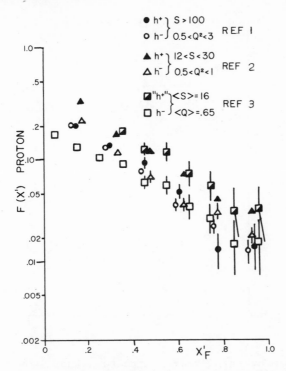

Fig. 9. Inclusive hadron structure function plotted for three
         lepton scattering experiments. The points labeled "h$^+$"
         were obtained from Ref. 3 by multiplying their reported
         negative spectrum by their measured charge asymmetry.

     If the low energy and high energy data are charge averaged,
a much more universal curve emerges as is shown in Fig. 10. If
there is any systematic s dependence, it would be characterized by
a decrease in F(x') in the large x' region. There is one low
energy point at x' = 0.2 which lies substantially above the
corresponding high energy point, but this x region could be contaminated with protons from the target hemisphere. At any rate,
there is a mutual consistency among the points indicating approximate scaling in Feynman x independent of s. We are encouraged to
expand our comparisons.

     Figure 11 compares the inclusive charged hadron yields from
several processes. The CHIO data are compared to e^+e^- annihilation
data from SPEAR and TASSO, deeply inelastic neutrino scattering in
neon in the Fermilab 15 Foot Bubble Chamber, and finally, with
inclusive hadron spectra from π^+ collisions with neon in the 15
Foot Chamber. The agreement among the various data sets is
startling. If the behavior of the π^+Ne is included, the results
are truly remarkable!

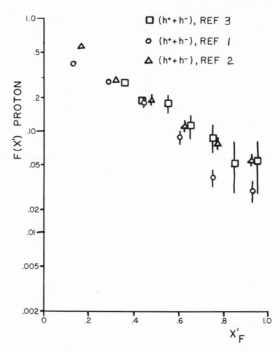

Fig. 10.   Charge averaged hadron structure functions for the data
           of Fig. 9.

        The idea of a quark instantaneously transported to a distant
point in momentum space by absorption of a high Q^2 photon, a Z^0 or
W boson, or by the materialization of a massive time-like photon is
believable, but how do we reconcile this with the same fragmentation
function appearing when a low t collision occurs between a π^+ and
a neon nucleus? I can't explain it, but there must be some
analogous process by which quarks can exchange one or more gluons
such that the struck partons fragment in virtually the same way
as isolated quarks in lepton scattering. Existence of the leading
particle effect in hadron inclusive spectra from hadron-hadron
collisions indicates a basic difference in the low t events, of
course, but I find that the striking similarity of the non-leading
inclusive hadron spectrum is very impressive and calls for a
simple explanation. Perhaps the pretty model of Louis Osborne of
MIT, discussed elsewhere in this volume is a way to understand it.

        Before leaving the subject of inclusive hadron structure
functions, let me note the overall normalization difference
apparent in comparison of the SPEAR e^+e^- spectrum with that of
CHIO muon scattering. Note that the e^+e^- spectrum has the same
overall <u>shape</u> as the μp, but that it is smaller by a factor 1.4

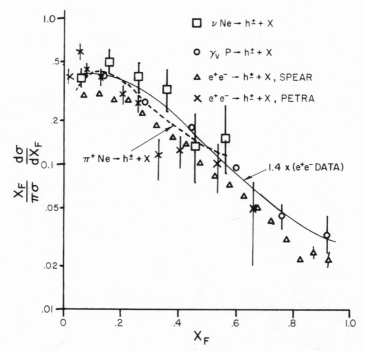

Fig. 11.   Inclusive structure function for charged hadrons from
           several scattering processes. The low energy e^+e^- data
           come from Ref. 4, the neutrino and π^+ data from Ref. 5.
           The high energy e^+e^- data are from G. Wolf presented at
           this seminar.

in overall <u>normalization</u>. The smooth curve shown was numerically
integrated for both the CHIO and SPEAR data with the results:

$$E_{chgd}(CHIO) = 0.58 \pm 0.03;$$

$$E_{chgd}(SPEAR) = 0.24 \pm 0.03.$$

The small fraction of energy carried by charged particles in the
e^+e^- case gave rise to the notion of an "energy crisis". The
abundance of neutrals has subsequently been partially explained by
resonance production. It is unlikely that such an explanation will
be adequate for explaining a similar result in the PETRA data if
the energy crisis persists, as it gives evidence of doing, in the
preliminary TASSO spectra shown by G. Wolf at this seminar.

CHARGE RATIOS AND ASYMMETRIES

Some attention has been paid to the electric charge ratio of inclusive hadrons. In the early days of the parton model, Feynman had proposed that the average quark quantum numbers would be preserved by the leading hadron from a deeply inelastic lepton scattering. If the scattering takes place at small values of ω (large x), one expects to see the charged ratio result dominated by the valence quarks. If it takes place at large ω (wee x), there should be no net charge asymmetry, corresponding to the charge symmetric sea of quark-antiquark pairs.

The observed results for protons and deuterons are shown in Fig. 12. As expected, the large ω points show charge symmetry for both large (leading hadron) and small (gluon bremsstrahlung cascade) values of Feynman x corresponding to the charge symmetry of the struck quarks and antiquarks. At small values of ω the leading hadrons should show a significant charge asymmetry, larger for protons than for neutrons. The deuteron should be a linear

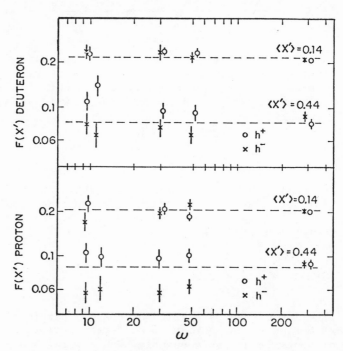

Fig. 12. Charge dependent hadron structure functions for inclusive hadrons from CHIO data. The data has been separated into samples for large and small Feynman x and is plotted explicitly as a function of Bjorken ω.

superposition of neutron and proton. The result shown in Fig. 12 exhibits such a charge asymmetry in a qualitative manner, but the statistics are poor for any numerical comparison to the quark model expectations from the Feynman conjecture.

Figure 13 shows the charge ratios explicitly, including a derived value for the neutron charge asymmetry. Through the data points is drawn the predicted curve from a simple quark model due to Dakin and Feldman[9]. There is fair qualitative agreement between the prediction and the data, but much better statistical precision will be needed to make a really rigorous test of the idea of charge asymmetry and its connection to quark fragmentation.

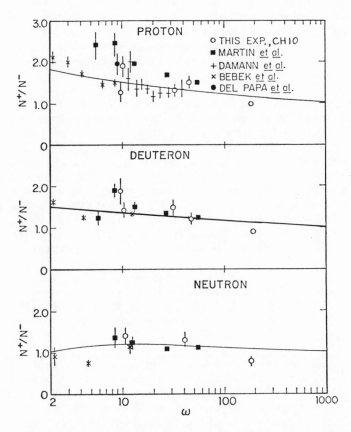

Fig. 13.   Explicit electric charge ratios for inclusive hadrons from deeply inelastic lepton scattering.

TRANSVERSE MOMENTUM DISTRIBUTIONS

Figure 14 shows the data on transverse momentum distributions of inclusive charged hadrons from the CHIO muon scattering experiment. The data are divided into broad categories by Q^2 and ω. The data were further cut on Feynman x to represent the hadrons in the jet portion of the cascade. Figure 15 shows another binning of the transverse momentum data. In this case, the invariant phase space momentum distribution is plotted and the results shown explicitly as a function of Feynman x. In this plot, the value of ω is high, the value of Q^2 is low, and we are sampling the sea quarks.

Since the data of Figs. 14 and 15 are difficult to integrate visually, and since there are obvious regularities, we show in Fig. 16 the transverse momentum distribution averaged over a broad interval of Feynman x and summed over all $Q^2 > 0.3$ $(GeV/c)^2$. Superimposed on the distributions for positive and negative hadrons from the CHIO experiment, we have plotted the data from the e^+e^- experiment at SPEAR. Over two decades on a semilog plot, we see a strikingly consistent behavior, even more congruent for the two data sets than the longitudinal structure function data were.

Inspired by the high quality of the data, I have ventured to fit the distribution to an analytical function based upon an exponential drop-off of the hadrons in $m = \sqrt{P^2 + M_0^2}$ measured from the virtual photon axis. The fit is obviously very good and therefore suggests we look for an explanation for the constant M_0. One candidate would be an intrinsic transverse momentum of the quarks in the nucleon prior to the scattering. A second possibility would be more mundane, namely, that M_0 is just the average mass of a primordial meson or cluster found in the early evolution of the struck quark. Further investigations using various probes of the nucleon will be needed to find the proper interpretation.

For purposes of comparison to other published data, it is useful to present the averaged values of $P_\perp$ and $P_\perp^2$ versus various other kinematic parameters. This is done in Figs. 17, 18 and 19 where $\langle P_\perp \rangle$ is displayed as a function of Q^2, Feynman x and ln s, respectively.

In Fig. 17, we observe that, contrary to many expectations, the average value of transverse momentum shows no clear Q^2 dependence. This result is somewhat unexpected by theorists who predicted an increase in $\langle P_\perp \rangle$ as Q^2 increases. It is consistent, however, with the universal quark fragmentation hypothesis, and represents a non-trivial support for it. The increase of $\langle P_\perp \rangle$ with ln s shown in Fig. 19 is also consistent with the picture of a hot quark de-exiting by a cascade process until it establishes

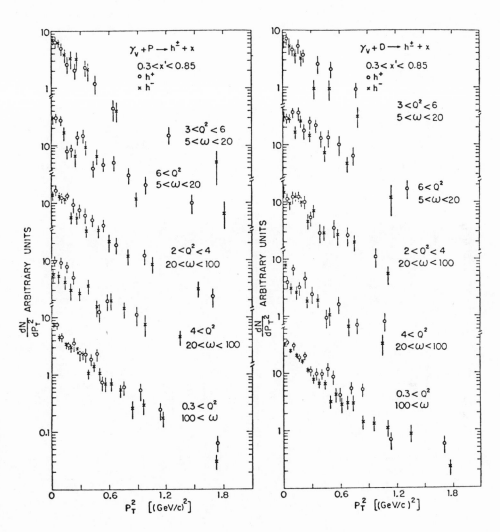

Fig. 14. CHIO data for transverse momentum of inclusive charged
          hadrons from deeply inelastic muon scattering. Data are
          selected by intervals in Q^2 and Bjorken ω.

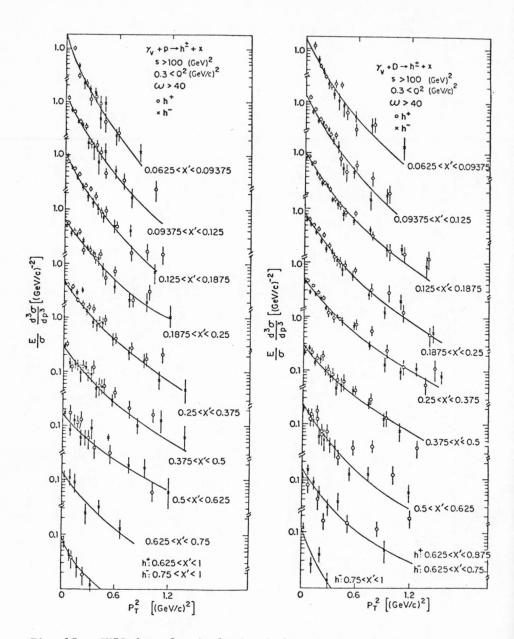

Fig. 15.   CHIO data for inclusive hadron transverse momenta are
           shown for various Feynman x intervals.

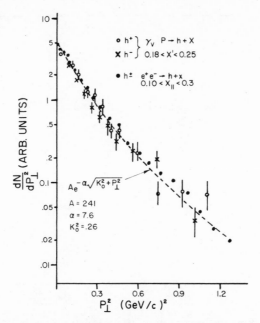

Fig. 16.   Transverse momentum behavior for inclusive hadrons from
           deeply inelastic muon scattering and e^+e^- annihilation
           processes.

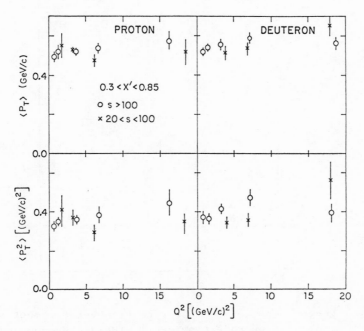

Fig. 17.   Average values for transverse momentum as a function of
           Q^2 for CHIO data on inclusive hadrons.

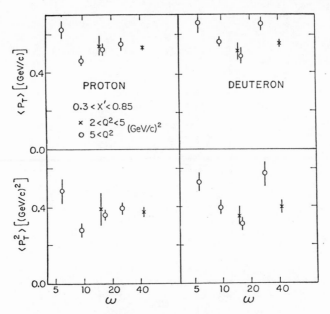

Fig. 18.   Bjorken ω dependence of average transverse momenta of
inclusive hadrons from CHIO muon scattering data.

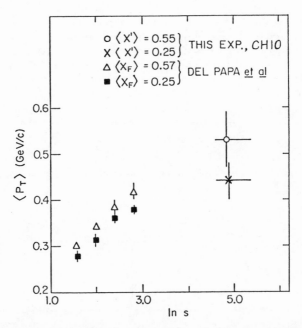

Fig. 19.   Energy dependence of average transverse momentum of
inclusive hadrons from deeply inelastic muon scattering
experiments.

contact with the spectator diquark system thereby preserving quark confinement. The logarithmic growth of the mean transverse momentum with s is just the behavior we would expect of such a statistical process.

The variation of mean transverse momentum with Feynman x is familiar from strong interactions where it has been called the "seagull effect". It has a largely kinematic origin and is connected with conservation principles.

CONCLUSIONS

We see from the above brief review of inclusive hadron properties that there are several striking regularities, not just in the deeply inelastic muon scattering data, but also among several other types of experiments. We should expect these regularities in closely analogous situations such as neutrino scattering (when the appropriate averages have been taken over the additive quark quantum numbers), but when we also see them in more distantly related experiments such as the e^+e^- annihilation and hadron nucleus scattering, we are clearly seeing evidence of a basic (hopefully universal) process in nature.

That process could well be the universal quark fragmentation principle noted above. I personally feel that the concepts and the available data at present put us in the same circumstances vis à vis quark fragmentation as we were in with quark constituents and deeply inelastic lepton scattering in the early 1970's. I hope the evolution and understanding of quark fragmentation will develop in the same satisfying way in the next few years as did the quark constituent model in the 1970's. If it does, I am sure that deeply inelastic scattering of muons and neutrinos along with e^+e^- annihilation will continue to play the major experimental roles.

REFERENCES

1. W. A. Loomis, B. A. Gordon, F. M. Pipkin, S. H. Pordes,
       W. D. Shambroom, L. J. Verhey, Richard Wilson, H. L.
       Anderson, R. M. Fine, R. H. Heisterberg, H. S. Matis, L. Mo,
       L. C. Myrianthopoulos, S. C. Wright, W. R. Francis, R. G.
       Hicks, T. B. W. Kirk, V. K. Bharadwaj, N. E. Booth,
       G. I. Kirkbride, T. W. Quirk, A. Skuja, M. A. Staton, and
       W. S. C. Williams, Phys. Rev. D (to be published).
2. J. T. Dakin, et al., Phys. Rev. D10 (1974) 1401.
3. C. del Papa, et al., Phys. Rev. D15 (1977) 2485.
4. G. Hanson, SLAC-PUB-2118 (1978).
5. H. Rudnicka, et al., VTL-PUB-53, University of Wisconsin (1978).

6. J. F. Martin, et al., Phys. Letts. 65B (1976) 483.
7. I. Daman, et al., Nucl. Phys. B54 (1973) 381.
8. D. J. Bebek, et al., Phys. Rev. 16D (1977).
9. J. T. Dakin and G. J. Feldman, Phys. Rev. 8D (1973) 2862.

FIRST RESULTS FROM THE BERKELEY-FERMILAB-PRINCETON

MULTIMUON SPECTROMETER*

Mark Strovink

Department of Physics and Lawrence Berkeley Laboratory
University of California, Berkeley, California 94720

INTRODUCTION

For the purpose of performing a highly sensitive study of muon-induced final states containing 1, 2, 3, 4, or 5 muons, data have been collected with a new spectrometer in the Fermilab muon beam. Examples of rare multimuon events and first results on J/ψ(3100) production by muons are presented in this report.

In early 1973 we proposed[1] to study multimuon final states at Fermilab using a spectrometer with target thickness and acceptance far exceeding that envisaged in other muon experiments. Of interest were a search for the heavy neutral muon (M^0) predicted by various theories of weak interactions with spontaneously broken gauge symmetry (Fig. 1(a)), and the study of deep-inelastic virtual Compton scattering (Fig. 1(b)). By 1975, observations[2] of the J/ψ(3100) had led us to expect[3] hidden and open charm to be manifested in a variety of muon-produced multimuon final states -- ψ production with $\mu^+\mu^-$ decay (Fig. 1(c)), and charmed hadron pair production with one or two semileptonic decays into a muon (Fig. 1(d)). The discovery[4] of scale non-invariance in lepton-nucleon scattering in the Fermilab muon beam emphasized the need to extend measurements of nucleon structure functions to higher Q^2.

---

*
Research performed by A.R. Clark, K.J. Johnson, L.T. Kerth, S.C. Loken, T.W. Markiewicz, P.D. Meyers, W.H. Smith, M. Strovink, and W.A. Wenzel (Berkeley); R.P. Johnson, C. Moore, M. Mugge, and R.E. Shafer (Fermilab); G.D. Gollin, F.C. Shoemaker, and P. Surko (Princeton).

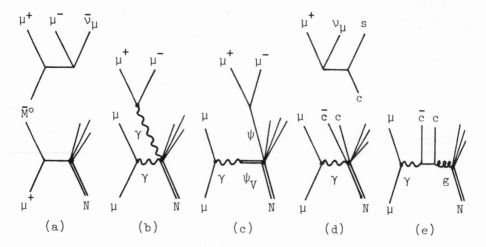

Fig. 1. Feynman diagrams.(a) $\bar{M}^0$ production and decay; (b) virtual
Compton scattering; (c) ψ production; (d) $c\bar{c}$ production and
decay; (e) "photon-gluon-fusion" (additional gluon exchange
is necessary to conserve color).

THE MULTIMUON SPECTROMETER

In the Fermilab muon beam the desired luminosity ($\gtrsim 10^6$ nb^{-1}
per experiment) could be achieved with a massive target (~ 5 kg/cm^2).
High acceptance over the full target length made necessary a
spectrometer magnet integral with the target. Its steel plates were
desirable also as hadron absorbers for calorimetry and muon identification. The earliest computer simulations of multimuon final
states underscored the necessity for full acceptance in the forward
direction, with no blind "beam hole". Inability to determine the
momentum of muons scattered or produced near $0°$ would remove vital
analysis constraints; inability to find all the final-state muons
would alter drastically the interpretation of many events. A dipole field configuration, requiring only one pair of coils for the
full magnet, was most compatible with high forward acceptance. Proportional and drift chambers could withstand the full beam flux at
Fermilab (typically 2×10^6 muons per 1-sec spill without deadening
in the beam area. With this design, the spectrometer optimized and
extended the utility of the existing beam.

Construction of the apparatus depicted in Fig. 2 was completed
in 1977. It consists of 18 25-ton modules each containing 5 10-cm-
thick steel plates, 5 calorimeter scintillators (omitted in modules
16-18), and a pair of proportional (PC) and drift chambers (DC).
Banks of 12 trigger scintillators (S_1-S_{12}) are located in even modules 4-18. The fiducial volume, 1.8×1 m^2 in area, extends 16 m

in the beam direction. Within the central 1.4 x 1 m^2 area of each
magnet plate, the 19.7 kgauss field is uniform to 3% and mapped to
0.2%. Located upstream of module 1 are one additional PC and DC,
63 beam scintillators, 8 beam PC's, and 94 scintillators sensitive
to accidental beam and halo muons.

Figure 3(a) is an exploded view of the detectors in a gap between modules. The 2 cm drift chamber cell and specially designed
readout electronics make possible a system efficiency exceeding 98%
during high-rate ($\lesssim 10^7$ Hz) conditions[5]. Two-fold DC ambiguities are
resolved by the PC anode wires, spaced at $\Delta x = 3$ mm. Coordinates
at 30° (u) and 90° (y) to the bend direction (x) are registered in
the proportional chambers by means of 5-mm wide cathode strips.
Each strip is connected to one input of a differential amplifier in
the network shown in Fig. 3(b). Although spread over many cathode
strips, the induced charge produces a count only in the one or two
electronics channels closest to the peak, even when the pulse
height far exceeds threshold. Through a specially stabilized amplifier, each calorimeter scintillator feeds two ADC's, which together
operate over a range of 0.03 to 1500 equivalent minimum-ionizing
particles. The resolution on hadron energy E_{had}, calibrated using
inelastic muon scattering, is $1.5E_{had}^{\frac{1}{2}}$ (GeV).

The spectrometer was triggered in parallel by ≥ 1, ≥ 2, or
≥ 3 muons in the final state. The required signatures in the scin-

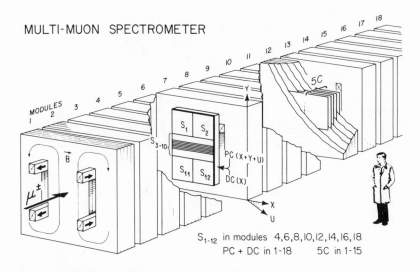

MULTI-MUON SPECTROMETER

S_{1-12} in modules 4,6,8,10,12,14,16,18
PC + DC in 1-18   5C in 1-15

Fig. 2.   Schematic view of apparatus. S_1 - S_{12} are trigger scintil-
          lators. PC and DC are proportional and drift chambers. The
          scintillators labelled 5C perform hadron calorimetry.

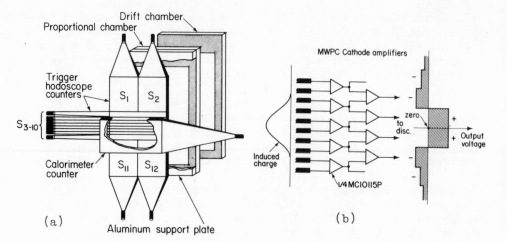

Fig. 3. Apparatus detail. (a) exploded view of detectors within a typical gap between magnet modules; (b) network of differential amplifiers sensing center-of-charge distribution induced on proportional-chamber cathode strips.

tillator hodoscopes and calorimeter counters are listed in Table 1. The multimuon triggers could not be vetoed by hodoscope and calorimeter information; the $\geq 1\mu$ and $\geq 3\mu$ triggers were affected in no way by the calorimeter signals. Under some conditions, when more than one count was required from a scintillator hodoscope, two of the contributing hodoscope elements were required not to be adjacent. During the run the instantaneous beam intensity varied from 0.03 to 0.11 muon per RF bucket. Each bucket was separated from the

Table 1.   Trigger requirements for $\geq 1\mu$, $\geq 2\mu$, and $\geq 3\mu$ final states

Final state	Requirements in each of 3 consecutive trigger banks	Requirement in calorimeter
$\geq 1\mu$	$A \cdot \bar{B}^{*}$	no requirement
$\geq 2\mu$	≥ 2 in A or B	≥ 20 GeV deposited ≥ 2 modules upstream
$\geq 3\mu$	≥ 3 in A or B	no requirement

*"B" refers to S_3–S_{10} in Figs. 2 and 3(a); "A" refers to S_1, S_2, S_{11}, and S_{12}.

next by 19 nsec. The trigger was vetoed by halo muons in the same
RF bucket, or beam muons in the same or adjacent buckets. The dead-
time thereby induced ranged up to $\sim$ 50%. As many as $\sim$ 60 events/1-sec
spill were recorded on a PDP-15 computer with read-in deadtime $\leq$ 15%.

EVENT RECONSTRUCTION AND SIMULATION

Beam muons were momentum-analyzed by systems of proportional
chambers and scintillator hodoscopes interspersed between magnets
producing two separate beam deflections. Pulse heights from calorimeter counters within the spectrometer provided a tentative longitudinal vertex position, using algorithms which depended on the type
of trigger. The beam track then was traced forward to this vertex
using the PC and DC hits. Outgoing tracks were recognized initially
at their downstream end. Hits were added extending the tracks upstream to the vertex, making adequate allowance for Coulomb scattering and momentum uncertainty. In order not to interfere with
rejection of halo tracks or later use of outgoing tracks to pin-
point the vertex, the transverse vertex position was not allowed
to influence this upstream projection. At least 4 PC hits in two
views and 3 hits in the third view were required for each accepted
track. The small electromagnetic showers found along high-energy
muon tracks in iron, due mainly to direct production of electron
pairs, contributed extra hits in the wire chambers which were not
completely rejected at this stage. After the full track was identified, it was possible to apply a complex momentum-fitting algorithm
capable of solving for the Coulomb-scattering angle in each magnet
module, yielding a rigorous χ^2 for the track. By iteration, this
algorithm identified and suppressed the false extra hits.

The beam and secondary tracks next were examined for consistency with a common vertex. The vertex position was moved by iteration in 3 dimensions to minimize the overall χ^2 while including all
associated tracks. This iteration provided an opportunity to final-
ize the inventory of hits on the tracks. After the vertex was fixed,
the coordinates and momentum of each track were redetermined subject
to the condition that it intersect the vertex point.

For analysis of 1μ final states (inelastic scattering) and some
3μ final states (ψ production), the events were subjected to a one
constraint fit demanding equality between the beam energy at the
interaction point and the sum of muon and hadron shower energies in
the final state. Using error matrices produced by the fits to individual tracks, the constraint perturbed all components of each
track momentum. The resulting momentum resolution was 7%-12% (typically 8%) per track. At the ψ mass, the dimuon mass resolution
was 9%. The uncertainty in Q^2 typically was 10%, but was bounded
below by $\sim$ 0.15 $(GeV/c)^2$ because of track angle uncertainty. The
strong features of the analysis described above -- rigorous criteria
for accepting hits on tracks, precise track definition close to the

common vertex, and energy-conserving 1C fit -- establish our confidence in its results.

The acceptance and resolution of the spectrometer were modeled
by a complete Monte Carlo simulation. Coordinates of randomly
sampled beam muons were used to represent the beam. Simulated muons
underwent single and multiple Coulomb scattering, bremsstrahlung, and
other energy-loss straggling in the steel magnet plates. Their
trajectories were deflected in each plate by the precisely mapped
magnetic field. Simulated interactions occurred between muons and
nucleons in non-degenerate Fermi motion, or coherently between muons
and Fe nuclei. At low momentum transfer the effects of nuclear
scattering were taken into account. Coherent and elastic processes
were attenuated by the appropriate form-factors even for forward
scattering (at $|t|_{min}$). Detector resolutions and efficiencies were
included throughout. Monte Carlo events were output on the same
magnetic tape format as raw data, and were reconstructed, momentum-
fit and histogrammed by the same programs.

COPIOUS AND RARE EVENTS

Data were accumulated during the first half of 1978 using
$\sim 4 \times 10^{11}$ (gated) 213 GeV muons, of which $\sim 90\%$ were μ^+. Approximately 5×10^9 100 GeV μ^+ were also used. The extent of the data is
shown in Table 2. Events with 2 or 3 muons in the final state, or
with $Q^2 > 50$ $(GeV/c)^2$, are not rare in this experiment; at least
10^5 of each category are on tape. "Rare" events thereby are re-
defined as those having ≥ 4 muons or 2 missing leptons in the final
state.

Identification of rare events begins in the programs which
reconstruct muon tracks and fit their momenta. Events satisfying
normal analysis criteria which possess unusual characteristics are
saved on microfilm containing tabulated data and computer-generated

Table 2. Extent of data

Final state	Cuts	Total on tape
1μ	$Q^2 > 10$ $Q^2 > 50$ $Q^2 > 100$	10^6 10^5 10^4
2μ	$E_{slow} \gtrsim 10$ GeV	2×10^5
3μ	$E_{slow} \gtrsim 10$ GeV ψ (background subtracted)	2×10^5 10^5

track pictures. A double scan by physicists of the microfilm
identifies a small sample of candidates for which are generated
$\sim$ 1 m^2 pictures containing all raw wire chamber hits, resolved to
better than 1 mm in real transverse coordinates. With the high-
resolution pictures, raw chamber hits are reconstructed by hand
into tracks and the vertex position determined. The track reconstruction program then is forced to fit the event using the
hand-selected information. To be accepted as a rare event, the
result of this hand-forced fit is required to differ in no significant respect from that of the original reconstruction. Close
inspection of each high-resolution picture insures that additional
tracks crossing as few as 3 chambers have not been missed, and that
distinct tracks separated along their full length by as little as
5 mm have not been combined. A particular concern, that two interactions not mistakenly be superimposed, is satisfied by four precautions: a) the trigger demands only one beam track within a 57 nsec
window centered on the event; b) all tracks are required to emanate
from a tightly difined common vertex; c) all tracks are required to
register in the appropriate fine-grained hodoscope scintillators,
sensitive within a $\pm$ 10 nsec window; d) adjacent drift and proportional chamber hits are required to register at a level rejecting
tracks out of time by more than $\sim$ 50 nsec. The accepted tracks
satisfy a tight χ^2 cut separately in both orthogonal views. At
least three hits in the third view unambiguously link the two projections. Each accepted track, passing smoothly through > 12
absorption lengths of steel, can be interpreted only as a muon.
The sign of each muon's charge is at least 8 standard deviations
from the reversed value.

Table 3 presents the properties of four rare events found in an
initial scan of 20% of the data. They consist of one 3μ event with
two missing μ^- or ν_μ, one 4μ event with large pair masses, and two
5μ events. The efficiency of the initial scan exceeded 50%, with
the possible exception of the 3μ event type. Although model-dependent, the detection efficiencies for these events may be estimated
to lie typically in the 10%–20% range. Each event therefore represents a cross-section of order 3 x 10^{-38} cm^2/nucleon.

3μ Event (Two Missing Leptons)

Four events of this type have been produced by neutrino interactions in the CDHS apparatus[6], at a rate consistent with π/K decay
contamination of dimuon events. The small pair masses and transverse momenta in event 851-5726 favor such an interpretation. The
size of the corresponding dimuon sample places a bound on the
probability of π/K decay into a detectable muon ($\geq$10 GeV). This
probability is less than 10^{-4} per hadron shower in the apparatus.

5μ Events

No 5-muon final state (or 4-muon final state with other than

μ or ν_μ incident) has been observed in any previous experiment.
It is natural to try to interpret events 1208-3386 and 851-11418 as
QED phenomena ("muon pentads"). Their negligible hadronic and missing energies support such an interpretation. Event 1208-3386 pos-
sesses values of Q^2, pair masses, and daughter transverse momenta
typical of muon tridents, which are more abundant by a factor consistent with α^{-2}. However, event 851-11418 has kinematic properties
which are puzzling. To explore these properties, all lowest-order
QED diagrams have been examined. One diagram has been found to
minimize the product of the denominators in the lepton and photon
propagators. It is a Bethe-Heitler-like graph with the incident
muon scattered into track 2 and coupled to track 5, and the target
coupled to track 3. Tracks 4 and 6 form a pair radiated by track 5.
The unseen $p_\perp$ is 1.8 ± 0.4; for the above choice of scattered muon,
Q^2 is 3.5 ± 0.6, and the daughter (unscattered) muons have a combined
mass $M_{3456} = 3.5 ± 0.3$. Of the muon tridents observed in the same

Table 3. Rare Events

Event	Scattered Muon	Energies (GeV)	Masses (GeV/c^2)	Unseen $p_\perp$ to γ_V (GeV/c)
851-5726 $\mu^- \to \mu^- \mu^- \mu^+$ 1 2 3 4	2 Q^2=0.1±0.1 ν =160±6	E_3= 19± 2 E_4= 11± 2 E_{had}=103±15 E_{miss}= 27±17	M_{34}=0.5±0.1	0.3±0.1
1191-5809 $\mu^+ \to \mu^+ \mu^+ \mu^- \mu^-$ 1 2 3 4 5	2 Q^2=0.3±0.2 ν =158±7	E_3= 26± 3 E_4= 18± 2 E_5= 25± 4 E_{had}> 57±11 E_{miss}< 31±14	M_{34}=3.0±0.3 M_{35}=3.2±0.3 M_{345}=4.6±0.3	2.0±0.2
1208-3386 $\mu^+ \to \mu^+ \mu^- \mu^- \mu^+ \mu^+$ 1 2 3 4 5 6	2 Q^2=0.2±0.2 ν =149±9	E_3= 50± 5 E_4= 27± 3 E_5= 61± 6 E_6= 10± 2 E_{had}= 6± 3 E_{miss}= -4±13	M_{35}=1.3±0.2 M_{36}=0.3±0.1 M_{45}=0.4±0.1 M_{46}=0.5±0.1 M_{3456}=2.0±0.2	0.1±0.3
851-11418 $\mu^- \to \mu^- \mu^- \mu^+ \mu^+ \mu^-$ 1 2 3 4 5 6	Q^2=3.5±0.6 ν = 61±12	E_3= 13± 2 E_4= 19± 2 E_5= 15± 2 E_6= 10± 2 E_{had}= 5± 3 E_{miss}= -1±13	M_{34}=2.3±0.2 M_{35}=2.0±0.2 M_{46}=0.5±0.1 M_{56}=0.3±0.1 M_{3456}=3.5±0.3	1.8±0.4

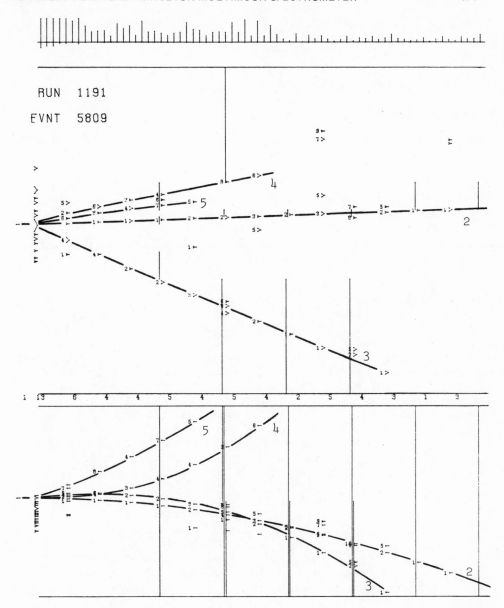

RUN 1191

EVNT 5809

Fig. 4. Computer-generated picture of proportional-chamber hits in event 1191-5809. Top frame: elevation view; bottom frame: plan view. Short vertical lines at the top are calorimeter counter pulse heights. Long vertical lines are projections of trigger counters which were tagged. Heavy broken lines are tracings of the computer-reconstructed trajectories. Large (typed) numbers identify each track.

data set, fewer than 200 have $Q^2 > 2.5$ and combined daughter muon mass > 2.8; fewer than 20% of these have unseen p > 1.4. A parent sample of that size would not be expected to yield one event with an additional energetic muon pair. The event requires a more plausible explanation.

High-Mass 4μ Event

     One neutrino-induced four-lepton final state has been observed by each of three groups: CDHS[7], BFHWW[8], and HPWFOR[9]. The respective authors expect $\sim$ 1/5 of an event as background in the CDHS and HPWFOR samples, and $\sim 10^{-3}$ events in the BFHWW sample. The muon-induced 4μ event 1191-5809 is pictured in Fig. 4. Its kinematics are summarized in Table 3 and compared with those of the neutrino-induced events in Table 4. The muon-induced event is qualitatively different from the others. The softest lepton has at least 4x the energy and the lightest μ⁺μ⁻ daughter pair has at least 4x the mass of any neutrino-induced counterpart.

     A relatively model-independent limit can be placed on this event's most obvious potential background -- single muon production due to any process, in *random* association with μ⁺μ⁻ pair production due to any process within the same diagram. Choosing the leading secondary (track 2) as the scattered muon produces the smallest Q^2 and determines the virtual photon direction, represented by the central axis of Fig. 5. The μ⁺μ⁻ pair is interpreted as that formed by tracks 3 and 4, because it is no more massive than that formed by tracks 3 and 5, but has only half the transverse momentum. Let

N_0 = calculated number of μ scatters with y > 1/2 corresponding to the sensitivity of this data set;

Table 4. Comparison of event 1191-5809 to published neutrino-induced 4-lepton events

Group	Final state leptons	Smallest lepton energy (GeV)	Smallest μ⁺μ⁻ or e⁺e⁻ pair mass (GeV/c)²
CDHS	μ⁺μ⁻μ⁺μ⁻	4.5	0.4
BFHWW	μ⁺e⁻e⁺e⁻	0.9	0.8
HPWFOR	μ⁻μ⁺μ⁺μ⁻?	3.0	0.5
This experiment	μ⁺μ⁺μ⁻μ⁻	18.3	3.0

N_1 = observed number of 2μ final states in which y > 1/2 and the non-leading μ has E > 15 and p > 1.7 (Fig. 5);
N_2 = observed mumber of 3μ final states with E_{pair} > 30, $(ν - E_{pair})$ > 100 GeV and M_{pair} > 2.75.

The number N_B of background events is then

$$N_B < ε_B N_1 N_2 / (ε_1 ε_2 N_0),$$

where the ε's are detection efficiencies. Using $ε_B / ε_1 ε_2 < 10$, $N_1 < 200$, $N_2 < 13$ and $N_0 = 3.7 \times 10^7$, one obtains $N_B < 7 \times 10^{-4}$.

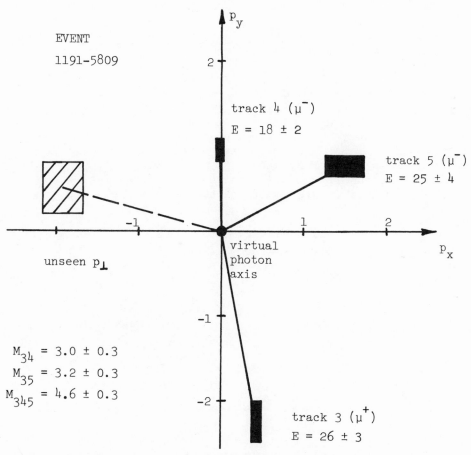

EVENT
1191-5809

track 4 (μ⁻)
E = 18 ± 2

track 5 (μ⁻)
E = 25 ± 4

unseen $p_⊥$

virtual
photon
axis

M_{34} = 3.0 ± 0.3
M_{35} = 3.2 ± 0.3
M_{345} = 4.6 ± 0.3

track 3 (μ⁺)
E = 26 ± 3

Fig. 5. Momenta transverse to the direction of the virtual photon in event 1191-5809. Dark boxes: secondary muons other than the scattered (spectator) muon. Hatched box: unseen transverse momentum carried off by neutrinos and/or hadrons. Box sizes indicate measurement errors.

Table 3 and Fig. 5 constrain kinematically the origin of this
event. Rigorously, the parent system for the 3 daughter muons has
a mass between 4.3 and 17.6 GeV (limits include one standard deviation in measuring error). If emitted at 90° in a frame travelling
in the virtual photon direction, tracks 3, 4, and 5 require that
frame to have γ = 11.3, 19.1 and 14.9 for an average of 15.1. If
this frame is the rest frame of a parent system carrying all of the
virtual photon energy, the parent mass is 158/15.1 = 10.5 GeV. If
the daughter muons are descendants of a heavy quark pair $Q\bar{Q}$ produced in the virtual photon direction, $M_Q > 4$ GeV. However, such
cascades should produce a larger number of events not containing
the analog of track 3. Both M_{34} and M_{35} are consistent with the ψ
mass. If this ψ was produced by Drell-Yan fusion of charmed quarks,
the third muon is most likely a descendant of the unfused member
of the projectile charmed quark pair. This intepretation requires
the charmed quarks to possess considerable transverse momentum. In
any case, charmed-quark fusion is constrained by these data to account for no more than $\sim 10^{-2}$ of ψ photoproduction.

ONE-MUON AND TWO-MUON FINAL STATES -- ANALYSIS STATUS

The extent of data with one and two muons in the final state
has been mentioned in Table 2. The acceptance for muon scattering
is shown in Fig. 6. Four years have elapsed since the original observation of scale-noninvariance[4] in muon scattering. Systematic
errors in the current analysis are steadily decreasing toward a
level appropriate to new results in this mature field.

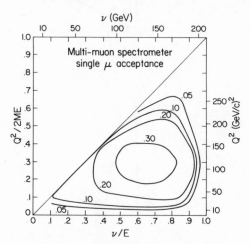

Fig. 6.   Calculated detection efficiency vs. Q^2 and ν for inelastic
          muon scattering. An average over the full target length
          is shown. If the target is restricted to the few modules
          furthest upstream, the low-Q^2 acceptance is much more uni-
          form.

The signatures for $\bar{M}^0$ production and decay in 2μ final states
(Fig. 1(a)) are sufficiently unique that useful signal levels/
limits can be determined even in the presence of copious background
from charm production (Fig. 1(d)) and other sources.   Such limits
will be reported on the basis of a larger sample than has been analyzed at present. With the hadron calorimeter calibrated using inelastic scattering, substantial average missing energy is observed
in the 2μ sample (Fig. 7(a)). Two-muon final states associated
with backgrounds to charm production -- π/K decay, and μ⁺μ⁻ pair
production with one muon undetected -- also are expected to exhibit
exhibit missing energy. Figure 7(b) shows the missing energy associated with 3μ final states produced by highly inelastic muon interactions. When the μ⁺μ⁻ pair mass lies between the ψ region and the
small values expected[10] from Bethe-Heitler and bremmstrahlung processes, the missing energy takes on a modest average value. In the
3μ channel, non-negligible backgrounds to pair-production of heavy
quarks are expected to conserve visible energy. In comparison to
the 2μ value, the average value of the 3μ missing energy suggests
the presence of significant visible-energy-conserving background
even in the intermediate-mass sample. Therefore it is clear that
inelastic 3μ final states cannot naively be interpreted as arising

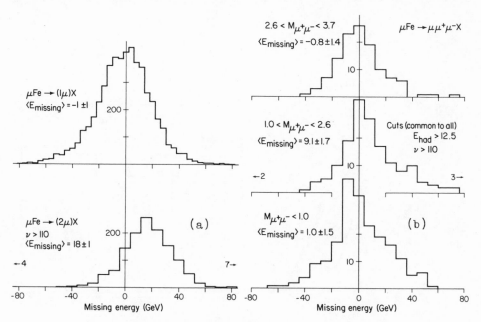

Fig. 7.   Difference between beam energy and the sum of secondary
          muon and hadron shower energies at the vertex.   (a) 1μ
          (top) and 2μ final states; (b) 3μ final states in three
          regions of μ⁺μ⁻ pair mass. The 1μ result is used to
          calibrate the calorimeter; the 2μ and 3μ results are
          sensitive to the indicated ν cut.

largely from charmed quark production and decay. When properly corrected for backgrounds, the same c$\bar{c}$ production cross-section should
be measurable in final states containing three muons, opposite-sign
muon pairs, and same-sign pairs.

J/ψ(3100) PRODUCTION BY MUONS

Previous experiments at Fermilab[11,12], SLAC[13], and Cornell[14]
have measured ψ photoproduction. No previous experiment has detected ψ production by spacelike photons. Reported here are
1000 ± 80 $\mu^+\mu^-$ pairs from ψ decay (Fig. 1(c)), drawn from 16834 muon
interactions producing 3 fully-reconstructed muon tracks in the
final state. These represent 12% of the data on tape.

Figure 8(a) displays the spectrum of $\mu^+\mu^-$ pair masses before
and after continuum subtraction. The smoothness of the continuum
emphasizes the uniformity of acceptance in the ψ mass region. Only
one choice per event of $\mu^+\mu^-$ pairing is plotted. The unpaired
(scattered) muon is taken to be the more energetic if the candidates differ in energy by more than a factor of two. Otherwise, it

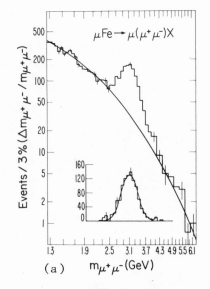
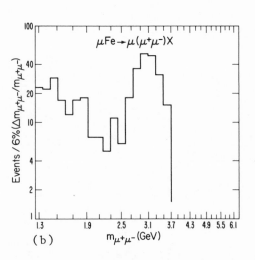

(a)

(b)

Fig. 8.   Invariant mass spectrum of muoproduced $\mu^+\mu^-$. (a) All
          data, before and after subtraction of the continuum (re-
          presentative fit is shown). The Gaussian curve is cen-
          tered at 3.1 GeV with a 9% rms width. (b) Data with (12 ± 5)
          < E_{had} < (36 ± 9).

is taken to be the candidate making the smallest angle relative to the beam direction. The effects of this and many alternate pairing algorithms have been studied extensively. Used on Monte Carlo events, this algorithm retains 92% of the ψ's in the mass peak and distributes the rest smoothly between 0.7 and 6 GeV. There is little tendency for mispaired events to dominate particular kinematic regions, such as high mass, Q^2, or ν. The same is not true for a number of other algorithms with similar overall probability of success in identifying the correct pair. Under less favorable experimental conditions, inability to reconstruct all 3 final state muons in effect can impose a pairing choice which is far from optimum. It is seen in Fig. 8(a) that the centroid of the ψ peak is consistent with 3.1 GeV, and its width with 9% rms. The latter is the average resolution calculated from first principles for each event, taking detector resolution and Coulomb scattering into account. It is also the width of the mass peak for Monte Carlo events. The effects on this mass spectrum of a $\psi' \rightarrow \mu^+\mu^-$ component at $\sim$ 5% of the ψ rate occur at the same subtle level introduced by possible variations in resolution shape, energy calibration, and continuum parametrization. A more promising method for studying the ψ' utilizes its cascade to hadrons $+ \psi \rightarrow \mu^+\mu^-$. The ratio of hadron energy (detected in the calorimeter) to ψ energy is expected to peak at $\sim$ 0.2. Figure 8(b) shows the enhancement in the ψ peak relative to continuum when a moderate energy deposit in the calorimeter is required.

The data were absolutely normalized using optimum experimental conditions; low beam intensity and an interaction region (first 8 magnet modules) allowing very long tracks downstream*. The effects of acceptance were removed by a Monte Carlo parametrization (described below) fitting all characteristics of the ψ sample. Allowing for the 7% $\psi \rightarrow \mu^+\mu^-$ branching ratio, the total cross-section i

σ/nucleon (μ Fe $\rightarrow \mu\psi$X) = 0.76 $\pm$ 0.22 nb.

Monte Carlo corrections for nuclear coherence, shadowing, and $|t|_{min}$ effects yield

$\sigma(\mu N \rightarrow \mu\psi X)$ = 0.67 $\pm$ 0.20 nb.

Essentially all the error is due to normalization uncertainty. A calculation[10] using the "photon-gluon-fusion" diagram (Fig. 1(e)) has yielded a cross-section consistent with this result.

In order to reproduce the experimental ratio of coherent to incoherent ψ production from Fe nuclei and to parametrize $|t|_{min}$ effects, the Monte Carlo simulation assumes the t dependence of the

_____
*
If normalized using the full data sample, the cross-sections are reduced by less than one standard deviation.

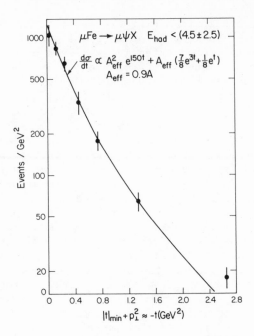

Fig. 9.   Comparison of experimental t distribution to a Monte
          Carlo calculation (solid line) using the indicated para-
          metrization of $d\sigma/dt$. The (unresolved) coherent compo-
          nent steepens the observed slope below $|t| \approx 0.5$.

cross section to be

$$d\sigma/dt \ (\gamma \ Fe \rightarrow \psi X) = G(t) \ d\sigma/dt \ (\gamma N \rightarrow \psi N) \ (t = 0)$$
$$G(t) = A_e{}^2 \exp(\alpha t) + A_e[(1-\varepsilon\delta)\exp(\beta t) + \varepsilon\delta\exp(\delta t)]. \quad (1)$$

The t resolution of the spectrometer is such that a δ-function at
$t = 0$ is smeared into $\sim \exp(5t)$. Therefore, data from other photon-
nucleus experiments[15], mainly at lower energies, are averaged to
set the coherent slope α to 150 $(GeV/c)^{-2}$. The shadowing factor
A_e is taken to be 0.9 x (A = 55.85) based on electron-nucleus scattering data[16] at similar average Q^2. We have used $\beta = 3 \ (GeV/c)^{-2}$,
$\delta = 1 \ (GeV/c)^{-2}$, and $\varepsilon = 1/8$. These choices are consistent with
high-energy ψ photoproduction data[11] and with the experimental t
distribution, as shown in Fig. 9. In this and in Fig. 10, each
data point is the result of a continuum subtraction like that in
Fig. 8(a); the errors include uncertainties due to variation in
the form and in the parameters used to fit the continuum. We have
investigated the effects of varying α, A_e, β, δ, and ε over a
range allowed by these data and (in the case of α and A_e) by other
measurements. The effects on the shape of the Q^2 distribution are
negligible. The effects on the E_γ distribution are summarized in
Table 5. Variation of any one parameter "tilts" the distribution
by less than $\sim 5\%$.

Table 5. Percent reduction in $d\sigma/dt(t = 0)$ for ψ production by
virtual photons, induced by variations in nuclear and
nucleon parameters α $(GeV/c)^{-2}$, A_e, ε, β $(GeV/c)^{-2}$, and
δ $(GeV/c)^{-2}$.

Parameter	α	A_e	ε	β	δ
Best value	150	50.27	1/8	3	1
Varied value	135	55.85	1/5	2.5	0.5
$<E_\gamma>$ (GeV)					
34	3	11	10	12	5
56	5	12	9	10	4
77	5	13	8	9	3
106	5	13	7	8	3
140	5	14	7	8	3

Figure 10 presents the basic ψ production results, in the form
of cross-sections for the equivalent flux[17] of transversely polarized virtual photons. Since the Q^2 averaged best-fit physical
value of σ_L/σ_T in these data is zero, we neglect any longitudinal
cross-section at the present statistical level. In order to suppress contamination from $\psi' \to \psi$ + hadrons and from other inelastic
processes, the energy deposit in the calorimeter is required not
to exceed an amount consistent with elastic ψ production. The E_γ
dependence in Fig. 10(a) is obtained assuming that the photon
cross-section has the Q^2-dependence given by the solid line in
Fig. 10(b), and vice versa. In order to make comparison with SLAC
data[13] accumulated mainly at $t \approx 0$, we express the observed cross-
section for ψ photoproduction in terms of $d\sigma/dt(t=0)$. This is done
by assuming in the Monte Carlo that $d\sigma/dt(t = 0)$ has the value indicated by the solid line in Fig. 10(a). The cross-section over all
t is obtained using Eq. (1). The quoted values of $d\sigma/dt(t = 0)$ are
equal to the solid line multiplied by the ratio of subtracted real
data to Monte Carlo data within each E_γ bin. This procedure corrects
precisely for acceptance, nuclear coherence, shadowing, and $|t|_{min}$
effects. It should be remembered, however, that the measured
quantity is the cross-section rather than its intercept at $t=0$;
the intercept could be affected by variations of the t slope over
the energy range. Significant variations of this sort are not observed when the SLAC[13] and Fermilab[11] ψ-photoproduction data are
compared.

Over the range of $<E_\gamma>$ (34 to 140 GeV) the ψ-production cross-
section is observed in Fig. 10(a) to rise by a factor close to two.
While a QCD calculation[18] using photon-gluon-fusion also predicts a
significant rise in this energy region (hatched band), that rise is
seen upon close examination to be significantly larger than can be
accomodated by these data. Both the calculation and the data evaluate σ rather than $d\sigma/dt$ ($t = 0$); each in the same way has been

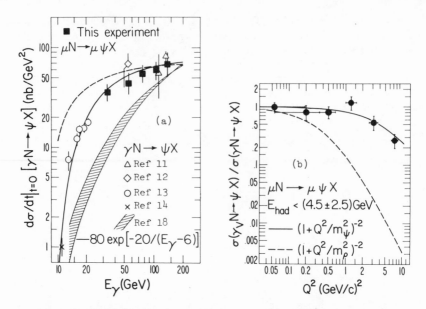

Fig. 10. Cross sections for ψ production by the equivalent flux of
         virtual photons. (a) E_γ-dependence at t = 0. The muo-
         production points have an overall ± 30% normalization un-
         certainty (not shown). (b) Q^2-dependence, normalized to
         1 at the lowest Q^2 point. Horizontal error flags show
         typical Q^2 resolution.

converted to $d\sigma/dt$ (t = 0) for display in Fig. 10(a). Within vector-
meson dominance (VMD), the ratio of the solid line to the broken
line ($p_{c.m.}^\psi{}^2 / p_{c.m.}^\gamma{}^2$) gives the energy-dependence of the square
of the ψ-nucleon total cross section. Taking together all data in
Fig. 10(a), the threshold behavior of ψ photoproduction now has
been measured more precisely than, for example, that of ϕ photoproduction.

Figure 10(b) establishes the shallow Q^2 dependence of ψ photoproduction. The data have been normalized to the lowest Q^2 point
as an approximation to $Q^2 = 0$. The result fits a propagator-depen-
dence $(1 + Q^2/M^2)^{-2}$ with M = 2.7 ± 0.5 GeV. Within VMD, consistency
of M with the ψ mass implies that no additional Q^2 dependence is
needed to parametrize the $\gamma-\psi$ coupling. The kinematics of photon-
gluon fusion also produce a Q^2 dependence similar to that observed,
although the QCD calculations[18] are not presented in a form easy to
compare with Fig. 10(b). If these calculations can possess fundamental validity, the broad range of parameters measurable in ψ
production by muons can provide many of the crucial tests.

    I am pleased to acknowledge the considerable efforts made by
Fermilab in behalf of this experiment, and in particular the cooperation of its Neutrino Department. I am very grateful for the hard
work of my experimental colleagues and the collaborating support
groups. In particular, Mr. T. Markiewicz contributed directly and
substantially to the analysis of ψ production. I wish especially
to acknowledge the basic contributions made by Mr. J. Caron to the
analysis of multimuon final states in two experiments, both as a pro-
grammer and as a physicist. This work was supported by the High
Energy Physics Division of the U.S. Department of Energy under
contract Nos. W-7405-Eng-48, EY-76-C-02-3072, and EY-76-C-02-3000.

REFERENCES

1. R. Cester, C.M. Hoffman, M. Strovink, and F.C. Shoemaker,
       Fermilab Proposal 203 (1973, unpublished).
2. Reviewed by S.C.C. Ting, Rev. Mod. Phys. 49, 235 (1977);
       B. Richter, Rev. Mod. Phys. 49, 251 (1977).
3. A.R. Clark, E.S. Groves, L.T. Kerth, S.C. Loken, M. Strovink,
       W.A. Wenzel, R. Cester, F.C. Shoemaker, P. Surko, M.S.
       Witherell, and R.P. Johnson, Fermilab Proposal 391 (1975,
       unpublished).
4. Y. Watanabe, L.N. Hand, S. Herb, A. Russell, C. Chang, K.W.
       Chen, D.J. Fox, A. Kotlewski, P.F. Kunz, S.C. Loken, M.
       Strovink, and W. Vernon, Phys. Rev. Lett. 35, 898 (1975);
       C. Chang, K.W. Chen, D.J. Fox, A. Kotlewski, P.F. Kunz,
       L.N. Hand, S. Herb, A. Russell, Y. Watanabe, S.C. Loken,
       M. Strovink, and W. Vernon, Phys. Rev. Lett. 35, 901 (1975).
5. G. Gollin, M.V. Isaila, F.C. Shoemaker, and P. Surko, IEEE
       Trans. Nuc. Sci. NS-26, 59 (1979).
6. T. Hansl et al., Phys. Lett. 77B, 114 (1978).
7. M. Holder et al., Phys. Lett. 73B, 105 (1978).
8. R.J. Loveless et al., Phys. Lett. 78B, 505 (1978).
9. A. Benvenuti et al., Phys. Rev. Lett. 42, 1024 (1979).
10. V. Barger, W.Y. Keung, and R.J.N. Phillips, Univ. of Wisconsin
       preprint COO-881-83 (1979).
11. B. Knapp et al., Phys. Rev. Lett. 34, 1040 (1975); W.Y. Lee,
       in Proc. Int. Symp. on Lepton and Photon Interactions at
       High Energies (DESY, Hamburg, 1977);
       M. Binkley, private communication.
12. T. Nash et al., Phys. Rev. Lett. 36, 1233 (1976).
13. U. Camerini et al., Phys. Rev. Lett. 35, 483 (1975).
14. B. Gittelman et al., Phys. Rev. Lett. 36, 1616 (1975).
15. See, for example, A. Silverman, in Proc. Int. Symp. on Elec-
       tron and Photon Interactions at High Energies (Daresebury,
       1969), Table 2.

16. W.R. Ditzler et al., Phys. Lett. <u>57B</u>, 201 (1975).

17. L.N. Hand, Phys. Rev. <u>129</u>, 1834 (1963).

18. M. Glück and E. Reya, Phys. Lett. <u>79B</u>, 453 (1978);
        M. Glück and E. Reya, DESY preprint 79/05 (1979). In
        Fig. 10(a) we have multiplied their result for σ by 2.4
        to obtain $d\sigma/dt$ (t = 0).

DEEP INELASTIC MUON NUCLEON SCATTERING:

A PROBE OF HADRONIC STRUCTURE

H. E. Stier

University of Freiburg, Germany, Physics Faculty

D7800 Freiburg, Hermann-Herder-Strasse 3

INTRODUCTION

Among the different ways of probing the nucleon structure, high energy muon-nucleon scattering plays an important role. As shown in Fig. 1a, the scattered muon interacts with the nucleon via the exchange of a virtual photon with the energy ν and the four momentum transfer squared q^2. The emission of the virtual photon from the pointlike muon is exactly calculable in Quantum-electrodynamics. But the interaction of this virtual photon with the nucleon will strongly depend on the structure of the nucleon. If the nucleon consists of partons, the photon may interact with a single parton, especially at high values of q^2, as shown in Fig. 1b.

The first observation of scaling[1,2], about ten years ago, was a hint on the quark parton substructure of the nucleon. The structure functions F_1 and F_2 which determine the inelastic charged lepton-nucleon scattering seemed to depend only on the dimensionless variable

$$x = \frac{q^2}{2M\nu} \quad .$$

The experimental observation could easily be explained by the naive quark-parton model, which predicted for the structure function F_2 the following relation:

$$F_2(x) = \sum_i e_i^2 \, xq_i(x) \tag{1}$$

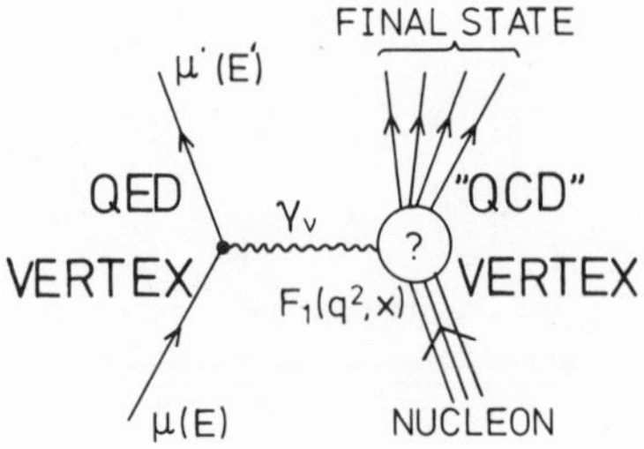

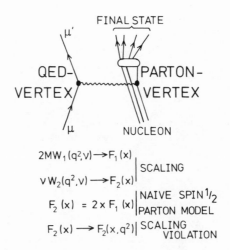

$$q^2 = 4EE' \sin^2 \frac{\theta}{2}$$

$$\nu = E - E'$$

$$x = \frac{q^2}{2M\nu}$$

$$\frac{d^2\sigma}{dq^2 d\nu} \propto \frac{4\pi\alpha^2}{q^4} [W_2(q^2,\nu) + W_1(q^2,\nu)]$$

Fig. 1a.   Kinematics of deep inelastic muon nucleon scattering
in the form factor picture

$$2MW_1(q^2,\nu) \rightarrow F_1(x) \Big| \text{SCALING}$$

$$\nu W_2(q^2,\nu) \rightarrow F_2(x) \Big|$$

$$F_2(x) = 2x F_1(x) \Big| \begin{matrix} \text{NAIVE SPIN}^1/_2 \\ \text{PARTON MODEL} \end{matrix}$$

$$F_2(x) \rightarrow F_2(x,q^2) \Big| \begin{matrix} \text{SCALING} \\ \text{VIOLATION} \end{matrix}$$

Fig. 1b.   Inelastic muon nucleon scattering and scaling as
seen in the quark-parton picture

x gets the physical interpretation of being the fractional longitudinal momentum inside the nucleon, e_i is the charge of the ith constituent and q_1 - the probability of the ith quark to have the fraction
x of the total longitudinal momentum. $F_2(x)$ is thus the fraction of
total momentum inside the nucleon carried by charged constituents
weighted by the squares of the constituent charges.

The scaling violation, observed for the first time in 1974[3,4]
as a weak but significant dependence of F_2 on x and q^2, showed
the necessity of more accurate experimental determination of F_2 up
to high values of q^2 and over a large range of x values between
0 and 1. It has been interpreted[5] as showing the influence of
gluon emission inside the nucleon.

The field theory of quarks and gluons (QCD) predicts asymptotic
freedom for the quarks. The interactions that bind the quarks
together, i.e., the interactions of quarks and gluons, will show up
if the time of the current quark interaction is short enough to be
sensitive to interactions of quarks and gluons on still shorter time
scales. If at fixed values of x, the variable q^2 increases, the
resolving power of the current probe increases. It can now show
whether the quark has just radiated a gluon and whether it, therefore,
interacts with a lower momentum. The q^2 dependence of the scaling
violation can answer questions about the gluon itself. The precise
inside the nucleon and also about the glue itself. The precise
determination of the scaling violation is, therefore, very important.

EXPERIMENTAL APPARATUS

A precise determination of the nucleon structure functions was
one of the main aims of the physics program of the European Muon
Collaboration (EMC). Twelve different institutions (CERN - Hamburg -
(DESY) - Freiburg - Kiel - Lancaster - Annecy (LAPP) - Liverpool -
Oxford - Rutherford (RHEL) - Sheffield - Turin - Wuppertal) joined
their efforts to prepare the experimental apparatus for muon nucleon
scattering experiments at energies up to 300 GeV, known as the NA2
experimental program[6].

As a first step towards these experiments a suitable muon beam
had to be built at CERN. The extracted 400 GeV proton beam hitting
a 50 cm long Be target produced pions and kaons. Their decay muons
were used for the muon beam. To increase the beam intensity, compared
to another high energy muon beam at FNAL, the principle of strong
focussing was used to focus kaons and pions in a 600 m long
decay region so that a maximum muon intensity could be obtained in
the beam direction.

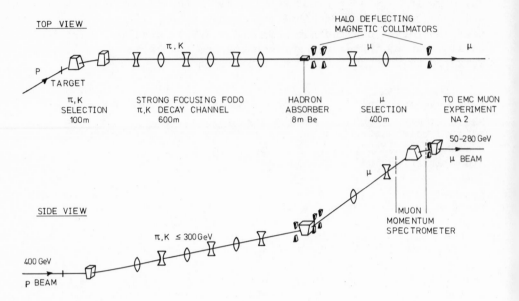

Fig. 2 Schematic layout of the CERN–SPS muon beam

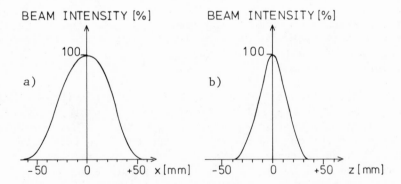

Fig. 3 Horizontal and vertical beam distributions at
the entrance of the NA2 experiment

Remaining hadrons were absorbed in an absorber consisting of
8 m Be and 2 m polyethelene. Besides a high beam intensity, a minimized
muon halo has been aimed for. For this purpose, movable magnetized
iron scrapers along the beam have been used to deflect halo muons
out of the primary beam direction. Magnetization and position of
these scrapers have been changed to minimize the muon halo. The main
muon beam properties are given in the table. A schematic layout of
the beam is shown in Fig. 2.

The horizontal and vertical beam distributions at the entrance
of the NA2 experiment are given in Figs. 3a and b. The beam momentum distribution is shown in Fig. 4. The accepted momentum band
is Δ p/p = ± 5%. With the aid of beam momentum spectrometer in the
beam (see Fig. 2), the momentum of the incoming muon is known to
better than Δ p/p = 0.5%. In this way, the kinematics of the muon
nucleon scattering process is very well defined by knowing the
accurate values of the incoming muon momentum.

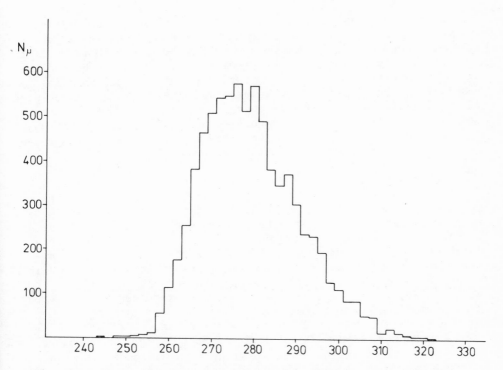

Fig. 4   Beam momentum distribution at an energy of 280 GeV.

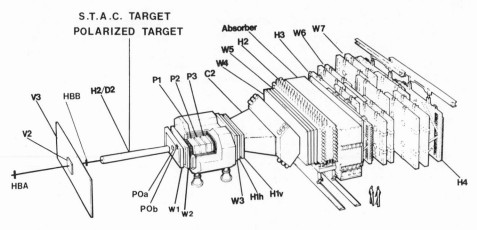

Fig. 5   Artists view of the NA2 experiment

Figure 5 shows an artist's view of the NA2 experimental
apparatus, the so-called EMC Forward Spectrometer. The position
and the direction of the incoming muon are determined in two scintillation counter beam hodoscopes (BHa and BHb) just in front of the
6 m long liquid hydrogen or deuterium target, which can be replaced
by an iron scintillator sandwich target (STAC Target: Sampling Total
Absorption Counter Target). The beam hodoscopes contain three planes
of 60 elements each. Muons in the beam, as well as muons scattered
under very small angles, are detected in 12 planes of proportional
chambers (Pϕa and Pϕb) which have a 15 cm ϕ sensitive area and can
stand intensities up to 10^8 particles/second.

Muons scattered at larger angles are detected in 16 planes of
driftchambers (W1, W2) in front of the spectrometer magnet. These
chambers have dead areas in the beam region, which are covered by
the P$\emptyset$ chambers. The spectrometer, therefore, has no acceptance hole
in the forward direction; the whole area is covered by detectors.
Three proportional chambers in the gap of the 5.25 T m forward
spectrometer magnet serve to connect the particle tracks found in
front of the magnet with the corresponding tracks in drift
chambers W3, 4 and 5 behind the magnet to determine the scattered
muon or produced hadron momenta.

A Čerenkov counter C2, with N_2 or N_e filling, can separate kaons
from pions within limited momentum ranges. A hadron calorimeter H2
is used to determine hadron energies and to separate electrons from
hadrons. Two meters of magnetized iron stop all hadron tracks.
Only muons penetrate this hadron absorber. Their tracks are
detected in large area drift chambers W6, and 7.

Scintillation counter hodoscopes H1, H3 and H4 are used to trigger the apparatus. Computer programmable coincidence matrices reduce the trigger rates by accepting only tracks pointing to the target region and by triggering on events with scattering angles and scattered particle momenta larger than a minimum value (Fig. 6). Typical values of the minimum angle are 0.5°, 1° and 2° depending on the primary beam energy, and about 10 GeV for the minimum muon momentum. A schematic view of the basic trigger is shown in Fig. 6.

The scintillation counter hodoscope H5 which covers the beam area, is used to monitor the incoming muon beam intensity which can additionally and independently be measured in the beam hodoscopes BHa and BHb. None of these hodoscopes has been used in the trigger.

Three sets of scintillation counter hodoscopes (V1, V2, V3) veto events with incoming halo muons. Due to their unknown and mostly lower momenta, halo muons would produce unwanted triggers.

The counter information is read out via a computer system consisting of four interconnected PDP 11/70 computers[7]. The readout is fast enough to allow up to 150 events to be read out during a spill time of about one second.

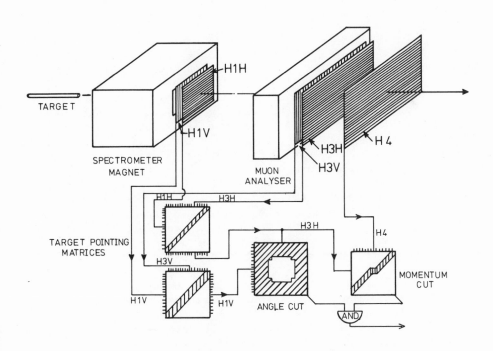

Fig. 6   Schematic view of the trigger system

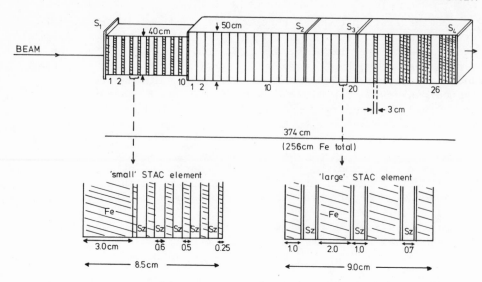

Fig. 7   The heavy target STAC

EXPERIMENTAL RESULTS

        Data taking started in August 1978. This report contains the
first very preliminary results on the determination of structure
functions with a Sampling Total Absorption Counter Target (STAC
Target). It consists of iron scintillator sandwiches and is used
as a calorimeter to determine the hadronic and electromagnetic
energy deposit by pulse height measurements in the scintillator
elements. As shown in Fig. 7 it has a small upstream part and a
larger downstream part. The total target length is 3.74 m and it
contains 2.56 m iron in total. Each of the 36 sandwich elements
is viewed by a simple phototube.

        Figure 8 shows the correlation between the energy deposit E_{STAC}
measured in the STAC and the energy loss ν of the scattered muon
as determined in the magnetic Forward Spectrometer. The linear
correlation, as well as the resolution of about ± 30 GeV, agree very
well with a Monte Carlo simulation of the experimental apparatus.

        A typical event distribution in the q^2-ν plane is shown in
Fig. 9 for a minimum muon scattering angle of 2° in the trigger. This
minimum angle is smeared out by the length of the target. This
distribution should be compared with the total kinematic range
accessible in this experiment as it is shown in Fig. 10. Even if
only ≈ 10 % of all data taken with the STAC are shown, there are
events up to $q^2 = 200$ (GeV^2/c^2).

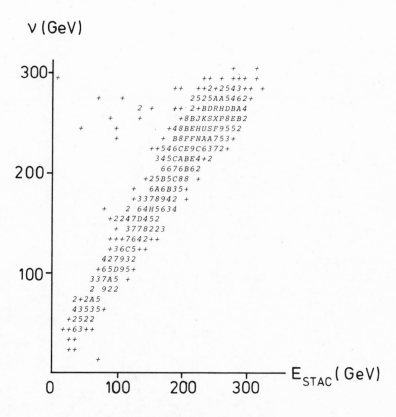

Fig. 8   Correlation between E_{STAC} and ν

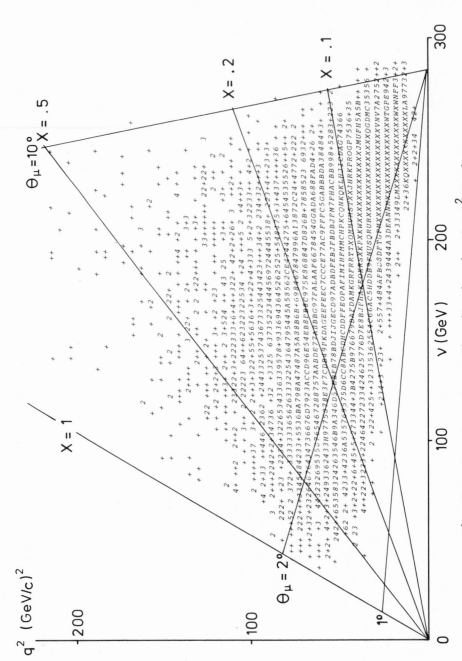

Fig. 9   Raw data event distribution in the $q^2-\nu$ plane

After correction with the acceptance and application of the necessary radiative corrections, the structure functions F_2 (x,q^2), as shown in Fig. 11 are obtained. They are average values of all protons and neutrons in the iron nucleus. The data are preliminary because all necessary corrections have not yet been applied to the raw data. In addition, the study of systematic errors is still going on, so that statistical errors only, are shown in Fig. 11. A rough estimation of the systematic uncertainty gives a value of about 20 % due to the different corrections not yet sufficiently studied and applied. The dotted lines in Fig. 11 show the prediction of Buras and Gaemers[8] for the q^2 dependence of the structure function $F_2(x,q^2)$, as given by asymptotically free gauge theories in a parametrization of parton distributions. Within the present measurement uncertainty, these QCD predictions agree very well with the experimental results. As soon as all our STAC data is analysed, we shall try to optimize the parametrization according to our results, which has not yet been done in the comparison shown in Fig. 11.

Figure 11 also contains the final result of the CDHS neutrino experiment[9] on $F_2(q^2)$ for the highest x regions, multiplied by 5/18. According to Equation (1) the structure functions of muon proton scattering $F_2^{\mu p}$ and muon neutron scattering $F_2^{\mu n}$ which both determine muon nucleus scattering, are as follows, if isospin in-

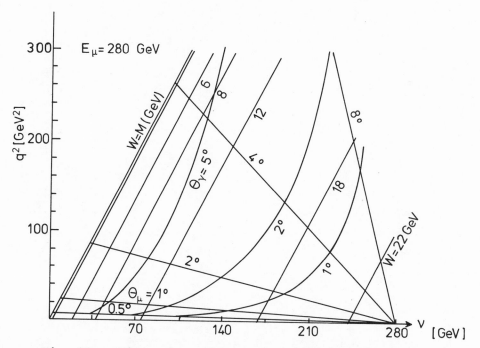

Fig. 10   Kinematic range of the scattering experiment

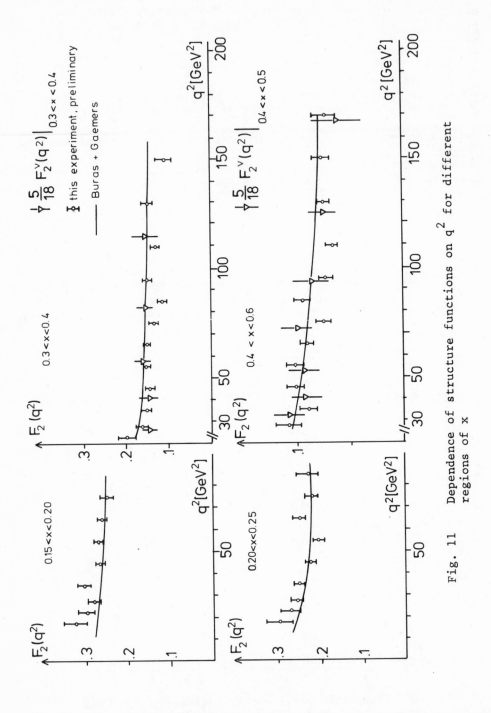

Fig. 11   Dependence of structure functions on q^2 for different regions of x

variance is assumed and ν,d,s are the quark distribution functions
inside the nucleon:

$$F_2^{\mu p} = x \left[\frac{4}{9} (u+\bar{u}) + \frac{1}{9} (d+\bar{d}+s+\bar{s}) \right] \tag{2}$$

$$F_2^{\mu n} = x \left[\frac{4}{9} (d+\bar{d}) + \frac{1}{9} (u+\bar{u}+s+\bar{s}) \right] \tag{3}$$

In the case of neutrino nucleus scattering the structure functions
$F_2^{\nu p}$ and $F_2^{\nu n}$ determine the cross section:

$$F_2^{\nu p} = 2 x (d+\bar{u}) \tag{4}$$

$$F_2^{\nu n} = 2 x (u+\bar{d}) \tag{5}$$

From equations (2) to (5) we have, if s quarks inside the nucleon
are neglected:

$$F_2^{\nu p} + F_2^{\nu n} = 2 x (d + u + \bar{d} + \bar{u})$$

$$= \frac{18}{5} x \left[\frac{5}{9} (d + u + \bar{d} + \bar{u}) \right] = \frac{18}{5} (F_2^{\mu p} + F_2^{\mu n}) \tag{6}$$

The experimental proof of this ratio between neutrino and muon nucleon
structure functions, as shown in Fig. 10, is evidence for the quark
charges + 2/3 and - 1/3 that determine the structure function ratio.
Neutrino and muon scattering are probing the same parton structure
inside the nucleon. The comparison of both methods of probing the
hadron structure in Fig. 10 shows that, especially in the high q^2
region, muon-nucleon scattering gives smaller error bars than neutrino
nucleon scattering thus allowing a more precise determination of the
q^2 dependence of the scaling violation.

The final data analysis will also show whether q^2 and x are the
important variables to describe the structure function behaviour or
whether the total hadronic energy W in the center-of-mass system
is an important variable as described in Ref. 10.

A second important part of the EMC physics program is muon scattering on hydrogen and deuterium targets to determine the proton and
neutron structure functions minimizing the possible nuclear effects. In
addition, in this case, the hadronic final states can be studied, at
least in the forward direction, because the hadrons can leave the low
density target in contrast to the heavy iron target measurements.
Data taking with 6 m long hydrogen and deuterium targets had started
at 280 GeV muon energy immediately after the first STAC measurements.

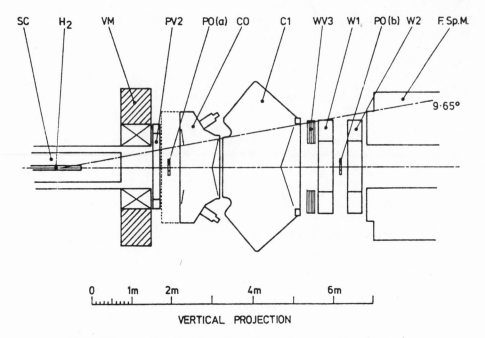

Fig. 12a   The Vertex Detector System in side view

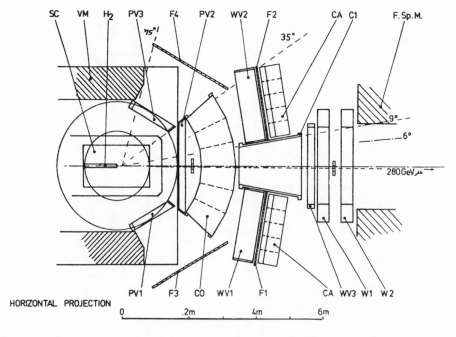

Fig. 12b   The Vertex Detector System in top view

FUTURE EXPERIMENTS

The first extension of the EMC Forward Spectrometer is a 450 channel lead glass detector to be added in the forward direction. With the aid of this detector the production of photons in coincidence with the scattered muons can be studied. Neutral hadron, as well as direct photon production, can be investigated.

The second essential change of the present experimental set-up will be the use of a polarized proton target to determine the spin dependent functions which average to zero in unpolarized inelastic lepton scattering. In the quark parton model, these functions represent the difference between the parton structures in the two spin states of the nucleon and contain, therefore, important information on the nucleon substructure. Measurements with the polarized target are planned for the beginning of 1980.

After those measurements, the angular acceptance of the Forward Spectrometer, which is $\pm 9^{\circ}$, will be increased to $\pm 60^{\circ}$ by the addition of a Vertex Detector System as shown in Fig. 12. This system will essentially improve the possibility of studying complete final states in deep inelastic lepton nucleon scattering and is known as experiment NA9 [11]. A streamer chamber in a C-shaped vertex magnet will be used in connection with proportional chambers PV1 - 3, wire chambers WV1 - 3 and scintillation counter hodoscopes F1 - F 4 to detect the final state particles. Three additional Čerenkov counters C 0, C 1 and C A will facilitate the π,k,p separation in different particle momentum ranges. The study of new quantum numbers like charm, connected with complicated decay final states, as well as the test of different QCD predictions for final states, like jet structure, angular distributions and inclusive hadron distributions, makes the Vertex Detector System an additional interesting part of the CERN muon scattering facility.

Table 1.   Muon Beam Properties

μ momentum range	50 - 280 GeV/c
parent particle (π, K) momentum range	50 - 300 GeV/c
μ momentum spread $\Delta p/p$	$\pm$ 5%
accuracy of μ momentum measurement $\Delta p/p$	$\pm$ 0.5%
μ^{+} number per incident proton	1.6×10^{-6} at 280 GeV
typical running intensity per SPS pulse	up to 4×10^{7} μ^{+}
spill length	~ 1 sec
muon halo	~ 7%

REFERENCES

1. W.K.H. Panofsky, Proceedings of the 14th International Conference
    on High Energy Physics, Vienna, 1968, p. 23.
2. M. Breidenbach et al., Phys. Rev. Lett. 23:935 (1969).
3. D.J. Fox et al., Phys. Rev. Lett. 33:1504 (1974).
4. E.M. Riordan et al., SLAC-PUB-1634 (1975).
5. H. Georgi and H.D. Politzer, Phys. Rev. D9:416 (1974).
6. CERN-SPS Proposal, SPSC/P.18 (1974).
7. D. Botterill, B. Charles and V. White, CERN-EP/79-52, submitted
    to Nucl. Instr. and Methods.
8. A.J. Buras and K.J.F. Gaemers, Nucl. Phys. B132:249 (1978).
9. J.G.H. de Groot et al., Phys. Lett. 82B:456 (1979) and Z. für
    Physik C1:143 (1979).
10. R.C. Bell et al., Phys. Rev. Lett. 42:866 (1979).
11. CERN-SPS Proposal, SPSC/P.18/Add.1 (1977) and SPSC/P.18/Add.2
    (1979).

STUDIES OF MUON INDUCED MULTIMUON FINAL

STATES IN THE EMC EXPERIMENT NA2

J.J. Aubert - LAPP, Annecy-le-Vieux, FRANCE

European Muon Collaboration: Annecy (LAPP), CERN,
Lancaster, Liverpool, Oxford, Rutherford, Sheffield,
DESY, Freiberg, Kiel, Wuppertal and Torino

The European Muon Collaboration has set-up a rather complete
spectrometer, mainly designed for doing deep inelastic scattering
experiments, measuring the so-called single arm experiment and the
produced hadronic final state. Even if the set-up has not been
optimised towards detecting the multimuon final state we will show
what has been achieved.

I. PHYSICS INTEREST IN MULTIMUON FINAL STATES

In physics, during the last decade, the leptonic final state
has been a powerful tool for new discoveries (J, Υ, heavy lepton...).
We will not speculate here on these new objects even though one
wants to keep an eye open for possible strange final states. The
discussion will be restricted to more conventional production of
heavy vectors mesons and their corresponding scalar mesons (charm,
bottom...).

I.1 Production of heavy vector mesons

J production. Since up to this time there has been no observation of J muoproduction, one does not know how heavy mesons are
produced. There are predictions from VDM[1] which relate the virtual
photon cross-section to the photon cross-section, namely:

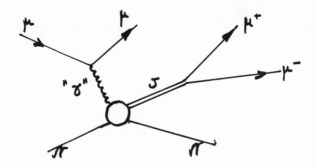

$$\frac{d^5\sigma_{"\gamma"}}{dE'd\cos\theta d\ dtd\phi_J} = \Gamma\ \left|\frac{k^o_c}{k_c}\right|\ \frac{M_J^{\ 4}}{(M_J^{\ 2} + Q^2)^2}\ \frac{d^2\sigma\gamma}{dtd\phi_J}$$

where E', θ,   are the energy, the scattered angle and the azimuth
                angle of the scattered muon;

Q^2, ν are the usual variables defining the virtual photon;

$t=-(P_N-P_{N'})^2$ is the momentum transfered to the nucleon (nucleus);

ϕ_J is the angle between the hadronic plane and the lep-
                tonic plane;

M_J is the mass of the vector meson;

Γ is the flux of the virtual photon;

$k^o_c = k_c$ at $Q^2=0$ is the known kinematic factor;

$\sigma\gamma$ is the cross-section of J production with real pho-
                tons where one neglects the contribution from longi-
                tudinal polarization of the virtual photon.

One obvious problem is to determine what the mass scale which
governs this process is: is it the mass of the vector meson as ad-
vised by VDM, or some universal mass around the proton mass?

For the energy dependence of J photoproduction, one would
expect a diffractive production, i.e., a flat behaviour far above
threshold production. M. Glück and C. Reya[2], using combined par-
tonic semi-local duality ideas, combined with QCD, derive absolute
predictions for J photoproduction cross-sections. They get a
rather flat cross-section only above a centre-of-mass energy of
20 GeV/c. One notices that this prediction concerns inelastic
production.

Production of heavier vector mesons. T can be photoproduced
and observed through its $\mu^+\mu^-$ decay. The signature is clean but
the luminosity which can be reached with present muon beams makes
its observation difficult.

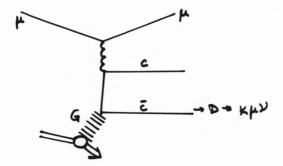

Fig. 1   Dimuon production by charmed particles.

I.2   Charm production

Charmed mesons (baryons) will be produced in pairs (Fig. 1) with a rather high cross-section[3] and they can decay through their leptonic channels. Experimentally, a rather clear signature would be a dimuon signal with missing energy (ν). The main background comes from π, K decays and can be distinguished. One will then be able to calculate the charm contribution to the structure function F_2.

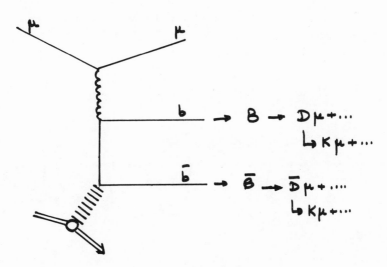

Fig. 2   Multimuon production induced by cascade decay of B B.

I.3 Bottom production

Bottom mesons (baryons) will also be produced in pairs (Fig. 2) in muoproduction. Their semileptonic cascade decays through charm and strange particles will end up with multimuon final states (up to five muons). The expected cross-section and branching ratio are large enough to expect events with more than three muons in the final state.

The problem will be to distinguish them from the background. It is, in principle, feasible, since B decays will induce large p_T muons but[4] background fake by charm decays and π, K decays might also extend to large p_T depending on their own p_T production.

I.4 Background

All Bethe-Heitler trident production will act as a large source of background (Fig. 3). In principle, they can be accurately estimated.

Hadron decays. Hadrons produced in μp interactions will decay and create multimuon final states. Their production rate can be estimated and is important mainly for low energy muons and for small angle muons. In the case of a heavy target, the rate is greatly decreased.

II. EXPERIMENTAL SET-UP

The experimental set-up[5] is discussed in detail in the contribution of H. Stier[6] (Fig. 4), so we will only describe the part which is specifically relevant to the multimuon experiment.

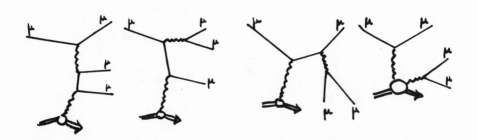

Fig. 3 Trident production : Feynman diagram

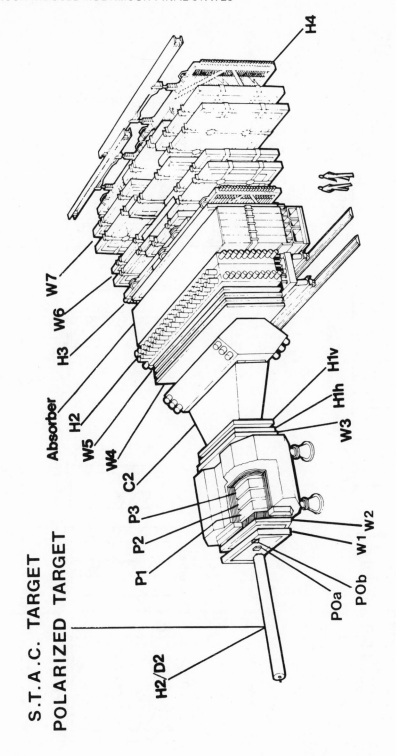

Fig. 4. EMC experimental set-up, forward spectrometer

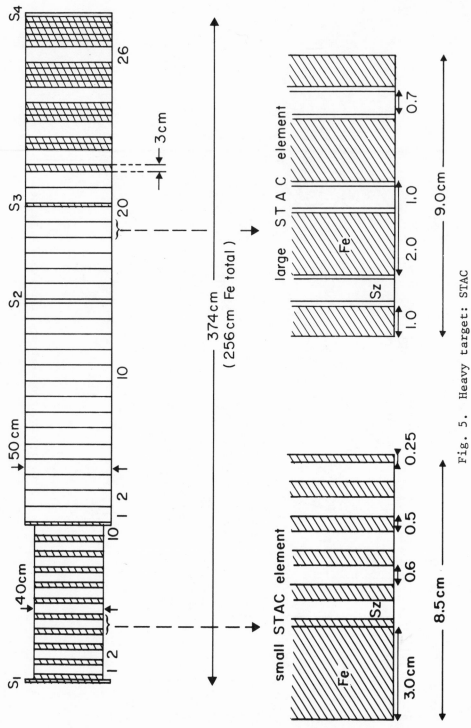

Fig. 5.   Heavy target: STAC

II.1 Heavy target STAC

A sandwich of iron and scintillator plate (Fig. 5) constitutes a heavy target and a calorimeter. The total iron thickness is 2,000g/cm^2. The target is sensitive to electromagnetic and hadronic showers.

Figure 6 shows the resolution of this target as a function of the energy.

There are several reasons for such a target choice for multimuon physics:
- the luminosity is higher than with the standard 6 m H_2 target;
- the background from π, K decay is reduced;
- apart from the missing neutrino energy, the forward spectrometer and the target give an over-all constraint for the total energy. Figure 7 shows the hadronic energy E_x for the deep inelastic event $\mu p \rightarrow \mu + X$, measured with the STAC and with the spectrometer;
- the vertex of the interaction can be determined directly from the amplitude measurement inside the target. This measurement cross checks the software determination of the vertex. Figure 8 shows the two measurements for the deep inelastic event $\mu p \rightarrow \mu x$.

The disadvantages are two: first, coherent electromagnetic production is enhanced, and second, angular resolution deteriorates because of the multiple scattering.

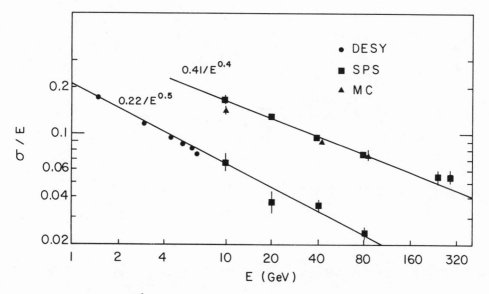

Fig. 6. STAC energy resolution

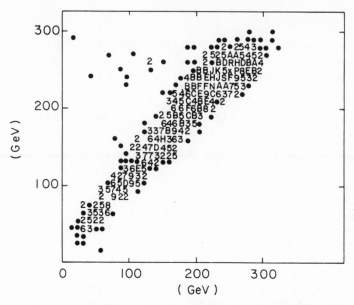

Fig. 7.   Hadronic Energy measured with the STAC on the horizontal
coordinate, compared with the forward spectrometer
determination on the vertical coordinate.

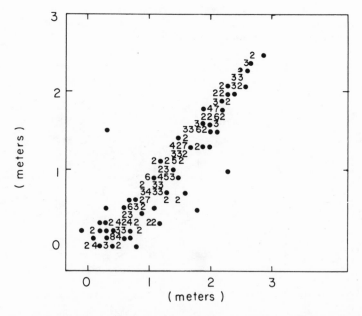

Fig. 8.   Vertex determined with the target on the horizontal
coordinate versus the vertex determined with the
forward spectrometer on the vertical coordinate

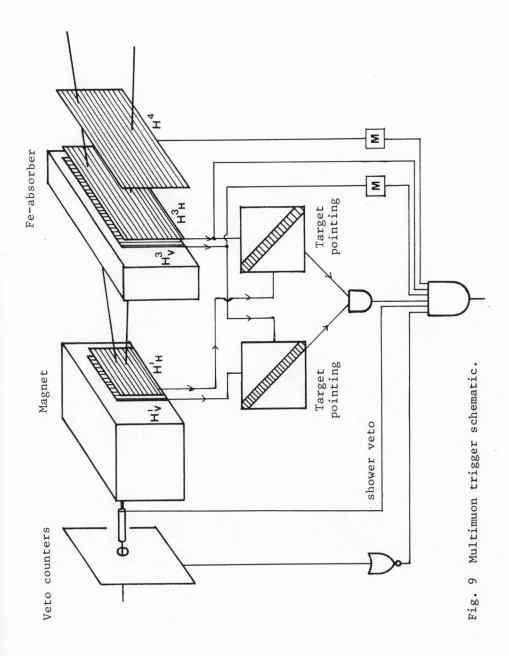

Fig. 9 Multimuon trigger schematic.

II.2 Trigger

Because of the high luminosity of the experiment (10^7 muons per burst at 280 GeV) and the high rate of coherent trident production, it is not possible to record all multimuon events during the data taking. We have then to make some choice and so have defined three triggers (Fig. 9):

trigger 3: a high mass dimuon trigger (mass cut around 2.4 GeV/c^2) designed to look for heavy vector meson J and above;

trigger 4: a lower mass dimuon trigger (rate is very high and it has been scaled down during data taking);

trigger 12: larger or equal to a three muon trigger with a cut against trident (i.e., requesting a minimum p_T for at least one muon).

We have to keep in mind that our hodoscope system is blind around the beam, so high energy, small angle muons are undetected through the trigger system. The trigger rates obtained are as follows:

- trigger 3 : 2.0×10^{-6}
- trigger 4 : 5.6×10^{-6}
- trigger 12: 1.2×10^{-6}

III. DATA AND STATUS OF ANALYSIS

Data were taken in August 1978 with the heavy target with 280 GeV/c incident muon (30% of the statistics) and in November and December 1978 with 250 GeV/c incident muon (70% of the statistics). Up to now we have analysed 10% of the whole sample and we present preliminary results on that analysis. Figure 10 shows one reconstructed trimuon.

III.1 Dimuon signal

Dimuon events are processed through the whole reconstruction analysis program chain and we retain only events in which the two muons have been recognized all the way along the forward spectrometer. From a sub-sample of the analyzed events we get 921 $\mu^+\mu^+$ and 6,230 $\mu^+\mu^-$ events.

There is, in many cases, a third muon which has stayed mainly inside the beam (for $\theta\mu < 10$ mrd and $P\mu > 100$ GeV/c) so it has not been reconstructed. We eliminated most of these false dimuon events by requiring that it was not a trigger 12, that there was no hit in H5 (hodoscope counter inside the beam at the end of the experiment) and that there was no muon reconstructed in the set of proportional

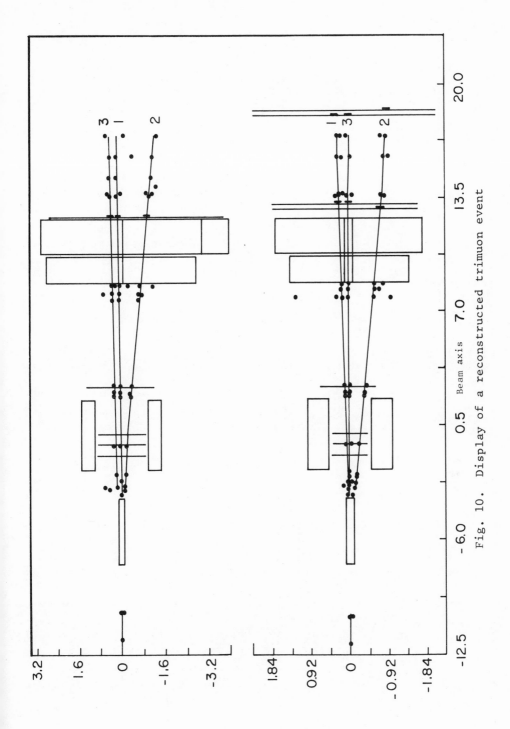

Fig. 10. Display of a reconstructed trimuon event

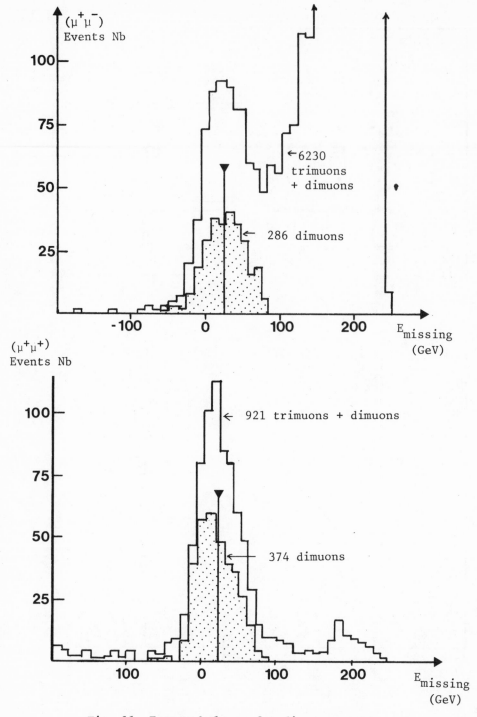

Fig. 11   Energy balance for dimuon events.

chambers P_O inside the beam before the magnet. Provided our detectors P_O and H5 are efficient, only very low energy muons will escape detection.

After those cuts, we were left with 535 $\mu^+\mu^+$ and 959 $\mu^+\mu^-$. We then asked for a good vertex χ^2, $P\mu^+ > 20$ GeV, with no shower escaping from the STAC and an energy loss smaller than 90 GeV. We ended up with 374 $\mu^+\mu^+$ and 286 $\mu^+\mu^-$ events. A visual scan of those events shows that they are mostly clean dimuon events.

We show in Fig. 11 the energy balance $E_{beam} - E_{\mu 1} - E_{\mu 2} - E_{STAC}$. The upper line is the sample before removing trimuons; the dashed area contains the dimuons left after cuts. About 20 GeV missing energy is clearly visible for both samples. If such events are unlikely to be related to π/K decays, only a detailed study with Monte Carlo calculation of cascades will prove this statement.

III.2   J production

With the sample of completely reconstructed dimuons, one can calculate the invariant mass $\mu^+\mu^-$. Figure 12 is clearly evidence for the muoproduction of J.

The distribution of those events as a function of ν (energy of the virtual photon) is shown in Fig. 13. (This result is not corrected for acceptance.)

If one defines $z = E_J/\nu$, elastic production on nucleons (coherent production on nucleus) will show up as a peak at $z = 1$. The histogram of z is shown in Fig. 14 for events in the J mass. One can see the elastic peak. Events with $z \ll 1$ can be either inelastic production, production of $\psi_{3.7}^!$ decaying into $J + \pi + \pi$, or radiative tail. One then selects the J events with $z > .9$ and studies their Q^2 dependence. For that, we calculate the acceptance inside the spectrometer as a function of ν and Q^2 and we integrate over the t distribution (t being the momentum transfered to the nucleon-nucleus) since the acceptance weakly depends on t, (the effect of coherent production, and t_{min} effect has been incorporated).

Figure 15 shows the Q^2 distribution of J events. The data are in good agreement with a VDM propagator $(M^2_J/M_{J2} + Q^2)^2$ and disagree with a mass propagator aroung 1 GeV/c.

Further work is needed to appreciate the full significance of this distribution.

I would like to thank all my colleagues from EMC for their contribution to this experiment.

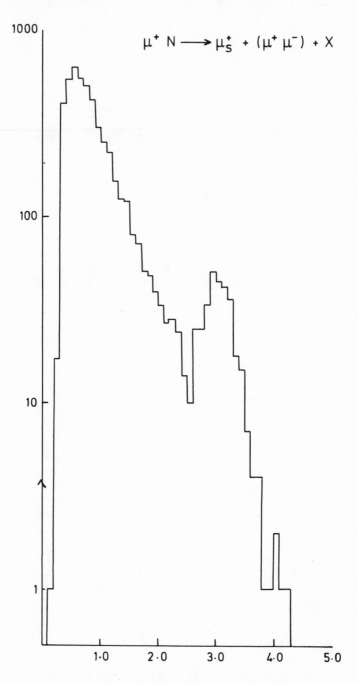

Fig. 12   Dimuon invariant mass spectrum.

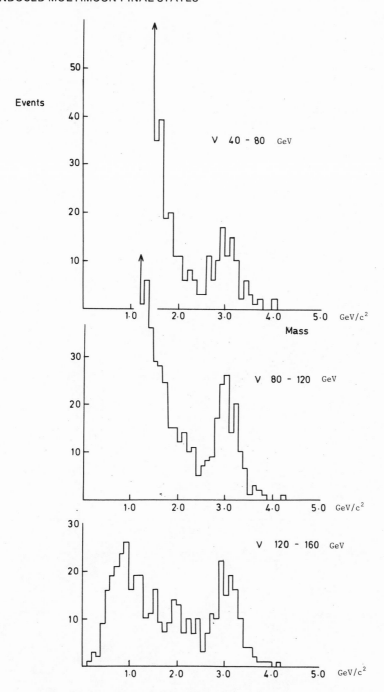

Fig. 13   Muoproduction of J for different values of ν.

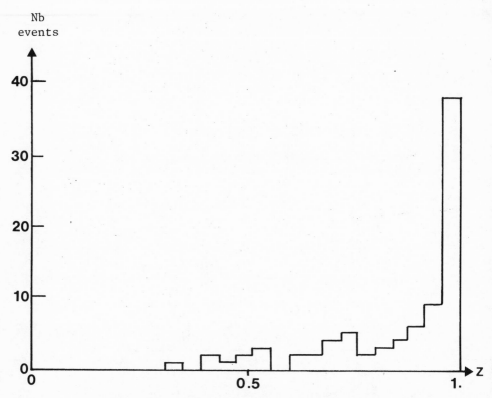

Fig. 14   z distribution of J events.

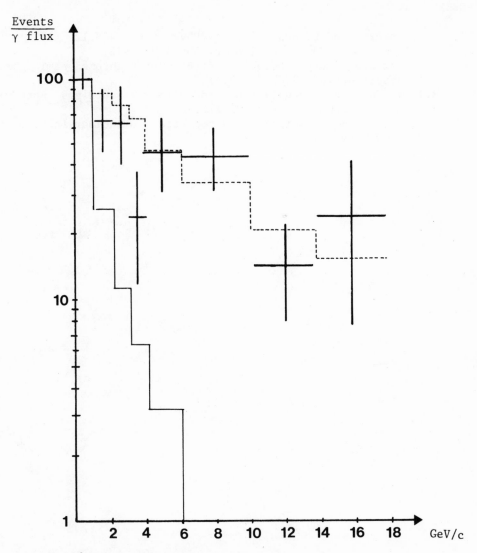

Fig. 15 Q^2 distribution of J events:
- the experimental points shown are the number of events in a bin, divided by the virtual photon flux for this bin. The scale is chosen arbitrarily to normalize for the first bin to 100.
- the dotted line is the result of the VDM prediction for a propagator $|M_J{}^2/(M_J{}^2+Q^2)|^2$ folded with the acceptance and set-up resolution.
- the plain line shows the VDM prediction for a ρ propagator.

REFERENCES

1. J.J. Sakurai and D. Schildknecht, Phys. Lett. 40B:121 (1972).
2. M. Glück and E. Reya, Phys. Lett. 79B:453 (1978).
3. M. Glück and E. Reya, Deep inelastic quantum chromodynamic charm leptoproduction, DESY 79-05 (1979)
4. R.J. N. Phillips, On b flavour production by neutrinos, Rutherfor Lab. preprint, RL 79-05 T 237.
5. EMC forward spectrometer, to be submitted to Nuclear Instruments and Methods (1979).
6. H. Stier, see contribution in these proceedings.

DEEP INELASTIC MUON SCATTERING AND RECENT RESULTS AT HIGH q^2 [*]

K. W. Chen[†]

Physics Department
Michigan State University
East Lansing, Michigan 48824

INTRODUCTION

I would like to present some new results from the MSU-Fermilab collaboration on deep inelastic muon scattering. Our group has
accumulated both single muon and multimuon data. Since analysis
of our single muon scattering data is at a more advanced state and
the preliminary conclusion is somewhat interesting, I will concentrate on that measurement.

Deep inelastic lepton scattering probes the structure of matter at large momentum transfers. The expression for scattering of
an incoming lepton by the nucleon is simple, permitting one to examine the internal structure with precision. Figure 1 shows the Feynman diagram and the definition of the kinematic variable for deep
inelastic scattering. The differential cross-section for scattering
is given by

$$\frac{d^2\sigma}{dE'd\Omega}(q^2,\nu) = \frac{\alpha^2\cos^2\theta/2}{4E_o^2\sin^4\theta/2}[W_2(q^2,\nu) + 2\tan^2\theta/2\ W_1(q^2,\nu)] \tag{1}$$

The function $\nu W_2(q^2,\nu)$ and its moments are related to the momentum distributions of the constituents of the nucleon presumed to be quarks.

[*]Work supported by theNational Science Foundation Grant GP 29050
[†]Other members of the collaboration include: B.C. Ball, D. Bauer
 J. Kiley, I. Kostoulas, A. Kotlewski, L. Litt, P.F. Schewe and
 A. Van Ginneken.

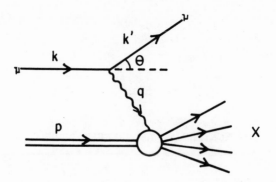

Fig.1. Feynman diagram for deep inelastic lepton scattering $ep \to ex$.
$p=(m,0,0,0)=$ proton at rest in lab frame
$k=(E_0,0,0,E_0)=$ incident muon
$k'=(E',0,E' \sin\theta, E' \cos\theta)=$ scattered muon
$q=(\nu,0,-E' \sin\theta, E_0 - E' \cos\theta)=$ virtual photon
$\nu = q \cdot P/m = E_0 - E' =$ energy transfer
$q^2 = (k-k')^2 = 4E_0 E' \sin^2\theta/2 =$ momentum transfer squared
$W^2 = M^2 = 2m\nu + m^2 - q^2 =$ hadron final state mass squared
$x = 1/\omega = q^2/2m\nu =$ Bjorken scaling variable
elastic scattering: $2m\nu/q^2 = \omega = 1$
inelastic scattering: $2m\nu/q^2 = \omega = 1/x > 1$

Before we describe the current experimental situation regarding the knowledge of $\mu p \to \mu X$, we comment that other lepton induced
reactions also study the structure of nucleons. For space-like momentum transfers ($q^2 > 0$), the reactions $\mu p \to \mu X$ and $ep \to eX$ are equivalent from μ-e universality. The electron-positron annihilation
reaction $e^+ e^- \to X$ is for time-like momentum transfers, $q^2 < 0$. For
neutrino scattering one probes the hadron's weak current via the
vector intermediate boson W. The Feynman diagram for $e^+ e^-$ and neutrino induced reactions is shown in Figure 2 and the respective
cross-sections are given by

$$\frac{d^2\sigma}{dE'd\Omega}(e^+ e^- \to X) = \frac{2\alpha^2}{q^4} m^2 \sqrt{\nu^2/q^2 - 1}$$

$$\{2W_1^{e^+ e^-} + \frac{2m}{q^2}(1 - \frac{q^2}{\nu^2}) \frac{\nu W_2^{e^+ e^-}}{2m} \sin^2\theta/2\} \qquad (2)$$

$$\frac{d\sigma}{dxdy}(\nu p \to \mu X) = \frac{G^2}{2\pi} s [F_2(1-y) + F_1 xy^2 \pm y(1-y/2)xF_3] \qquad (3)$$

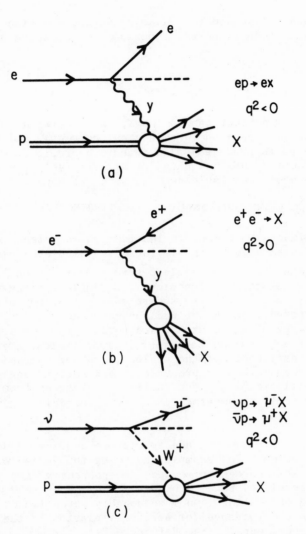

Fig.2. Feynman diagrams for other kinds of lepton-photon scattering
       a) $ep \to eX$ $q^2 < 0$
       b) $e^+e^- \to X$ $q^2 > 0$
       c) $\left.\begin{array}{l}\nu p \to \mu^- X \\ \bar{\nu} p \to \mu^+ X\end{array}\right\}$ $q^2 < 0$

where $y = \nu/E_o$ and F_3 is a third structure function necessitated by parity violation in the weak interaction, and s is the total center-of-mass energy.

The space-like and time-like structure functions are related to the annihilation structure functions by "crossing symmetry"

$$W_1^{e^+e^-}(q^2,\nu) = -W_1^{ep}(q^2,-\nu) \tag{4}$$

$$\nu W^{e^+e^-}(q^2,\nu) = -\nu W_2^{ep}(q^2,-\nu) \tag{5}$$

Electroproduction reactions $\mu p \to \mu h X$, $ep \to eX$ (with polarized beam and target), $\nu p \to \nu p$ (weak neutral current), and $pp \to \mu\mu X$ (massive lepton pairs produced) also measure hadronic structure. All of these interactions are now being studied actively. Now we shall review some newly available data in μp scattering.

PREVIOUS SINGLE-ARM DEEP INELASTIC SCATTERING EXPERIMENT

Experiments[1-3] on deep inelastic muon and electron scattering have established significant deviations from Bjorken scaling[4] up to $q^2 = 50$ (GeV/c)^2. The nucleon structure function, νW_2, can no longer be expressed as a function of a single scaling variable, say x ($\equiv q^2/2m\nu$, where q^2 is the square of the four momentum transfer of the scattered muon, M the nucleon mass and $\nu \equiv E-E'$ the difference in energy of incident and scattered μ). Theoretical interpretations[5] of non-scaling behaviour include field theoretic arguments, composite constituents of the nucleon, and new hadronic degrees of freedom in deep inelastic processes. However, a convincing explanation is still lacking. Nonetheless, deviations from scaling are likely to reveal fundamental features of particle structure and dynamics.

Attempts first were made to restore scaling by defining new scaling variables. For example, by using the variable $\omega' = \omega + M^2/q^2$ some of the scale-breaking tendencies apparently disappeared. But the violations persisted to even higher values of q^2 (≈ 40(GeV/c)^2), and the breaking of scaling is now reasonably established. Figures 3, 4 and 5 show the evidence for scaling violation in the μFe, $\mu p-\mu d$ and ep-ed experiments. (See Reference 2)).

Experimentally observed scaling violation has been an important development in the study of constituent theories of the nucleon. For example, the field theory which seeks to explain how these violations come about is known as Quantum Chromodynamics (QCD). It tries to explain scaling violations by assuming gluon-quark interactions including the gluon radiative correction terms. Much hope is expressed in what might be the correct field theory to explain strong interactions and even possibly unite this theory with the weak and electromagnetic interactions as well.

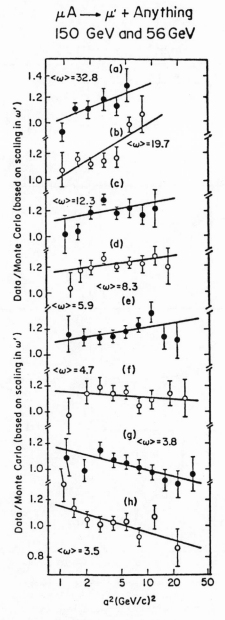

$\mu A \longrightarrow \mu' + $ Anything

150 GeV and 56 GeV

Fig. 3.   Early Cornell-Michigan State-Univ. of California Collaboration μ-Fe data (FNAL E-26).   Data/Monte Carlo ratios vs. q²
for various values of ω are shown.   The ratio has a negative
slope for large values of ω≥ 5.   Scaling violations of up to
25% are observed (C. Chang et al, 1975).

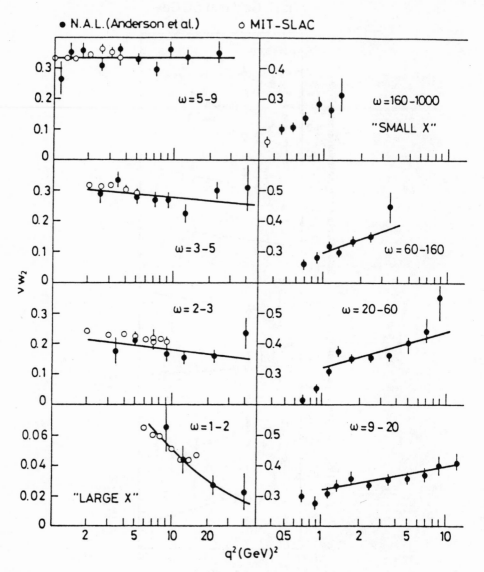

Fig.4. Early FNAL (Anderson et al., FNAL Exp. E-98) hydrogen data at
147 and 96 GeV. νW_2 vs q^2 for various ω bins. Open circles
are the MIT-SLAC data. The solid lines are fits to the
empirical formula $\nu W_2(q^2,\omega) = \nu W_2(q^2,\omega)(q^2/q_o^2)^b$, (1976)

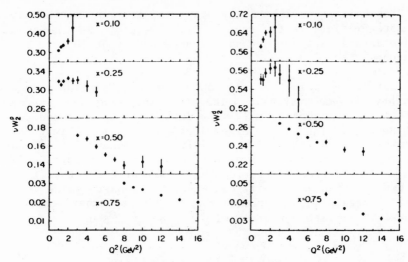

Fig. 5a.   SLAC ep and ed data. Separated values of $2MW_1^p$ and $2MW_1^d$ plotted versus Q^2 for selected fixed values of x. The error bars shown represent only the random errors in these quantities, (1975).

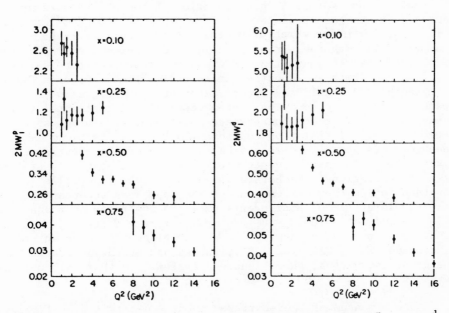

Fig. 5b.   SLAC ep and ed data. Separated values of νW_2^p and νW_2^d plotted versus Q^2 for selected fixed values of x. The error bars shown **represent** only the random errors in these quantities, (1975).

Before one could be fully convinced of such an explanation
many more critical tests, such as comparison of the theory with more
precise data in a wider dynamical range than that covered thus far,
are needed.

During the past year, new data from two Fermilab muon experiments have become available. The μ-Fe experiment was repeated by the
MSU-Fermilab collaboration[6]. The data has been extended to q^2 =
= 120 GeV^2/c^2. The second experiment (μ-p) is the Chicago-Harvard-
Illinois-Oxford Collaboration (CHIO). Preliminary data of both experiments have been briefly presented at the 1977 Hamburg Conference
The data show better statistical and systematic errors. The
μ-Fe experiment uses a heavy target with a density which is several
orders of magnitude higher than the μ-hydrogen experiment. One thus
can explore kinematic regions with higher q^2 and ν.

MUON SCATTERING AT 270 GeV

The present experiment, carried out at Fermilab, increases
significantly the statistical certainty and the kinematic range
over which deep inelastic muon scattering (μN → μX) has thus far
been explored. For 2 x 10^{10} incident muons with an energy of 270
GeV (μ$^+$ and μ$^-$), 10^6 deep inelastic events above q^2 = 5 (GeV/c)^2 are
recorded. Results from a large fraction of the μ$^+$ exposure are reported. The target consists of a 7.4 m long (4,260 g/cm^2) iron-scin-
tillator calorimeter which also measures the final state hadron
energy. Following the target is a 745 g/cm^2 thick steel hadron shield
and a spectrometer consisting of eight fully powered toroidal magnets
(4,973 g/cm^2 thick and about 90 cm in radius). As shown in Figure 6,
both the hadron shield and spectrometer are interleaved with wire
spark chambers. In addition, three vertical and horizontal trigger
banks of scintillation counters are positioned within the spectrometer. Three scintillation counters (15.9 cm radius) centred on
the beam axis and placed in the back section of the spectrometer form
a veto which eliminates events with a penetrating particle at a small
angle. Proportional chambers and beam halo veto (scintillation)
counters define the incident muon. The apparatus acceptance is relatively large (> 50%) for events having a scattered muon energy E' >
> 50 GeV and an angle 20 < θ < 80 mrad. Figure 7 shows curves of
equal apparatus acceptance in the q^2 - ν plane. Also shown are lines
of constant W^2 ($\equiv$ 2Mν - q^2 + M^2), the square of the centre-of-mass
energy of the virtual photon-nucleon system. In this talk only those
data with 40 $\leq$ W^2 $\leq$ 300 GeV^2 are presented.

The momentum of the scattered muon is determined by fitting
its trajectory through the magnetic spectrometer. The fitting algorithm is based on detailed magnetic field maps, known muon dE/dx
values and calibration of the spectrometer using the muon beam with

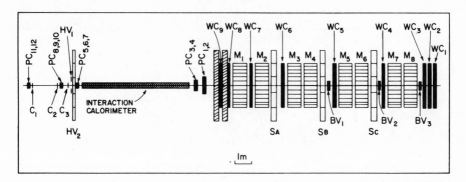

Fig. 6. Michigan State-FNAL. E-319 apparatus cross-section. PC
are proportional chambers HV and BV are halo and beam
particle veto counters respectively, WC are wire spark
chambers, M is a magnet and S_A, S_B, S_C form the trigger
bank.

known energies. Track finding efficiencies vary from 80% at low q^2
to 90% at high q^2 ($\sim$ 50 GeV^2/c^2). The resolution of the spectrometer is known to be about 2%, with its central value calibrated to
about 0.7%. The scattering angle is known to 0.4 mrad. The energy
of the incident muon is determined to within 0.75%. Muon tracks
with scattered energy $E' > 30$ GeV ($y < 0.88$) were recognized and
momentum fit making full allowance for Coulomb scattering, energy
loss, and bending in the iron magnets. Spark chambers were shielded
from the target by at least 640 g/cm^2 of iron. Intensities were
such that fewer than 30% of random triggers yielded one or more tracks.

The track finding inefficiency correction, based on studies
using auxiliary detectors and omitting various chambers, was less
than 7%. Absolute momentum calibration was based on these magnetic
field maps, dE'/dx measurements, steering beam muons into the spectrometer, and studying the endpoint of the E' spectrum. We ascribe
uncertainties of 1% to the relative momentum calibration and less
than 10% to the normalization of data.

RESULTS

The values of the structure function, F_2 per nucleon, are obtained by comparison of data with a Monte Carlo calculation based
on fits[7] of lower energy data in F_2^{ep} and F_2^{eD} and with nearly the
same result, using a QCD parametrization[8]. The calculation includes
effects of real incident beam distributions, Fermi motion of the
nucleons in the iron nucleus, radiative corrections and wide-angle
bremsstrahlung in simulating deep inelastic scattering. The iron
target nucleus was modeled as a collection of nucleons in Fermi mo-

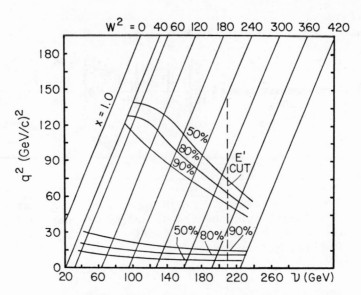

Fig. 7.   Contour plot of Michigan State-FNAL.   Experiment (E-319)
          apparatus acceptance as a function of q^2 and ν.   Also shown
          are lines of constant W^2 which are labelled at the top of
          the figure.   The E' cut requires the scattered muon to have
          total energy larger than 50 GeV.

tion.   Radiative corrections used **effective-radiators differing from**
the more exact correction by less than 3%.   The correction for coherent wide-angle bremsstrahlung rose from less than 3% for $\omega < 20$
to 15% at ω near 50.   The simulated muons suffered from μ-e scattering and bremsstrahlung in the iron, and measurement errors in-
curred in spark chambers.   Incident and scattered muons are traced
through the apparatus undergoing simulated magnetic deflection,
multiple Coulomb scattering, μ-e scattering, bremsstrahlung and
collision losses.   Further analysis treats data and Monte Carlo
events identically.   A number of tests establish the accuracy of
the simulation.   Shapes of real and simulated distributions in
azimuth and radius of tracks near fiducial boundaries agree well.
It accurately models differences between momentum reconstruction
algorithms which use different points on the muon trajectories.   The
data presented below show only statistical **errors and do not include**
the relative systematic and normalization uncertainties of under 10%.

Figure 8 (a-h) shows F_2 $(\bar{x}_i, q^2)$ versus q^2 for successive regions, where x is the weighted average of x for the ith data point.
Also shown is a QCD prediction [8] both for $F_2(\bar{x}_i, q^2)$ (the solid
curve, $\Lambda=0.5$) and for $F_2(\bar{x}, q^2)$ (the **dashed-dotted** curve, $\Lambda=0.5$;
dashed curve, $\Lambda = 0.4$) where x is the weighted average value for
the entire x region.   For $x \geq 0.4$ the data appear to agree with the

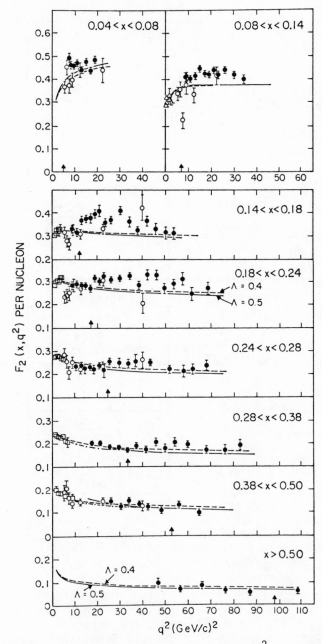

Fig.8. Nucleon structure function F_2 versus q^2. (a) 0.04<x<0.08, (b) 0.08<x<0.14, (c) 0.14<x<0.17, (d) 0.17<x<0.24, (e) 0.24< x<0.28, (f) 0.28<x<0.38, (g) 0.38<x<0.40, (h) x>0.5. Shown also are $F_2^{eD}/2$ (open square) of Ref.3, and $F_2^{\mu p}$ of Ref. 2 (open circle) after correcting for n - p diffence. The solid (dashed) lines are QCD predictions[8] in which Λ=0.5 (0.4) GeV/C is indicated by arrows. Errors are statistical only.

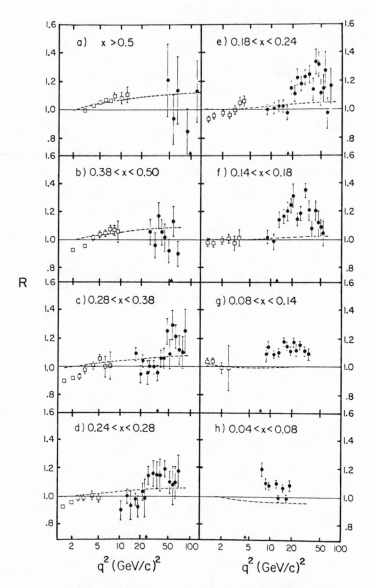

Fig. 9.   Ratio of observed to calculated (QCD) structure function vs.
          q^2:   (a) x>0.5, (b) 0.38<x<0.50, (c) 0.28<x<0.38, (d)
          0.24<x<0.28, (e) 0.18<x<0.24, (f) 0.14<x<0.18, (g) 0.08<x
          <0.14, (h) 0.04<x<0.08.   The QCD prediction scale parameter
          is Λ=0.5 GeV.   The value of W^2= 80 GeV^2 is indicated by
          arrows.   The dashed line is the ratio of the QCD prediction
          for Λ=0.4 to that for Λ=0.5 GeV/c.   Open squares show ratio
          for F_2^{eD} .

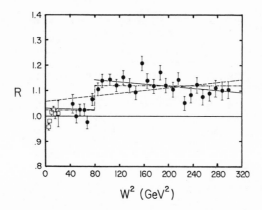

R

W² (GeV²)

Fig.10. R, ratio of observed to calculated (QCD) structure function,
versus W², the square of the **centre**-of-mass energy of the
virtual-nucleon system for all q²and x. The dashed line
shows the fit R=aW²+b for the entire W² region. Solid lines
show separate fits for the range 80<W² and W²> 80 GeV . The
dash-dotted line are fits to two separate constants. [See
Table 1]. $F_2^{eD}/2$ data with q²>2.0 (GeV/c)² (Ref. 3) are shown
(open square) but not fitted. Errors are statistical.

QCD curves within errors. However, for .28 < x < .38 and lower x
regions the data show a pattern of near agreement with QCD at lower
q², then a tendency to rise above QCD at moderate to higher q².
This behavior of F_2 versus q² for the various cuts in x is statistically consistent with that expected from a threshold behavior in
the variable W². The value of W² = 80 GeV² is indicated by an arrow.

Experimental structure functions compared to the QCD predic-
tion are shown in Fig. 9 (a-h). Values of R [$\equiv F_2(\bar{x}_i,q^2)/F_2(\bar{x}_i,q^2)$
QCD] are plotted versus ln q² for successive x cuts. The values
of R for SLAC-MIT data [*] calculated in the same manner, are also
shown. The ratio of the QCD prediction for Λ = 0.4 GeV/c to that
for Λ = 0.5 GeV/c is shown as a dashed line. In Fig. 9 (c-f) where
the data disagree with the QCD prediction, the difference due to
this change of Λ is not significant.

In Fig. 10, R is plotted as a function of W² between 40 < W² <
< 300 GeV². R (W²) is consistent with a rise of 12-15% beyond the
region of W² = (80 ± 10) GeV². Also shown are tests of a possible
W²-threshold effect. The data are fit to R = aW² + b, yielding
χ^2 = 3.57 per degree of freedom (DOF). More statistically significant fits are obtained with two separate straight lines for W² < 80
and for W² > 80 giving χ^2/DOF of 2.25 and **1.29, respectively.**

---

*) F_2 is extracted assuming R (=σ_s/σ_T)= 0.25. By assuming R = 0.44,
the variation of the **cross**-section is less than 2% for the range
of the data presented here.

TABLE I   Fits to R = aW^2 = b

W^2 Region	a $(\times 10^4)$	b	χ^2/DOF
All W^2 [1]	2.64 ± 0.66	1.058 ± 0.011	96.3/27
W^2 < 80 [2]	-0.72 ± 9.52	1.030 ± 0.058	11.3/5
W^2 > 80	-2.11 ± 0.91	1.159 ± 0.017	25.8/20
All W^2 [1]	0 [3]	1.097 ± 0.005	112.4/28
W^2 < 80 [2]	0 [3]	1.025 ± 0.010	11.3/6
W^2 > 80	0 [3]	1.21 ± 0.006	31.2/21
All W^2 [1]	0 [3]	1.00 [4]	485.4/29

1) $40 < W^2 < 300$ GeV^2. Data in **separate** band of q^2 are not fitted here.
2)   The choice of W^2 = 80 GeV^2 is dictated by the best possible fit.
     χ^2/DOF improves further when resolution effects are included.
3)   The slope is set to zero i.e., R = constant.
4)   Ratio of measured F_2 to QCD predictions.

Fitting R to one constant yields a poor fit with χ^2 = 17.3/DOF. Similarly, good fits are obtained using two separate constants (a step
function) yielding χ^2/DOF of 1.88 and 1.49 respectively. Fitting
R to QCD prediction[8] (a = 0, b = 1.00) yields χ^2 = 17.3/DOF. A
summary of these fits is given in Table 1.

      This observed behavior of F_2 in q^2 or W^2 cannot be accounted
for by known systematic effects which include uncertainties in measurement of E_0, E' and θ, variation in σ_s/σ_T * or in the application
of radiative corrections and corrections for wide-angle bremsstrah-
lung and Fermi motion *). The apparatus acceptance is a smooth function in W^2 and cannot accomodate a point-to-point rapid variation
of F^2 in the range $65 < W^2 < 95$ GeV^2. Although $F_2(x,q_2^2)$ can con-
ceivably be more closely fitted by readjustments of q_0, Λ, quark or
gluon distributions in QCD, the observed q^2 or W^2 variation is not
easily accomodated without major new assumptions since $F_2(q^2)|_x$
should be monotonic in q^2 or W^2, in contrast to the observed trend
in this experiment+).

---

*) Any unknown coherent **nuclear effect** cannot explain a rise in F_2
    in the range of x and q^2 of this experiment. In addition there
    is agreement with the available F_2^{up} data at large W^2.

+) In QCD, the moment integrals $M_n(q^2) = \int_0^1 dx\, x^n[F_2(s,q2)]$ vary as
    $a_n/(\ln q^2/\Lambda^2)^{b_n}$, where a_n, b_n and Λ are constants. For a fixed x,
    $\partial F_2(x,q^2)/\partial \ln q^2$ is constant. Thus, $F_2(q)$ varies monotonically
    in q^2.

COMPARISON WITH CHIO RESULTS

Our results significantly extend the range in q^2 over previous
measurement at SLAC-MIT[3] and Fermilab of νW_2 [1],[2]. This is
achieved chiefly by choosing the incident muon energy at 270 GeV,
1.83 times the energy chosen in the previous experiments. Figure 8
(a-e) shows values of the nucleon structure function $\nu W(q^2,x)$
as a function of x for various q^2 bands. The figure includes data
from SLAC-MIT measurements at lower energies and from the μN experiment at Fermilab. There is general agreement where the data sets
overlap ($5 < q^2 < 15$ and $15 < q^2 < 30$) between our data and earlier
measurements when due account is taken of systematic uncertainties
not shown in these data points. These amount to 10% in our data,
7% in the Chicago-Oxford-Illinois (CHIO) collaboration and .4% in
the SLAC-MIT data. (To compare with the proton data we have taken
into consideration the difference between νW_2^p and νW_2^n.) There are
no published data of νW_2 for $q^2 > 30$ GeV^2/c^2 at large ω. Our results
show the same pattern of scaling violation observed before where νW_2
decreases for $\omega > 3$ and increases for $\omega > 9$. The high beam energy
allows small values of x where the observed increase in νW_2 with q^2
cannot be removed by clairvoyant choice of scaling variables.

Our data are compared to a fit of the Fermilab μN experiment
(CHIO) expressing the observed scaling violations in a power-law
dependence of νW_2 on q^2.

$$\nu W_2(q^2,\omega) = \nu W_2(q_0^2,\omega)\ \left(\frac{q^2}{q_0^2}\right)^b \quad \text{where}$$

$$\nu W^2(q_0^2\omega) = \sum_{i=3}^{5} C_i\ \left(1-\frac{1}{\omega}\right)^i \tag{6}$$

with $a = 0.145 \pm 0.024$, $b = a\ \ell n\ \frac{1}{6}\ \omega$, $q_0^2 = 3$ GeV^2, $C_3 = 2.799 \pm 0.493$,
$C_4 = -4.048 \pm 1.134$ and $C_5 = 1.615 \pm 0.649$.

One observed characteristic of the new data is an increased
rise of νW_2 as $x \to 0$ for increasing q^2. For $q^2 \geq 30$ the data are
observed to rise more sharply than the fit, indicating some inade-
quacy of the CHIO fit. The implication of this observation is not
yet clear.

We also evaluated the integral of the structure function defined
by

$$I_2\ (q^2) = \int_0^1 \nu W_2\ (x,q^2)\ dx \tag{7}$$

in different bands of q^2. The fits of Ref 2 are used to evaluate
the integral from $x = 0.25$ to 1. This region is dominated by the
SLAC-MIT data. The maximum values of νW_2 at low q^2 are limited by
the requirement that $W > 2.0$ GeV. The value of the integral is well
determined by the polynomial fit. Summing over the present data
determines the integral from x_{min} to 0.25. The integral from x_{min}

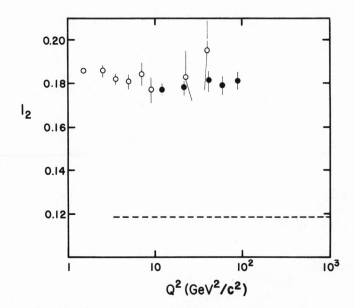

Fig. 11. The moment of nucleon structure function, I_2. The open
circle is the analysis of Anderson et al.[2,2] The closed
circle is for this experiment appropriately normalized,
$I_2 = \int_{x\,min}^{x\,max} F_2 dx$. X limits of this experiment are
$0.05 < x < 0.55$.

to $x = 0$ is evaluated by assuming it to be $x_{min} \cdot \nu W_2 (x_{min})$. The
structure function at a fixed q^2 is most sensitive to the assumed
form of R ($= \sigma_t/\sigma_s$) at large values of $y^{8)}$. We took this uncertainty
into account at this time by assigning the maximum error to those
points with y values exceeding 0.67, where $\Delta\sigma/\sigma \simeq 10\%$. Figure 11
shows $I_2^N (q^2)$ evaluated using this procedure. Also shown are $I_2^N(q^2)$
at lower q^2 value calculated as $1/A [Z I_2^p + (A - Z) I_2^n]$. We note
that values of $I_2^p = 0.333$ for scattering due to three valance quarks.
In the particular case of quantum chromodynamics, the four-quark,
three-color version in the asymptotic limit $I_2^p = I_2^n = 0.119$. The
present data shows that up to $q^2 = 120$ GeV^2/c^2 we are moving away
from this limit and not toward it.

SYSTEMATICS

There are many sources of systematics which could influence
the q^2 W^2 distribution. First the events could be subject to
inefficiencies, fitting or reconstruction bias. The Monte Carlo
program could be subject to inadequacies of input parameters in
simulation.

We made substantial tests of effects which could create a
12-15% rise as shown in the data. In no case did we find systematic
uncertainties drastic enough to give rise variation in q^2 or W^2
distributions.

Figure 12 shows the effects of uncertainties caused by a typical misrepresentation of our experiment in some critical effects.
The ratio of Monte Carlo cross section with or without the misrepresentation is shown. It is evident that the effect of these changes
are smooth and could not appear to give a 12-15% rapid variation to
$60 < W^2 < 95$ GeV^2.

CONCLUSIONS

It is too early to conclude that there is a major problem with QCD
or that we now have a new process $\sim 12\%$ at work in deep inelastic
scattering. At large x, (x > 0.4) the data agree very well with QCD,
andsmaller x, in some W^2 regions F_2 is higher than QCD trends. If one
were to shift the QCD prediction uniformly higher by 12-15%, in the
$W^2 > 80$ GeV^2 region, we could entertain a reasonably good fit. Then
one would have to explain the normalization problems at lower $W^2(q^2)$.
It is important to study carefully regions with moderate q^2; it is

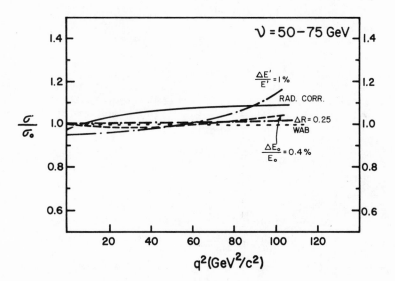

Fig. 12. Effect of systematic uncertainties on the measured cross-
section. σ is the cross-section after the uncertainties
of the corrections are applied from the original cross-
section σ_o. The uncertainties are those normally expected
in this experiment, which include radiative corrections,
$R = \sigma_S/\sigma_T$, error in momentum determination, wide-angle bremsstrahlung and error in incident muon momentum.

not simply important to push to highest q^2, since one could easily
gloss over the fine details. It is the over-all trend of $F_2(x,q^2)$
starting from q^2 = 1-5 GeV^2 up to 100-200 GeV^2 that could reveal
some interesting facets of the structure of nucleons.

Within the next year or so more data will become available, at
CERN, Fermilab and from our group. We look forward to a resolution
of the current mystery.

ACKNOWLEDGEMENT

We acknowledge the support of the Fermilab staff. We thank
K. Thorne and M. Ghods for their help in data reduction. One of us
(K.W.C.) would like to thank E. Lehman for useful discussions. This
research was supported in part by the National Science Foundation
under Grant No. 60950 and by the Department of Energy under Contract
No. E(11-1)-1764.

REFERENCES

1. C. Chang et al., Phys. Rev. Lett. <u>35</u>, 901 (1975).
2. H. L. Anderson et al., Phys. Rev. Lett. <u>38</u>, 1450 (1976); Phys.
     Rev. Lett. <u>41</u>, 615 (1978).
3. E. M. Riordan et al., SLAC Report No. SLAC-PUB-1634, 1975 (un-
     published).
4. J. D. Bjorken, Phys. Rev. <u>179</u>, 1547 (1969).
5. See, e.g., H. Cheng and T.T. Wu, Phys. Rev. Lett. <u>22</u>, (1969);
     K. Matsumoto, Progress Theor. Phys. <u>47</u>, 1975 (1972); M. S. Chan
     witz and S. D. Drell, Phys. Rev. Lett. <u>30</u>, 1975 (1973); H. D.
     Politzer, Physics Reports <u>14c</u> (1974) and references therein;
     M. Y. Han, Y. Nambu, Phys. Rev. <u>139</u>, B1006 (1965); J. C. Pati an
     A. Salam, Phys. Rev. Lett. <u>36</u>, 11 (1976).
6. For a schematic of the apparatus, see K. W. Chen, in "Proceedings
     of the 1977 International Symposium on Lepton and Photon
     Interactions at High Energies", edited by F. Gutbrod
     (DESY, Hamburg, Germany) p. 467.
7. S. Stein et al., Phys. Rev. <u>D12</u>, 1884 (1975).
8. See, e.g., A. Buras and B.G.F. Gaemers, Nuclear Phys. <u>B132</u> (1978)
     and private communication. The parton distributions for the
     proton with a scale parameter $\Lambda = 0.5$ GeV/c at q_o^2 = 2 (GeV/c)^2
     are $xS = (1-x)^8$, $xG = 2.41 (1-x)^5$, $xC = 0$, $xu_v = 3/B(0.7,3.6) x$
     $(1-x)^{2.6}$ and $xd_v = 1/B(.85,4.35) x^{0.85} (1-x)^{3.35}$.
     For the neutron, u_v and d_v distributions are interchanged. Λ =0.
     GeV/c is preferred by recent fits to lepton scattering data.
     [See for example, I. Hinchcliffe and C. H. Llewellyn-Smith, Nuc
     Phys. <u>B218</u>, 93 (1977)]

DEEP INELASTIC SCATTERING OF μ^+ AT 280 GeV/c ON CARBON

G. Smadja

Centre d'Etudes nucléaires, Saclay
Gif-sur-Yvette, France

CERN-Dubna-Munich-Saclay Collaboration*

INTRODUCTION

The NA4 spectrometer is characterized by its large luminosity, due to a target length of 40 m, and by its good acceptance to muons produced at large momentum transfers. At beam energies of 280 GeV/c, the toroidal magnetic field configuration confines deep inelastic events inside the iron up to $Q^2 = 270$ $(GeV/c)^2$, which allows us to cover a Q^2 range five times larger than previous experiments.

THE APPARATUS

The Magnetized Iron Toroid

The magnet, shown in Fig. 1, consists of 10 supermodules, 5.25 m each in length, with an outer diameter of 2.74 m and a central hole of 0.5 m diameter. The field B varies from 2.1 T at the centre to 1.7 T at the edge. In the toroidal field, the geometrical parameters

* CERN : A. Benvenuti, M. Bozzo, R. Brun, H. Gennow, M. Goossens, D. Reeder, C. Rubbia, D. Schinzel.

Dubna : D. Bardin, J. Cvach, I.A. Golutvin, I.M. Ivanchenko, Yu.T. Kiryushin, V.S. Kisselev, V.G. Krivokhizhin, V.V. Kukhtin, I. Manno, W.D. Nowak, I.A. Savin, D.A. Smolin, G. Vesztergombi, A.G. Volodko, A.V. Zarubin.

Munich : D. Jamnik, U. Meyer-Berkhout, F. Navach, A. Staude, K.M. Teichert, R. Tirler, R. Voss, Č. Zupančič.

Saclay : J. Feltesse, A. Lévêque, J. Maillard, J.M. Malasoma, J.F. Renardy, Y. Sacquin, M. Spiro, G. Smadja, M. Virchaux, P. Verrechia.

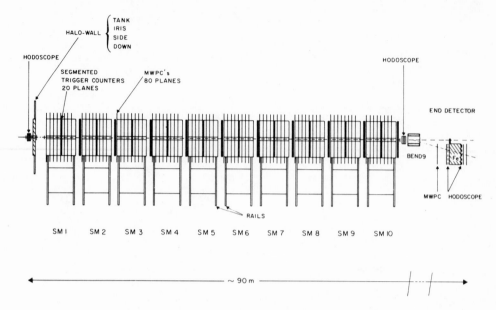

Fig. 1.   The NA4 spectrometer (top view).

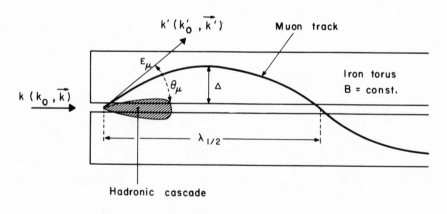

$$\lambda_{1/2} = 6.66 \ k'_T \ / \ B$$

$$\Delta = \frac{Mp}{0.3B} \ \frac{q^2}{q^2_{max}}$$

Fig. 2.   Kinematics of the toroidal field spectrometer. Here
$q^2 = (k'-k)^2$ and M_p is the proton mass. Units are m,
GeV/c, and T.

of a muon coming from the axial target are directly related to
$Q^2 = -(k - k')^2$, the 4-momentum transfer, and $k'_T = |\vec{k'}|\sin\theta$,
the transverse momentum, as shown in Fig. 2. The iron radii define
a maximal contained Q^2: $Q_c^2 = 0.6\ k_0 B\eta(r_{out}-r_{in})$, where $\eta = 0.66$ is
a reduction factor for B caused by air gaps. For $k_0 = 280$ GeV/c,
Q_c^2 is 270 $(\text{GeV/c})^2$.

The Chambers

Each one of the 10 supermodules contains 8 planes of multiwire
proportional chambers (2 MWPCs/plane), measuring alternately x and
y coordinates, as shown in Fig. 3. The main features of the readout are:

- a double delay stack of 450 nsec on each channel (every 4 mm),
  which avoids dead-time;

- a strobe width of 100 nsec.

Trigger Counters

Each supermodule contains two planes of liquid scintillator
trigger counters. A plane consists of 2 half planes of 7 rings
as shown in Fig. 4. The signals from the photomultipliers at
both ends of each ring go through a mean timer, which avoids the
time dependence upon the position of impact.

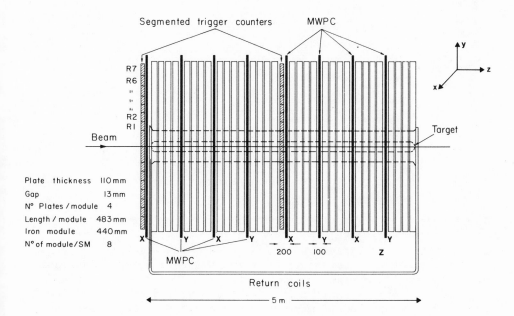

Fig. 3. Layout of one supermodule.

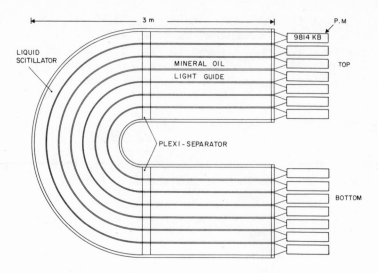

Fig. 4. A trigger counter.

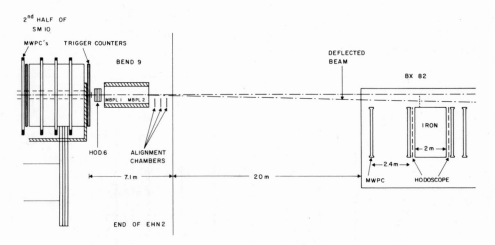

Fig. 5. The end detector.

The Target

Eight carbon targets, 5 m long each, with a density of 2 g/cm^3 are located at the centre of the first 8 supermodules. The choice of carbon yields a high luminosity: 5×10^{27} cm^{-2} per incoming muon and per nucleon. A small magnetized toroid, 1.5 cm thick, around the carbon, traps low-Q^2 muons down to the end of the apparatus, to improve the acceptance of the end detector. Liquid H$_2$ or D$_2$ targets will be installed in the future.

Table 1

Halo distribution normalized to beam (in %)

Front		Lateral	
14		5	
Central	Peripheral	Down	Side
8	6	2	3

The End Detector

The toroid is blind to muons with $Q^2 < 25$ $(GeV/c)^2$. A special set-up, shown in Fig. 5, with a bending magnet of 7 T·m (2.1 GeV/c), measures the momentum of the forward muons in low-Q^2 interactions $\left[Q^2 < 4 \ (GeV/c)^2\right]$. The end detector accepts particles in a momentum slice extending from 30 to 130 GeV/c, which can be changed if necessary.

Beam and Halo

The radial dispersion of the muon beam around the axis is 2.5 cm, and its momentum spread is $\sigma/k_0 = 4\%$. A set of 4 beam hodoscopes measures the momentum of each muon track with an accuracy of 1%, up to intensities of 10^9 μ/burst. A typical intensity at 280 GeV/c is 10^7 μ/burst, for an extracted beam of 7×10^{12} ppp.

The muon beam is surrounded by a halo with a main component, the front halo, which increases in the neighbourhood of the beam, and a smaller component, the lateral halo. Table 1 gives halo/beam ratios for the two components. An array of veto counters at the front of the spectrometer is used in anticoincidence and generates a dead-time of 25%.

THE TRIGGER

Trigger Scheme

The triggering scheme is shown in Fig. 6. The logic is distributed along the whole apparatus. Any coincidence of 4 consecutive planes will trigger the whole set-up via a fast "strobe" cable which runs along the spectrometer and preserves synchronization. We see in Fig. 7 that the acceptance of the 4-plane trigger extends from $Q^2 = 25$ $(GeV/c)^2$ to $Q^2 = 350$ $(GeV/c)^2$ over most of the range of $\nu = k_0 - k_0'$.

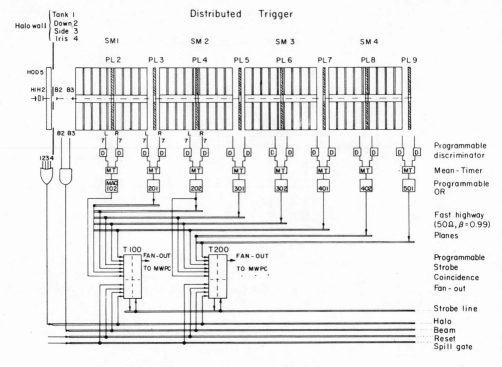

Fig. 6.   The trigger scheme.

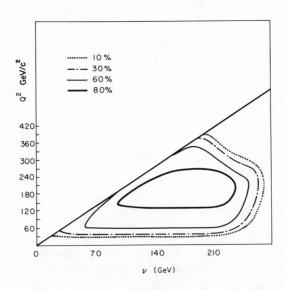

Fig. 7.   Acceptance contours for the 4-plane trigger.

Table 2

Characteristics of the four-plane trigger (in %)

$\dfrac{\text{Empty triggers}}{\text{Triggers}}$	$\dfrac{\text{Showers}}{\text{Triggers}}$	$\dfrac{\text{Halo events}}{\text{Triggers}}$	$\dfrac{\text{Good events}}{\text{Triggers}}$
14	38	34	14

The effective beam, or gated beam, is the fraction of beam left after taking into account the effect of the halo veto counters and computer dead-time.

The trigger rate of the 4-plane trigger is $5 \times 10^{-5} \times$ gated beam. A more selective trigger requires a minimal sagitta for the tracks. If, for instance, we do not use the inner rings, the trigger rate drops to 10^{-5}, and the Q^2 cut is around 50 $(\text{GeV/c})^2$.

Table 2 shows the main features of our triggers. Showers and empty triggers contain no track, but have or do not have a localized cloud of hits around the target; they are caused by electromagnetic and hadronic showers or low-Q^2 tracks, not seen in the chambers. Halo triggers are mostly lateral halo tracks not covered by the veto counters.

DATA ANALYSIS

The present analysis corresponds to a total luminosity of 5×10^{38} cm^{-2} for the dimuon events, representing half of the data collected in 1978. The deep inelastic interactions were studied on 10% of the data, corresponding to a luminosity of 1.1×10^{38} cm^{-2}. Furthermore, 10,000 μ^- events have been analysed to monitor the background of multimuon events in the deep inelastic sample. A more detailed analysis of multimuon events was presented by Feltesse[1].

Event Selection

The criteria are the following:

a) Track quality: at least 4 hits in one projection, 3 hits in the other one.

b) Halo rejection: a track is rejected if one finds more than 2 hits within a distance of 20 cm from the extrapolated track upstream of the vertex (Fig. 8).

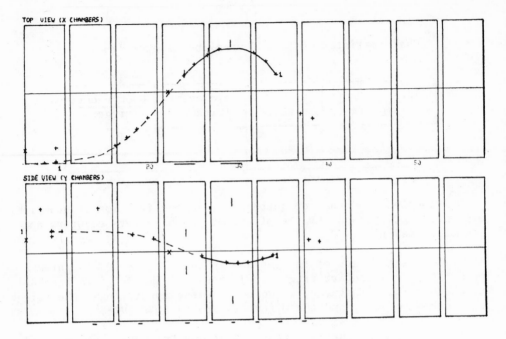

Fig. 8. Halo track simulating high-Q^2 scattering.

c) Rejection of accidentals: the electronic information concerning
    the trigger must be compatible with the position of the track.

All the events satisfying the previous criteria have been scanned in the present analysis.   We expect to bring the scanned fraction down to 10% in the future. The events in the spectrometer are
clean, with a spray of hits around the vertex as traces of the hadronic shower. We show a typical high-Q^2 event (Fig. 9a) and one
event with $Q^2 = 300$ $(GeV/c)^2 > Q_c^2$ which escapes the toroid (Fig. 9b).

Accuracy of the measurements

<u>Momentum resolution</u>. The momentum resolution has been found
to be 8% by 3 independent methods:

- Reprocessing of events generated by a Monte Carlo program with
  multiple scattering included.

- Measurement of the spread of the beam deflected into the spectrometer.

- Splitting tracks and plotting the difference of the 2 values of
  the momentum. The width gives the resolution and the mean the
  energy loss in iron (2.31 ± 0.04 GeV/m at 100 GeV/c, in agreement
  with Richard-Serre[2]).

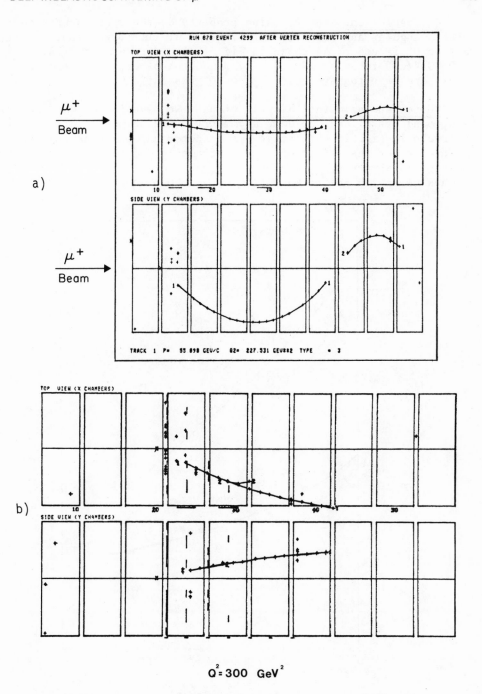

$Q^2 = 300 \ GeV^2$

Fig. 9.   Typical events.   a) High Q^2.   b) Very high Q^2.

Q^2 resolution. The focusing property of the spectrometer correlates angular and momentum measurements in such a way that the accuracy on Q^2 is good. As shown in Fig. 10 it improves from 10% at Q^2 = = 50 $(GeV/c)^2$ to 5% at 300 $(GeV/c)^2$. On the other hand, defocused μ^- tracks are seen to have quite a poor Q^2 resolution.

Vertex and azimuthal distributions. We have found that the halo background has a large up/down asymmetry. The azimuthal distribution of tracks in Fig. 11 shows that our detector has the expected isotropy (except for the effect of chamber frames), and that no halo contamination is apparent. The closest distance of approach to the target matches well with the radius of 6 cm, as shown in Fig. 12.

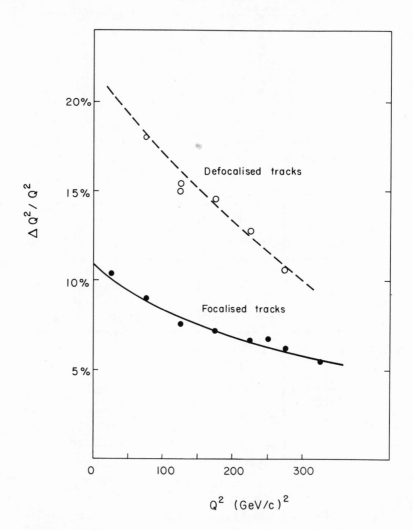

Fig. 10. Q^2 resolution for focused and defocused tracks.

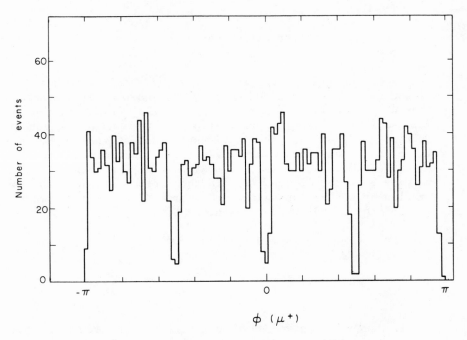

Fig. 11. Azimuthal distribution of μ^+ tracks.

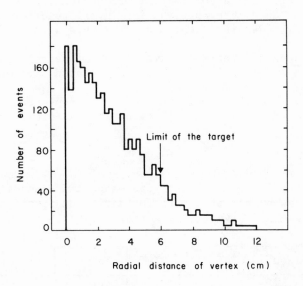

Fig. 12. Radial distribution of the vertex.

MULTIMUON EVENTS

Three main physical processes can generate multimuon events in the spectrometer.

i) <u>Vector mesons.</u> The μ cross-sections for ψ and T production on carbon can easily be computed from the known total photoproduction cross-sections on hydrogen within the virtual photon beam (Weizsacker-Williams) approximation[3]. The photoproduction cross-section of ψ on nucleons is given in Nash et al.[4] as 30 nb at large photon energies ($\gtrsim$ 55 GeV/c). The photoproduction cross-section of T on nucleons is computed in several papers[5] and can vary by a factor of ten. Within the virtual photon beam approximation of Ref. 3, using the propagator $M_V^4/(M_V^2 + Q^2)^2$, where M_V is the mass of the vector meson, we find

$$\sigma_\psi = 0.75 \times 10^{-33} \text{ cm}^2$$

$$\sigma_T = 0.05\text{-}0.5 \times 10^{-36} \text{ cm}^2 \ .$$

ii) <u>QED tridents.</u> Electrodynamic pair creation of muons can occur via the two graphs of Fig. 13. In fact, the Bethe-Heitler contribution dominates the Compton (time-like) process at all masses by a tenfold factor. Exchange terms must be included, but one configuration is always overwhelmingly larger in the present experiment. The absolute pair production cross-section computed by us on carbon is given in Fig. 14 together with the results of Mennessier[6], which agree within a factor of three.

iii) <u>Hadron decays.</u> There are two clases of hadron decays:
  - either a deep inelastic event associated with one leptonic decay,
  - or a low-Q^2 interaction and a double leptonic decay, the scattered μ^+ escaping detection in the toroid.

<u>Dimuons $\mu^+\mu^-$, $\mu^+\mu^+$, $\mu^-\mu^-$</u>

Data corresponding to a luminosity of 5×10^{38} cm^{-2} -- half of the 1978 statistics -- have been processed and scanned for multimuons. The raw number of events for different topologies are as follows (brackets designate the forward muon):

Topology	$\mu^+\mu^-$	$\mu^+\mu^-(\mu^+)$	$\mu^+\mu^+$	$\mu^+\mu^+(\mu^+)$	$\mu^-\mu^-$	$\mu^-\mu^-(\mu^+)$
Events	398	52	24	0	6	4

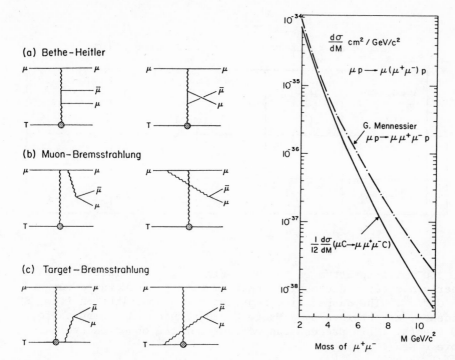

Fig. 13.   Feynman graphs for   Fig. 14.   Differential cross-
            QED tridents.   section for $\mu^+\mu^-$ pair
                                                   production on protons.

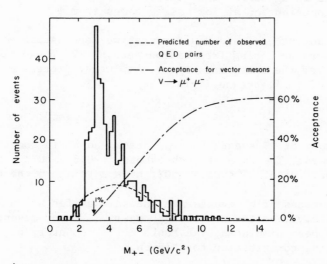

Fig. 15. $\mu^+\mu^-$ mass spectrum together with the contribution from
            QED pairs.

Table 3

Expected rates for vector meson production

	Acceptance (%)	Branching $\mu^+\mu^-$ (%)	σ (cm^2)	$\mu^+\mu^-$ events
ψ	0.7 → 1.5	7	0.7×10^{-33}	200–400
T	50	2	$(0.05 - 0.5) \times 10^{-36}$	0.3–3

The $\mu^+\mu^-$ mass spectrum is shown in Fig. 15. The distribution is clearly a steeply decreasing spectrum with an acceptance cut at 2.5 GeV/c^2. The dashed line represents the contribution from QED tridents, as obtained by a Monte Carlo program, and amounts to 200 events. This process can account for the observed events above 6 GeV/c^2.

We have computed the acceptance for $\mu^+\mu^-$ decay modes of ψ and T, taking the leptonic branching ratios from Boyarski et al.[7] and the PLUTO Collaboration[8], and assuming that the photoproduction on nucleons varies as e^{-3t}. The expected number of events, given in Table 3, is in agreement with the observed mass spectrum. The dominant contribution at low masses comes from ψ.

By studying $\mu^+\mu^+$ and $\mu^-\mu^-$ events, we have estimated the possible background coming from hadron decays. We found it to be less than 10% over the whole mass spectrum. Other processes, such as the decay of τ [9], are negligible.

μ^- Events

A sample of μ^- events (one μ^- track, no μ^+ in the toroid) has been analysed. The rate of such events is $0.6 \times 10^{-6} \times$ beam. A large proportion of μ^- events, 50%, have a track in the end detector.

The sources of μ^- events are the same as for all multimuons: hadronic decays, vector mesons, QED pairs. Vector mesons and QED pairs have been simulated by a Monte Carlo computation, and yield the following contribution to μ^- events:

$$\psi: \quad 10 \pm 5\% \text{ of } \mu^-$$
$$\text{QED:} \quad 15 \pm 5\% \text{ of } \mu^- .$$

We shall use the μ^- sample to monitor the backgrounds of deep inelastic events, as the background processes are nearly symmetric in
μ^+ and μ^-. The (Q^2, k_0') plot of μ^- events is shown in Fig. 16.
Almost all events are concentrated at low k_0', low Q^2, with only 4%
at $Q^2 > 50$ $(\text{GeV/c})^2$, $k_0' > 25$ GeV, so that

$$\frac{\mu^-}{\text{beam}} = 2.4 \times 10^{-8} \text{ when } \begin{cases} Q^2 > 50 \text{ } (\text{GeV/c})^2 \\ \\ k_0' > 25 \text{ GeV} \end{cases}$$

As $\mu^+/\text{beam} = 1.3 \times 10^{-6}$ in the same domain, this corresponds to a
2% contamination of decay μ^+ after correcting for acceptance.

Another estimate is obtained from the proportion of deep inelastic events associated with a track in the end detector, which is 0.1%.
These events, presumably not genuine deep inelastic events, can of
course be rejected. We know however from the μ^- sample that the acceptance of the end detector for the low-Q^2 interaction with "background" μ^- is 50%; we can assume that this acceptance is the same
for background μ^+, which leaves a contamination of 0.1%.

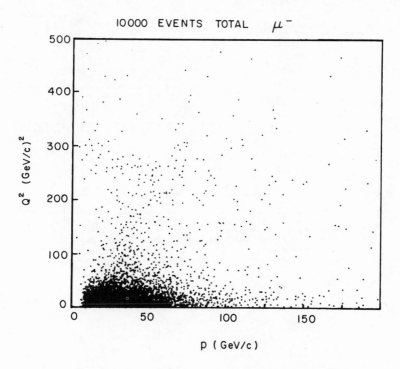

Fig. 16. (Q^2, k_0') plot of μ^- events.

PRELIMINARY RESULTS ON DEEP INELASTIC SCATTERING

This preliminary analysis is based on 16,000 scanned events, which correspond to 1.52×10^{10} incoming muons. The final sample of 8760 events is obtained after applying the cuts mentioned above, $k_0' > 25$ GeV, $Q^2 > 50$ $(\text{GeV/c})^2$, so that the contamination from multi-muon events is less than 2%.

The (Q^2, ν) plot is given in Fig. 17 before correcting for acceptance.

Q² Distribution

The raw Q^2 distribution above 50 $(\text{GeV/c})^2$ is given in Fig. 18. The comparison with SLAC parametrization[10], extrapolated from a Q^2 range ten times smaller, shows that the scaling violations are less than a factor of two in the Q^2 range considered. On the other hand, the agreement with FNAL parametrization, extrapolated from a fit below $Q^2 = 50$ $(\text{GeV/c})^2$, which includes scaling violations, is good.

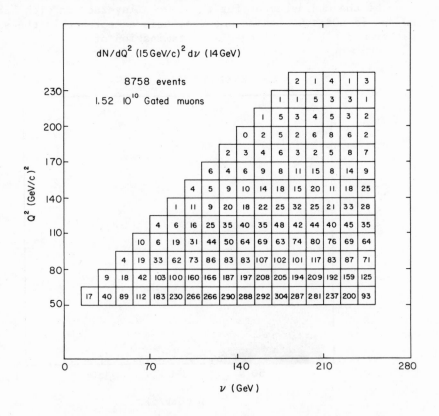

Fig. 17. (Q^2, ν) plot of deep inelastic events before acceptance corrections.

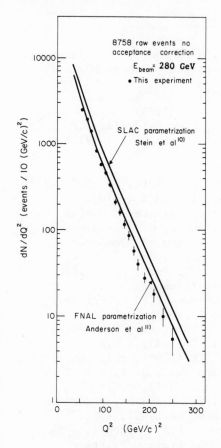

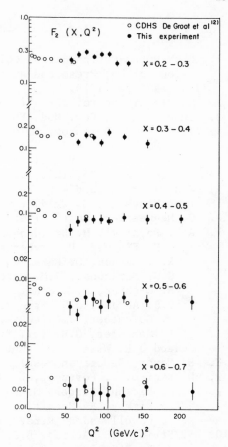

Fig. 18. Differential Q^2 distribution of deep inelas-
tic events.

Fig. 19. Comparison of ν and μ
structure functions
$F_2(x,Q^2)$.

Comparison with ν Experiments

In order to give a preliminary determination of the F_2 structure
function, we have assumed the ratio of longitudinal to transverse
photon cross-section on protons $R = 0.22$. We have corrected the
data for acceptance and measurement accuracy, but the acceptance map
is not yet final, and the normalization is therefore subject to a
10% uncertainty. A systematic effect, smaller than present statistical errors, could remain. Although our present data do not allow
a detailed comparison to be made, our determination of F_2 is in
agreement with the neutrino F_2 functions of De Groot et al.[12] (multiplied by $5/18$), as shown in Fig. 19.

REFERENCES

1. CERN-Dubna-Munich-Saclay Collab., High energy muon interactions
    on carbon, Presented at the 14th Rencontre de Moriond,
    Les Arcs, 1979.
2. C. Richard-Serre, CERN 71-18 (1971).
3. Yung Su Tsaï, Rev. Mod. Phys. 46, 815 (1974).
4. T. Nash, A. Belousov, B. Govorkov, D.O. Caldwell, J.P. Cumalat,
    A.M. Eisner, R.J. Morrison, F.V. Murphy, S.J. Yellin,
    P.J. Davis, R.M. Egloff, G. Luste and J.D. Prentice, Phys.
    Rev. Lett. 36, 1233 (1976).
5. G.J. Aubrecht and W.W. Wada, Phys. Rev. Lett. 39, 978 (1977).
    L.M. Jones and H. Wyld, Phys. Rev. 17, 2332 (1978).
6. G. Mennessier (Montpellier, France), private communication.
7. A.M. Boyarski, M. Breidenbach, F. Bulos, G.J. Feldman,
    G.E. Fischer, D. Fryberger, G. Hanson, B. Jean-Marie,
    R.R. Larsen, D. Lüke, V. Lüth, H.L. Lynch, D. Lyon,
    C.C. Morehouse, J.M. Paterson, M.L. Perl, P. Rapidis,
    B. Richter, R.F. Schwitters, W. Tanenbaum, F. Vannucci,
    G.S. Abrams, D.D. Briggs, W. Chinowsky, C.E. Friedberg,
    G. Goldhaber, J.A. Kadyk, A.M. Litke, B.A. Lulu, F.M. Pierre,
    B. Sadoulet, G.H. Trilling, J.S. Whitaker, F. Winkelmann
    and J.E. Wiss, Phys. Rev. Lett. 34, 1357 (1975).
8. H. Meyer, Report on results of PLUTO (DESY), this conference.
9. W. Bacino, T. Ferguson, L. Nodulman, W. Slater, H.E. Ticho,
    A. Diamant-Berger, M. Faessler, A. Hall, G. Irwin,
    J. Kirkby, F. Merrit, S. Wojcicki, R. Burns, P. Condon,
    P. Cowell and J. Kirz, Phys. Rev. Lett. 41, 13 (1978).
10. S. Stein, W.B. Atwood, E.D. Bloom, R.L.A. Cottrell,
    H. De Staebler, C.L. Jordan, H.G. Piel, C.Y. Prescot,
    R. Siemann and R.E. Taylor, Phys. Rev. D 12, 1184 (1975).
11. H.L. Anderson, V.K. Bharadwaj, N.E. Booth, R.M. Fine,
    W.R. Francis, B.A. Gordon, R.H. Heisterberg, R.G. Hicks,
    T.B.W. Kirk, G.I. Kirkbride, W.A. Loomis, H.S. Matis,
    L.W. Mo, L.C. Myrianthopoulos, D.M. Pipkin, S.H. Pordes,
    T.W. Quirk, W.D. Shambroom, A. Skuja, M.A. Staton,
    W.S.C. Williams, L.J. Verhey, Richard Wilson and
    S.C. Wright, Phys. Rev. Lett. 38, 1450 (1977).
    R.C. Ball, D. Bauer, C. Chang, K.W. Chen, S. Hansen, J. Kiley,
    I. Kostoulas, A. Kotlewski, L. Litt, P.F. Schewe and
    A. Van Ginneken, Michigan State University preprint
    MSU-CSL-60, 1978.
12. J.G.H. De Groot, T. Hansl, M. Holder, J. Knobloch, J. May,
    H.P. Paar, P. Palazzi, A. Para, F. Ranjard, D. Schlatter,
    J. Steinberger, H. Suter, W. von Rüden, H. Wahl, S. Whitaker,
    E.G.H. Williams, F. Eisele, K. Kleinknecht, H. Lierl,
    G. Spahn, H.J. Willutzki, W. Dorth, F. Dydak, C. Geweniger,
    V. Hepp, K. Tittel, J. Wotschack, P. Bloch, B. Devaux,
    S. Loucatos, J. Maillard, J.P. Merlo, B. Peyaud, J. Rander,
    A. Savoy-Navarro, R. Turlay and F.L. Navarria, Z. Phys. C 1,
    143 (1979).

THE PARTON MODEL AND ASYMPTOTIC FREEDOM REVISITED

Geoffrey B. West

Theoretical Division
Los Alamos Scientific Laboratory
University of California
Los Alamos, New Mexico 87545

INTRODUCTION

It is the intent of these lectures to give a retrospective review of some of the salient features of the parton model and its relationship to light cone dynamics and asymptotic freedom. There are, of course, many excellent review articles[1] covering these topics so in order to avoid excessive repetition I shall attempt to dwell on various aspects of the problem from a somewhat different perspective than usual.

The original "naive" parton model was designed to give a phenomenological understanding of scaling as observed in deep inelastic electron scattering. The heroic attempts to put these ideas on a firmer theoretical basis led, initially, to a recasting of the problem in terms of the behavior of the product of two electromagnetic currents near the light cone. When it was realized that a non-abelian gauge theory (and, in particular, quantum chromodynamics) exhibited asymptotic freedom and thus behaved in a quasi-free fashion near the light cone, the task of "understanding" the field theoretic origins of the parton model was basically complete. Indeed, this has spurred a whole new industry, so-called perturbative QCD,[2] which allows one to use one's early training in Feynman graphology to calculate deep inelastic-like processes. The main emphasis has naturally shifted from scaling itself to its breaking or approach.

In the problem of the deep inelastic structure functions, asymptotic freedom basically allows one to calculate only the radiative corrections due to the strong interactions. Indeed this is the reason that testing QCD, for example, is so difficult. One is focussing attention only upon a very small portion of the data; ultimately, one requires the theory to "predict" the structure functions themselves. Nevertheless, it is clear that so far, although there is no crucial test of the theory, there does not appear to be any contradiction. On the other hand, it is well-known that the general features of the structure functions themselves, as well as the behavior of the radiative corrections can be understood from some general physical ideas[3,4].

Of course, one of the great appeals of QCD is, that among its other virtues, it appears to exhibit both asymptotic freedom and infrared slavery (i.e., confinement). At the present time, the latter is neither established nor understood and indeed will presumably be crucial in allowing one to calculate the structure functions from theory in an a priori fashion. The manner in which asymptotic freedom passes over into infrared slavery will clearly be reflected in the manner in which, for example, the deep inelastic structure functions evolve into the total photoproduction cross-section as $q^2 \to 0$. From a parton model point of view the physics of the problem will presumably be reflected in the changeover from an incoherent state ($q^2 \to \infty$) to one of eventual total coherence ($q^2 \to 0$). One of the main purposes of these notes will be to examine these questions within a non-relativistic many-body field theory. We shall show how scaling evolves, even in a system with confining forces, and how the photoproduction limit is approached. We shall then move on to the covariant parton model and its relationship to the light cone[1,3,5] and point out what lessons can be inferred from the many-body discussions. Finally, we shall discuss one of the outstanding problems in particle physics, namely, the origin of the n-p mass difference and its relationship to some of the topics discussed above[6].

I. SUM RULES, CORRELATIONS AND THE APPROACH TO SCALING

In this first part we shall ignore spin completely and concentrate on the behavior of the structure functions in non-relativistic many body field theory. The structure function is defined as[3]

$$W(\nu, q^2) \equiv \frac{d^2\sigma/dE' \, d\Omega'}{(d\sigma/d\Omega')_{\text{Ruth}}} \tag{1}$$

where ν is the electron energy loss, q the three-momentum transferred and $(d\sigma/d\Omega')_{\text{Ruth}}$ the standard Rutherford cross-section. In first quantized form, W can be expressed as

$$W(\nu, q^2) = \sum_f |<\Psi_f| \sum_{i=1}^{N} Q_i e^{i\underline{q} \cdot \underline{r}_i} |\Psi_0>|^2 \delta[E_f - E_0 + \nu] \tag{2}$$

where $\Psi_{0(f)}$ is the initial (final) state of the target, Q_i the charge of the ith constituent and $E_{0(f)}$ the initial (final) electron energy. These are sums over the N constituents and all possible final states (f) of the target system. Notice that the Heisenberg equations of motion allow this to be re-expressed as a ground state expectation value:

$$W(\nu,q^2) = \int_{-\infty}^{\infty} \frac{dt}{2\pi} e^{i\nu t} <\Psi_0| \sum_{i,j} Q_i Q_j e^{i\underline{q}\cdot\underline{r}_j(t)} e^{-i\underline{q}\cdot\underline{r}_i(0)} |\Psi_0>$$

(3)

where the sum over all final states has been traded in for complete knowledge of the time development of the phase operator $e^{i\underline{q}\cdot\underline{r}_i}$. As a simplification of the notation, we shall suppress the time dependence of an operator when it is to be evaluated at $t = 0$, e.g., $e^{i\underline{q}\cdot\underline{r}_i} \equiv e^{i\underline{q}\cdot\underline{r}(0)}$.

In a second quantized language we can introduce charge density operators[7]

$$\rho_{\underline{q}} = \sum_{\underline{k}} a^+_{\underline{k}+\underline{q}} a_{\underline{k}}$$

(4)

to write the more familiar form

$$W(\nu,q^2) = \int_{-\infty}^{\infty} \frac{dt}{2\pi} e^{i\nu t} <\Psi_0| \rho^+_{\underline{q}}(t) \rho_{\underline{q}}(0) |\Psi_0>$$

(5)

where

$$\rho_{\underline{q}}(t) \equiv e^{iHt} \rho_{\underline{q}} e^{-iHt}$$

(6)

and $a_{\underline{k}}$, $a^+_{\underline{k}}$ are the standard creation-destruction operators. These satisfy the usual canonical commutation relations

$$[a_{\underline{k}}, a^+_{\underline{k}'}] = \delta_{\underline{k},\underline{k}'} \quad \text{etc.}$$

(7)

The Hamiltonian is assumed to be of the form

$$H[\underline{k}_i, \underline{r}_i] = \sum_i \frac{k_i^2}{2\mu} + \sum_{i \neq j} v(r_{ij}) \equiv H_0 + H'$$

(8)

where μ is the common constituent mass. In second quantized form these become

$$H_0 = \sum_{\underline{k}} \frac{k^2}{2\mu} a_{\underline{k}}^+ a_{\underline{k}} \tag{9}$$

and
$$H' = 1/2 \sum_{\underline{k}} V(\underline{k}) \rho_{\underline{k}}^+ \rho_{\underline{k}} \tag{10}$$

where
$$V(\underline{k}) \equiv \int d^3r \; e^{i\underline{k}\cdot\underline{r}} v(\underline{r}) \tag{11}$$

It is clear that these equations generate diagrams of the cat's ears type shown in Fig. 1 where the dotted lines represent the interaction $V(\underline{k})$. These graphs are generated by expanding $\rho_q(t)$ in the usual way as a power series in t. We shall show below that the leading term which corresponds to t = 0 gives rise to scaling. Relativistically, this will generalize to the relationship between scaling and the light cone behavior of the current product.

The elastic form factor can be expressed very simply in this formalism, namely as:

$$F(\underline{q}^2) = \langle\Psi_0| \sum_i Q_i e^{i\underline{q}\cdot\underline{r}_i} |\Psi_0\rangle \tag{12}$$

$$= \langle\Psi_0|\rho_{\underline{q}}|\Psi_0\rangle \tag{13}$$

Diagrammatically this corresponds to diagrams of the type given in Fig. 2. It is straightforward to show[3], using Eq. (12), that

$$F(\underline{q}^2) \rightarrow (1/\underline{q}^2)^2 \; , \; \text{or faster} \tag{14}$$

provided the single particle wave function is not singular at the origin.

An indication of a scaling phenomenon can be obtained by deriving a sum rule. This is done by integrating Eq.(3) over all ν for fixed $\underline{q}^2$: we obtain

$$\int_{\nu_0}^{\infty} d\nu \; W(\nu,\underline{q}^2) = \sum_i Q_i^2 + f(\underline{q}^2) \sum_{i \neq j} Q_i Q_j \tag{15}$$

where $\nu_0 = \underline{q}^2/2M$, the kinematic threshold for the process. In Eq. (15) we have separated the coherent from the incoherent contributions since the former are governed by the two particle correlation function

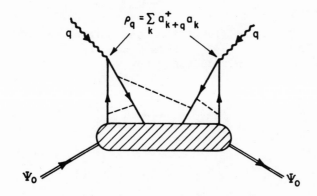

Fig. 1. Typical graph generated from Eq. (5).

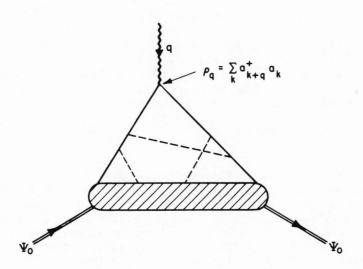

Fig. 2. Typical graph contributing to the elastic form factor.

$$f(\underline{q}^2) \equiv \langle \Psi_0 | e^{i\underline{q}\cdot\underline{r}_j} e^{-i\underline{q}\cdot\underline{r}_i} | \Psi_0 \rangle , \qquad (i \neq j) \qquad (16)$$

which drops off rapidly with q^2. Indeed it is clear that $f(q^2)$ has a rather similar behavior to the elastic form factor $F(q^2)$, Eq. (12). Both fall rapidly with q^2 and both are normalized to unity at $\underline{q}^2 = 0$. For large q^2 it is convenient to scale out the $\underline{q}^2$ dependence in ν by introducing a new variable $\xi = \underline{q}^2/2M\nu$ to give

$$I(\underline{q}^2) \equiv \int_0^1 \frac{d\xi}{\xi} \, W(\xi,\underline{q}^2) = \sum_i Q_i^2 + f(\underline{q}^2) \sum_{i \neq j} Q_i Q_j \qquad (17)$$

This implies that

$$I(q^2) \xrightarrow{q^2 \to \infty} \sum_i Q_i^2 \qquad [= N \qquad \text{if all } Q_i = 1] \qquad (18)$$

whereas

$$I(0) = \sum_{i,j} Q_i Q_j \qquad [= N^2 \qquad \text{if all } Q_i = 1] \qquad (19)$$

We shall refer to the fact that $I(q^2)$ becomes independent of q^2 for large $\underline{q}^2$ as "weak scaling"; one cannot, of course, deduce from this the "strong scaling" result that $\nu W(\xi,q^2)$ itself becomes independent of $\underline{q}^2$. To prove this via sum rules requires knowledge of all the moments of ξ; however, strong scaling is certainly the simplest way of satisfying the weak scaling. At this stage we therefore take as suggestive the scaling result,

$$\lim_{q^2 \to \infty} \nu W(\xi,\underline{q}^2) = F(\xi) \qquad (20)$$

Below we shall show how it follows dynamically from Eqs. (3) or (5). Before doing so, however, we wish to stress that a consequence of the argument given so far is that the approach to scaling is governed by correlations in the system as expressed by the correlation function $f(q^2)$. This is certainly true of weak scaling, and presumably will be at least partially valid for strong scaling. To what extent this is so will be examined below.

It is amusing to extend this physical picture to the nucleon and assume that the approach to scaling is dominated by quark correlations. In this regime radiative corrections will only be of secondary importance, although ultimately they, of course, dominate the corrections. Let us attempt to make some quantitative estimates. We shall consider the structure function[1,2] $F_2(q^2,x)$. Recall that gauge invariance requires F_2 to vanish at $q^2 = 0$;

more precisely

$$F_2(q^2,x) \xrightarrow{q^2 \to 0} \frac{q^2 \sigma_\gamma(\nu)}{4\pi^2 \alpha} \tag{21}$$

where σ_γ is the total photo-absorption cross-section. Enforcing this constraint leads to the analog of the weak scaling result, Eq. (17), namely

$$I_2(q^2) \equiv \int_0^1 F_2(q^2,x)\frac{dx}{x} = I_2(\infty)[1 - f(q^2)] \tag{22}$$

where $f(q^2)$ is interpreted as a quark correlation function. Note incidentally, that the non-relativistic version of this can be cast in the form of Eq. (22) by replacing q^2 with $\underline{q}^2$ and $I(q^2)$ by $\tilde{I}(q^2) \equiv I(0) - I(q^2)$. Recall that, roughly speaking, $f(q^2) \sim F^2(q^2) \sim (1 + q^2/m_0^2)^4$ with $M_0^2 \simeq 0.71$ GeV^2, so if $F(q^2)$ is smooth, as it is for the nucleon, we can expect the approach to scaling also to be smooth. Furthermore, the scale of this approach is governed roughly by M_0^2; since $M_0^2 \sim 0.71$ GeV^2 this gives a natural "explanation" for precocious scaling. Figure 3 shows the early SLAC data plotted in the form $I(q^2)$ vs. q^2 nicely illustrating the smoothness of the approach[8]. Indeed the curve is well fit by $\{1 - F^2(q^2)\}$ - see below. There are a few points worth emphasizing here:

a) One of the curious properties of the nucleon is that it is one of the few physical systems that exhibits a smooth elastic form factor. Almost all other systems have "edges" which lead to diffractive oscillatory behavior in their elastic form factors. For these we would predict that $I(q^2)$ approaches $I(\infty)$ in an oscillatory fashion. As an example of this consider $I(q^2)$ as measured by scattering thermal neutrons from liquid argon[9]. The data is shown in Fig. 4 and clearly shows evidence of an oscillatory behaviour. The smoothness of $I(q^2)$ for the nucleon is therefore not to be taken lightly.

b) The presence of $f(q^2)$ reflects a certain degree of coherenc in the system and, as such, is presumably related to the presence of vector mesons in the system. Furthermore, since $f(q^2)$ is not so dissimilar to $F(q^2)$ one can think of this contribution as related to the standard vector dominance (VD) contribution to $F_2(q^2,x)$[10]. Indeed, from our point of view the VD contribution is simply one model for $f(q^2)$*

---

* Although the $1/q^4$ asymptotic behavior of the nucleon form factor is "understood," the fact that it is so smooth and, in particular, is a dipole, is still a mystery.

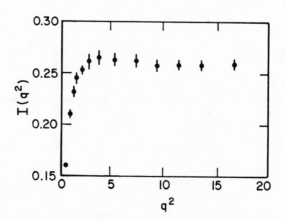

Fig. 3.   Plot of $I(q^2)$ vs. q^2 [8] (SLAC data) showing smoothness of
          approach to scaling, which, according to Eq. (22),
          reflects smoothness of the elastic form factor; this
          is to be contrasted with data of Fig. 2.

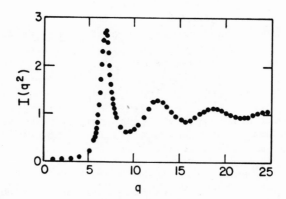

Fig. 4. $I(q^2)$ vs. $\underline{q}$ for thermal neutron scattering from argon [9],
          showing an <u>oscillatory</u> approach to scaling, reflecting
          an edge of the wave function; this is to be contrasted
          with the smooth approach shown in Fig. 3 for the nucleon.

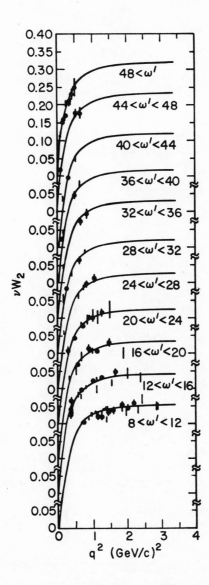

Fig. 5. $F_2(x,q^2)$ vs. q^2 for fixed values of ω'(1/x) showing
         smooth approach to scaling[11]. The solid lines are
         $[1 - F^2(q^2)]$ showing agreement with Eq. (22).

c) Just as strong scaling is suggested by weak scaling, so one is tempted to go further and introduce

$$\tilde{F}_2(x) \equiv \frac{F_2(q^2,x)}{1 - f(q^2)} \simeq \frac{F_2(q^2,x)}{1 - F^2(q^2)} \tag{23}$$

$\tilde{F}_2$ should scale much better than F_2 for small q^2. This is nicely illustrated in Fig. 5 where the dependence of $F_2(q^2,x)$ is well-fitted by $[1 - F^2(q^2)]$ for fixed x[11]. Taking this to its ultimate limit, let us examine $q^2 \to 0$ to illustrate how the deep inelastic region "joins on" to photoproduction. For simplicity, assume $\tilde{F}_2(0)$ to be constant, then Eq. (23) leads to

$$\sigma_\gamma(\gamma) \simeq \frac{8\pi^2\alpha}{M_0^2} \tilde{F}_2(0) \tag{24}$$

$$\simeq 100 \ \mu b$$

in remarkable agreement with experiment. Note that the scale that dominates the approach to scaling is here correlated to both the nucleon radius and the total photoabsorption cross-section.

Since the origin of the factor $[1 - f(q^2)]$ is so physical, and presumably related to the VD model, we feel that such contributions should be factored out of the data before attempts are made to compare it to the predictions of asymptotic freedom.

II. DYNAMICAL ORIGIN OF SCALING (WITH CONFINING FORCES)

In order to determine $W(\nu,q^2)$ we need to know the time development of the charge density operator $\rho_k(t)$, as given in Eqs. (5) and (6). First observe that $[H',\rho] = 0$ so we need only consider $[H_0,\rho]$. A straightforward calculation gives:

$$[H_0,\rho_q] = \sum_k \left\{ \frac{k^2 - (k + q)^2}{2\mu} \right\} a_k^+ a_{k+q} \tag{25}$$

The quantity in curly brackets is simply the difference in kinetic energies of the struck particle. In general, the sum on the right hand side of Eq. (25) cannot be further simplified; however, in the deep inelastic limit it can clearly be replaced by

$$[H_0,\rho_q] \to (- q^2/2\mu)\rho_q \tag{26}$$

We can now use the well-known operator identity: if $[A,B] = \alpha B$, then $e^A B e^{-A} = e^\alpha B$. Applying this to Eq. (26) allows us to solve for $\rho_{\underline{k}}(t)$:

$$\rho_{\underline{k}}(t) \to e^{-i(q^2/2\mu)t} \rho_{\underline{k}} \tag{27}$$

Substituting this into Eq. (15) straightforwardly gives

$$\nu W(\nu,\underline{q}^2) \to \delta(\xi - \mu/M) <\Psi_0 | \rho_{\underline{q}}^+ \rho_{\underline{q}} | \Psi_0> = \tag{28}$$

$$= \delta(\xi - \mu/M) [\sum_i Q_i^2 + f(\underline{q}^2) \sum_{i \neq j} Q_i Q_j] \tag{29}$$

i.e., νW does indeed scale, but to a delta function at the quasi-elastic peak! As already discussed, the approach to scaling is, to some extent, governed by $f(q^2)$; however, this derivation also shows that we require $\underline{q}^2 \gg <2\underline{q}\cdot\underline{k}>$ or, if $\underline{q} \equiv |q|\hat{\underline{z}}$, that $|\underline{q}| \gg 2 < k_z >$.

It should be emphasized that, although the derivation **involves** only H_0, it _is_ _valid_ _for_ _an_ _arbitrary_ _interaction_. To pursue this further let us temporarily set $H' = 0$ and consider the corresponding structure functions, $W_0(\nu,\underline{q}^2)$. **One easily finds that**

$$W_0(\nu,\underline{q}^2) = \sum_{\underline{k},\underline{k}'} <\Psi_0 | \delta \left[\nu - \frac{(\underline{k}+\underline{q})^2 - \underline{k}^2}{2\mu} \right] a_{\underline{k}}^+ a_{\underline{k}+q} a_{\underline{k}'}^+ a_{\underline{k}'+q} | \Psi_0> \tag{30}$$

This clearly generates the graphs shown in Fig. 6: the terms where $\underline{k} = \underline{k}'$ are the incoherent contributions corresponding to the box graph, whereas those with $\underline{k} \neq \underline{k}'$ are the coherent pieces. The delta function occuring here requires that

$$k_z = \frac{2\mu\nu - q^2}{2|\underline{q}|} \equiv \mu y \tag{31}$$

so, introducing a normalized single particle wave function $\psi(\underline{k})$ we have a new scaling law, namely

$$|\underline{q}| W(\nu,\underline{q}^2) \to P(y) \tag{32}$$

where

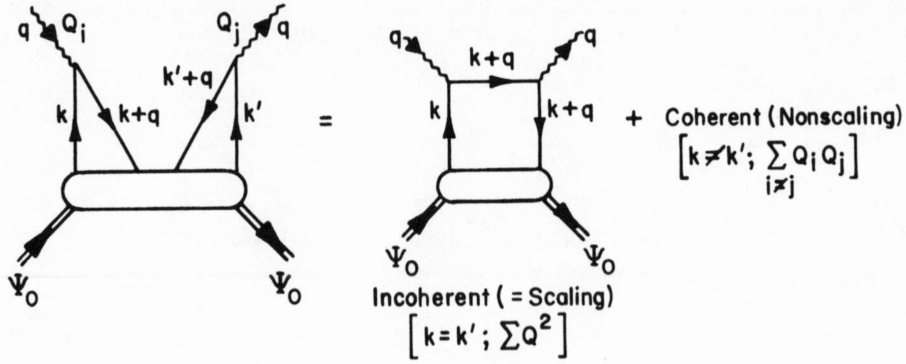

Fig. 6.   Graphs generated by H_0 showing decomposition into incoherent (scaling) and coherent pieces; see Eq. (30).

$$P(k_z) = \int \frac{d^2 k_\perp}{(2\pi)^3} \, |\psi(k_\perp, k_z)|^2 \tag{33}$$

Thus $|q|W$ measures the longitudinal momentum distribution of the
constituents.   This scaling law is thus physically closer in spirit
to Bjorken scaling than the ξ one described above.   There have been
some recent experiments[12] carried out at the Bates machine at MIT to
test this scaling law.   Figure 7 shows some preliminary data
which exhibits some scaling features.   Note incidentally, that
plotting the data this way automatically takes care fo the single
particle contribution, in much the same way that using a structure function takes care of the purely electromagnetic part of the
scattering.   At least, to this extent, y scaling is a useful way of
exhibiting data.

    In this derivation of y-scaling we explicitly ignored interac-
tions.   It turns out (in hindsight) that it is much simpler to deal
with them using a first quantized formalism.   So, let us return to
Eq. (3) and consider the operator

$$O(t) = e^{-i\underline{q}\cdot\underline{r}_j(0)} \, e^{i\underline{q}\cdot\underline{r}_i(t)} \tag{34}$$

Using the fact that $e^{i\underline{q}\cdot\underline{r}_i}$ is the unitary operator that shifts
the momentum of the ith constituent by $\underline{q}$ we can re-express (34) as

$$O(t) = e^{i\underline{q}\cdot(\underline{r}_i - \underline{r}_j) + i\underline{q}^2 t/2} \, e^{i[H - \underline{k}_i \cdot \underline{q}]t} \, e^{-iHt} \tag{35}$$

where for simplicity we have set $\mu = 1$. We now use the operator identity

$$e^{i(H+B)t} = e^{iHt} \, Te^{i \int_0^t dt' B(t')}$$
(36)

where T is the standard time ordering operator and B an arbitrary operator to write

$$0(t) = e^{i\underline{q}\cdot(\underline{r}_i - \underline{r}_j) + i\underline{q}^2 t/2\mu} \, Te^{-i\underline{q}/\mu \cdot \int_0^t \underline{k}_i(t')dt'}$$
(37)

Now, even in quantum mechanics, Newton's second law is valid(!), namely

$$\underline{k}_i(t) = \underline{k}_i(0) + \int_0^t F_i(t')dt'$$
(38)

where the force $\underline{F}_i \equiv -\nabla_i v$. Note that by $F_i(t')$ we mean only $\underline{r}_i$ is to be taken at $t = t'$; the remaining co-ordinates are at $t = 0$. Introducing dimensionless variables $\beta \equiv qt$ and $\beta' = qt'$ we finally arrive at

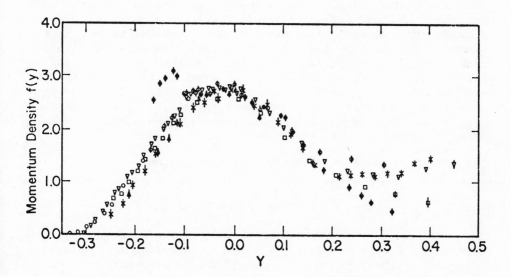

Fig. 7. Preliminary data illustrating y-scaling[13]; see Eq. (32).

$$qW(\nu,\underline{q}^2) = \langle\Psi_0| \sum_{ij} Q_iQ_j \acute{e}^{\,i\underline{q}\cdot(\underline{r}_i-\underline{r}_j)} \int_{-\infty}^{\infty} \frac{d\beta}{2\pi} e^{\,i\beta(y-P_{iz})}$$

$$\times\, T_\beta e^{\,-i\beta/q \int_0^\beta d\beta'\,(1 - \beta'/\beta)F_i(\beta'/q)} |\Psi_0\rangle \qquad (39)$$

First note that the factor $Q_iQ_j e^{\,i\underline{q}\cdot(\underline{r}_i-\underline{r}_j)}$, is the same as that in the sum rule, Eq. (15) and simply determines the degree of coherence. Since we have already **discussed this aspect of the** problem in some detail let us concentrate on the incoherent contribution (i = j). It is clear already from Eq. (39) that when q → ∞, the effects of F_i become irrelevent and qW scales to a function of y only. This result is valid even for a confining force. Indeed, <u>this</u> <u>demonstrates</u> <u>that</u> <u>even</u> <u>if</u> <u>the</u> <u>constituents</u> <u>are</u> <u>never</u> <u>free</u>, <u>they</u> <u>effectively</u> <u>obey</u> <u>free</u> <u>kinematics</u> <u>in</u> <u>the</u> <u>deep</u> <u>inelastic</u> <u>region</u>. More explicitly, note that

$$\int_0^\beta F_i(\beta'/q)d\beta' = v(\underline{r}_i - \beta\hat{\underline{z}}) - v(\beta\hat{\underline{z}}) \qquad (40)$$

Thus, for a short range interaction (unconfining), only small values of β and, therefore, only <u>very</u> small values of t(≡ β/q) are important when q → ∞. On the other hand, for a long-range con- fining force large values of β are clearly allowed. However, t is still limited by the other factors in the exponent. These larger values of t characteristic of a long-range force are presumably necessary for the system to readjust itself to a confining situation following the interaction. Finally, it is amusing to note that for a constant **force (i.e., a linearly rising potential) the** β **inte-** gration can be carried out explicity. One then finds that the approach to scaling can be described (up to correlations) by introducing a new scaling variable $y' \equiv y - (F/2q)^{1/2}$. This is clearly analogous to the Bloom-Gilman **variable. It is suggestive** that only with this confining potential can such a procedure be carried through.

III. LIGHT CONE ANALYSIS AND THE RENORMALIZATION GROUP EQUATION

In the previous Section we showed how a non-relativistic many-body field theory with confining forces manifests scaling. The extension to the relativistic domain is far from trivial, especially because we did not include radiative effects in our derivation even though the effects of the inter-particle potential were included to all orders. Let us begin this part of our discussion by considering the analog to W^0, the structure function generated by free field behavior of the currents. This is simply obtained by treating **Fig. (6) as Feyman graphs:** they clearly have the structure

$$W_2^0(q^2, \nu) = \int \frac{d^4k}{(2\pi)^4} |f(k)|^2 \delta[(k+q)^2 - \mu^2]\theta(\ell_0 + q_0) \tag{41}$$

where $f(k)$ (the nucleon "wave function") contains the unknown dynamics of the target itself. Note, furthermore, that correlations have been dropped. As before, scaling results from the quasi-free kinematic constraint contained in the delta-function which can be re-expressed as $k_- \equiv k_0 - k_z \simeq x - (k^2 - \mu^2)/2\nu$. The correction to x given by $(k^2 - \mu^2)/2\nu$ has been suggested as the origin of a new **Bloom-Gilman type of scaling variable**[13]. Unless $f(k)$ falls off sufficiently fast, scaling will not result; indeed this does not happen in perturbation theory and scaling is violated logarithmically. Only in asymptotically free theories where the effective coupling constant **vanishes for large** q^2 **can one justify a perturbative** expansion and, as a bonus, in **these theories** the scaling violations are only log log. Let us briefly review this.

Eq. (41) can be recast in co-ordinate space as:

$$W_2^0(q^2, \nu) = \int d^4z e^{iq \cdot z} \Delta(z, \mu^2) <p|\phi(z)\phi(0)|p> \tag{42}$$

where
$$<p|\phi(z)\phi(0)|p> = \int d^4k e^{ik \cdot z}[f(k)]^2 \quad , \tag{43}$$

is regular as $z \to 0$ (ϕ being the constituent field) whereas $\Delta(z, \mu^2)$ (its free causal propagator) becomes singular. Recall that when $\nu \to \infty$, $q_+ = q_0 + q_z \to 2\nu$, $q_- = q_0 - q_z \to x$, $z_- \to 0$ and therefore $z^2 = z_+ z_- - z_\perp^2 \to - z_\perp^2 \leqslant 0$. But causality requires $z^2 \geqslant 0$; hence $z^2 \to 0$ showing that the light cone is the relevant space-time region. It is convenient **to invert Eq. (43); in the** $q^2 \to \infty$ limit this gives rise to the moment equations

$$M_n(q^2) \equiv \int_0^1 F(x, q^2) x^{n-1} dx \tag{44}$$

$$= f_n(q^2) <p|0_n|p> \tag{45}$$

where $f_n(q^2)$ is the transform of $\Delta(z)$ and $<p|0_n|p>$ are q^2 **indepen-** dent coefficients derived from a Taylor series expansion of $<p|\phi(z)\phi(0)|p>$. The idea of the operator product expansion is that the factorization between q^2 dependent but target independent functions derived from light cone singularities and target dependent q^2 independent coefficients is a general property of any field theory. For the free field case, $f_n(q^2)$ is independent of q^2 and exact scaling follows.

This analysis shifts the focus of the problem from $F_2(q^2,x)$ to the $f_n(q^2)$. A crucial property of the $f_n(q^2)$ is that they are independent of the target mass, and therefore the infra-red properties of the theory. For large q^2, the explicit mass parameters of the theory become irrelevent; however, a hidden mass scale (M) is still present because we need it to normalize the meaning of our coupling constant g(M). Clearly, if we change M by letting $M \to \lambda M$, say, $f[q^2/M^2, g(M)]$ must remain invariant. This can be expressed in differential form as

$$\frac{\partial f}{\partial t} + \beta(g)\,\frac{\partial f}{\partial g} + \gamma(g)f = 0 \tag{46}$$

where $t \equiv -1/2\,\ell n q^2$, $\beta = \partial g/\partial \lambda$ and $\gamma(g)$ is the anomalous dimension of f; i.e., when $M \to \lambda M$, $f \to \lambda^\gamma f$. Although the solution to this is well-known it is of interest to derive a generalization of it via the method of characteristics

Introduce

$$k(g) = \int_{g_0}^{g} \frac{dg}{\beta(g\ )} \tag{47}$$

so that $k(g_0) = 0$, then Eq. (46) reads

$$\frac{\partial f}{\partial t} + \frac{\partial f}{\partial k} + \gamma(k)f = 0 \tag{48}$$

This is easily solved to give

$$f_n(t,g_0) = f_n\,[t + \phi(t)\, , \, g\{\phi(t)\}]\,e^{\int_0^t dt'\gamma_n[g(t')]} \tag{49}$$

where g(k), the running coupling constant, is obtained by inverting Eq. (47) and $\phi(t)$ is an arbitrary functions of t. The standard solution[1] has $\phi(t) = -t$. In QCD with an SU(3) color group one finds that to lowest order in g:

$$\beta(g) = -\frac{g^3}{16\pi^3}\,(11 - 2/3\,n_f) \equiv -bg^3 \tag{50}$$

$$\gamma_n(g) = \frac{g^2}{6\pi^2} \left[1 - \frac{2}{n(n+1)} + \sum_{M=2}^{n} \frac{1}{M} \right] \equiv a_n g^2 \tag{51}$$

leading, via Eq. (47), to

$$g(t) \rightarrow \frac{g_0^2}{1 + 2bg_0^2 t} \quad . \tag{52}$$

Note that $g(t)$ vanishes as $t \rightarrow \infty$ justifying perturbation theory. This is asymptotic freedom. With $\phi(t) = -t$, one finds

$$f_n(t,g_0) \xrightarrow{t \to \infty} \frac{f_n(0,0)}{(2bg_0^2 t)^{a_n/2b}} \tag{53}$$

giving a logarithmic fall-off for each moment. Note that $f_n(0,0)$ are basically the free field moments. Now, suppose we choose $\phi(t)$ to be a rapidly growing function so that $g\{\phi(t)\}$ falls <u>very</u> rapidly to zero so perturbation theory on the RHS should be <u>very</u> good. Then Eq. (40) naively leads to

$$f_n(t,g_0) \rightarrow \frac{f_n[\phi(t),0]}{[2bg_0^2\phi(t)]^{a_n/2b}} \tag{54}$$

which is most paradoxical since $f[\phi(t),0]$ is given by free field theory and is therefore t **independent**. We would **therefore** conclude the $f_n(t,g_0)$ falls like $[\phi(t)]^{-a_n/2b}$, i.e. in an <u>arbitrary</u> <u>fashion</u>! Equation (49) is, of course, valid for any value of g_0, in particular for g_0 very small where perturbation theory can be used. Let us first see what constraints this imposes if we assume that $f_n(t,g_0)$ can be expanded perturbatively: e.g.,

$$f_n(t,g_0) = A(t) + B(t)g_0^2 + \ldots \quad . \tag{55}$$

Using this in Eq. (49) gives

$$A(t) + B(t)g_0^2 + \ldots = A\{t + \phi(t)\} + \frac{B\{t + \phi(t)\}g_0^2}{1 + 2bg_0^2\phi(t)} + \ldots$$

$$\times \left[\frac{1}{1 + 2bg_0^2\phi(t) + \ldots} \right]^{a/2b} \quad . \tag{56}$$

Equating coefficients up to $O(g_0^2)$ shows that A must e independent of t and that

$$B[t + \phi(t)] = B(t) - aA\phi(t) \tag{57}$$

or

$$B(t) = B(0) - aAt \tag{58}$$

Thus $B(t)$ must grow with t and this is presumably the origin of the
"paradox". For it must be, that when we look at the large t behavior of Eq. (49), the rapid growth of $B\{t + \phi(t)\}$ must compensate the
rapid fall-off in the corresponding g to give Eq. (53). In other
words, in general, one can not replace $g\{\phi(t)\}$ by zero for a rapidly growing $\phi(t)$ and use free field theory even though g is becoming infinitesimal since the amplitude itself is growing rapidly.

IV. N-P MASS DIFFERENCE IN QCD

Finally, we turn to an old problem which is intimately related
to deep inelastic scattering, namely the n-p mass difference, Δm.
Let us briefly review the situation[6] and make some comments on the
present status of the problem. First, it should be remarked that
there are two issues to be dealt with: a) the finiteness of Δm
and b) its sign. The physics of the situation can be summarized
as follows: we know that for a point-like object Δm is logarithmically divergent. Thus, we can expect that if the $F_i(q^2,x)$ scale
exactly, Δm will remain divergent[14]. It has been suggested that
in a unified theory the weak interactions contribution will cancel
the electromagnetic leading to a convergent result[15]; indeed this
has been demonstrated in specific models where the nucleon is
treated as point-like ignoring the strong interactions[16]. Now, in
QCD scaling is broken logarithmically due to gluon contributions
[Eq. (53)] and there now exists the possibility that the electromagnetic contribution itself is convergent. We shall try to indicate when this is possible and how the weak interactions can give
the correct sign.

The divergent part of the one photon contribution to Δm can
be expressed as[15]

$$(\Delta m) = \frac{3}{4} \int_{q^2}^{\infty} \frac{dq^2}{q^2} \{q^2 T(q^2) + \int_0^1 [2 \times F_1(q^2,x) + F_2(q^2,x)] \, dx\} \tag{59}$$

where $T(q^2)$ is the subtraction at $\nu = 0$ in the transverse part
of the virtual Compton amplitude[1,3]. Of course, only the differences in these quantities is required. For exact scaling it is
clear that Δm diverges logarithmically. In QCD, on the other
hand, the q^2 behaviour of all terms in the integrand is, in

principle, known and is given by Eq. (53). For example,

$$\int_0^1 [2 \times F_1(q^2,x) + F_2(q^2,x)] \, dx \rightarrow \frac{1}{(\ln q^2/\Lambda^2)^p} \qquad (60)$$

where $\Lambda \lesssim 1$ GeV and $p = a_2/2b = 32/3(33 - 2n_f)$, n_f being the number of quark flavours. This will give a convergent contribution (albeit an exceedingly slow one) to Δm provided that $p > 1$ which leads to $n_f \geq 12$. Before examining $T(q^2)$, note that Δm arises from the trace of the Compton amplitude and is therefore sensitive to the scalar operators in the theory. For the self-mass (the singlet piece) the leading operator is the trace of the energy-momentum tensor (θ) which has no anomalous dimension, so $a_2 = 0$. Thus even in QCD, the self-mass of each nucleon is logarithmically divergent. Although θ has quark mass terms (m_i) built into it these do not change the form of Eq. (60); in fact, it has been conjectured that at large q^2 such terms simply serve to re-define a new scaling variable[13] in the way discussed in the many-body example of the last section. When we come to examine the subtraction contribution $q^2 T(q^2)$, however, we find that mass terms can contribute since they are weighted with their respective charges (Q_i). Thus, we find

$$q^2 T(q^2) \rightarrow \langle \Sigma Q_i^2 \, m_i \, \bar{\psi}_i \, \psi_i \rangle \qquad (61)$$

where only the nonsinglet piece is required. This, of course, gives a divergent contribution to Δm whose magnitude is roughly proportional to $[Q_u^2 \langle m_u \rangle - Q_d^2 \langle m_d \rangle]$, where the m's are current quark masses occurring in the Lagrangian. The physical origin of this contribution is clearly the different e.m. self-mass of quarks with different charges.

There are several attitudes to take towards this result if we demand a finite Δm:

a) We could demand that, for some unknown reason, $Q_u^2 m_u = Q_d^2 m_d$. However, then $m_u \neq m_d$ (as commonly believed) and the major contribution to the n-p mass difference would arise form this and would not be weak or electromagnetic in origin. To some extent, the fact that Δm is of $O(\alpha)$ is then an accident in the sense that the fact that m_i/M is $O(\alpha)$ is also accidental.

b) The m_i are all zero, the physical quark masses being derived purely from radiative corrections. Recall that the m_i in Eq. (61) are the parameters occurring in the fundamental Lagrangian and, as such, are only loosely connected with the physical masses.

c) The divergent piece coming from Eq. (61) is cancelled by an analogous piece arising from weak interactions[15,16] leaving only the continuum contribution, Eq. (60) to which we must now add the weak structure function pieces. Of course, such a cancellation can take place only in a unified theory where the strength of the weak and electromagnetic become identical at high energies.

Of these various scenarios, (c) has to us the nicest appeal. Let us follow its consequences a little further: with the divergences arising from the different self-energies of the quarks cancelled, the remaining pieces converge extremely slowly. Therefore the contributions from large q^2 can be important and, in particular, the weak contribution can be significant even in the convergent part. Symbolically the contribution to Δm coming from the scaling region reads:

$$(\Delta m)_\infty \sim A\, e^2 \int_{q_0^2}^{\infty} \frac{dq^2}{q^2} \left[\frac{1}{\ln\, q^2/\Lambda^2} \right]^p$$

$$+\, B\, e^2 \int_{q_0^2}^{\infty} \frac{dq^2}{q^2 + M_z^2} \left[\frac{1}{\ln\, q^2/\Lambda^2} \right]^p \tag{62}$$

where A and B are associated, via the light-cone expansion, with the e.m. and weak structure functions, respectively, q_0 represents the threshold for scaling and M_z the mass of the Z-boson that mediates the neutral weak current. Thus

$$(\Delta m)_\infty \sim \alpha \left[\frac{A}{\{\ln(q_0^2/\Lambda^2)\}^{p-1}} + \frac{B}{\{\ln(M_z^2/\Lambda^2)\}^{p-1}} \right] \tag{63}$$

where $p-1 = (6n_f - 67)/3(33 - 2n_f)$. Now, asymptotic freedom requires $n_f \leqslant 16$, whereas the convergence of Eq. (60) requires $n_f \geq 12$. In this range $(p-1)$ 1 and, in particular, for $n_f = 12$ $p - 1 = 5/27$. Thus, although $M_z^2 \gg q_0^2 > \Lambda^2$, $\{\ln(q_0^2/\Lambda^2)\}^{p-1} \simeq \{\ln(M_z^2/\Lambda^2)\}^{p-1}$; in other words, the weak and electromagnetic contributions are comparable and this is a consequence of QCD. Furthermore, in principle, it is even possible for the weak contribution to overcome the electromagnetic and to lead to a reversal of the sign. For this to occur one might expect $(p-1)$ to take on as small a value as possible in order to minimize the effects of M_z, especially since $|A| \sim |B|$. This minimum occurs at $n_f = 12$, which has the significance of being the smallest number of flavors consistent with the known facts and a purely vector-like theory.

Thus, to summarize: A possible scenario for a calculation of mass differences is that a unified theory ensures the cancellation of the logarithmic divergences, whereas asymptotic freedom allows the possibility that the weak contribution is comparable to the electromagnetic. For it to be big enough to reverse the sign suggests that $n_f = 12$. In any case absolute convergence requires $n_f \geqslant 12$.

REFERENCES AND FOOTNOTES

1. Suitable reviews and an extended list of references can be found in the 1976 Les Houches Session XXIX "Weak and Electromagnetic Interactions at High Energy" (ed. R. Balian and C. H. Llewellyn Smith). See also the other lectures in this volume and, in particular, H. D. Politzer, Phys. Rep. 14C, 129 (1974).
2. See, e.g., I. Hinchliffe and C. H. Llewellyn Smith, Nucl. Phys. B128, 93 (1977).
3. G. B. West, Phys. Rep. 18C, 265 (1975).
4. H. Harari, SLAC preprint.
5. Within the covariant parton model, there exists an attempt to calculate $\sigma_{\gamma p}$ using $F_2(x)$ as input: see G. B. West, Phys. Rev. Lett. 31, 798 (1973).
6. See, e.g., A. Zee, Phys. Rep. 3C, 129 (1972).
7. See, e.g., A. L. Fetter and J. D. Walecka, "Quantum Theory of Many-Particle Systems" (McGraw-Hill, N. Y., 1971).
8. Fig. 3 is old SLAC data taken from the talk by E. D. Bloom at the 1973 Electron-Photon Symposium held in Bonn.
9. J. L. Yarnell et al., Phys. Rev. A 7, 2130 (1972).
10. See, e.g., J. J. Sakurai, talk at Electron-Photon Symposium held at Liverpool in 1969.
11. S. Stein et al., Phys. Rev. D 12, 1884 (1974).
12. P. D. Zimmerman, private communication.
13. H. Georgi and H. D. Politzer, Phys. Rev. D 9, 416 (1974).
14. G. B. West, Phys. Rev. D 5, 1987 (1972).
15. S. Weinberg, Phys. Rev. Lett. 29, 388 (1972).
16. D. F. Freedman and W. Kummer, Phys. Rev. D 7, 1829 (1973).

THE HADRONIC FINAL STATE IN DEEP INELASTIC SCATTERING

P.V. Landshoff

Department of Applied Mathematics and Theoretical Physics
University of Cambridge
Silver Street, Cambridge   CB3 9EW

INTRODUCTION

I am going to talk about three approaches to deep inelastic lepton scattering:

1. The naive parton model[1]
2. The covariant parton model[2]
3. Quantum chromodynamics (QCD)

The covariant parton model is an attempt to formulate Feynman's naive parton model in field theoretic terms. Nowadays, one or other of these versions of the parton model is regarded as the non-perturbative element in the scattering cross-section; this is then corrected by perturbation theory calculations based on QCD. That is, the parton model is the zeroth-order contribution to QCD. The naive and covariant forms of the parton model make the same predictions for νW_2.

A major theoretical problem, which will recur throughout my discussion of the hadronic final state, is our continuing ignorance of the effects of quark or colour confinement on the final state. The principal difference between the naive and covariant parton models, for practical purposes, is in their attitude to this problem.

Suppose that a parton of four-momentum k absorbs a virtual photon or W-boson of four-momentum q, as shown in the standard parton-model diagram of figure 1. With x the usual Bjorken variable and M the mass of the target hadron, it follows from simple kinematics that at large q^2

$$k^2 = - \frac{xs_0 + k_T^2}{1 - x} + xM^2.$$

(1)

Here $\sqrt{s_0}$ is the invariant mass of the residual system of hadronic fragments remaining in the target particle after the parton k has been extracted from it. It is the invariant mass of this system before the occurrence of the final-state interaction necessary in order to bring about confinement; this final-state interaction has to be added as a subsequent stage in **Figure 1**. Thus $\sqrt{s_0}$ does not describe the invariant mass of a physical system of particles. Nevertheless, in the covariant parton model it is assumed that already $s_0 > 0$, as for a physical system of particles. Then it can be seen from (1) that the parton k cannot be "on shell"; indeed, for x close to 1 the value of k^2 even has to be large and negative. In the naive parton model, on the other hand, it is assumed that the confinement effects are such as to allow k always to be "on shell", $k^2 = m^2$; from (1) we see that then one has to give up the assumption that $s_0 > 0$, and for x close to 1 one even has s_0 large negative. In this case it must be assumed that the final-state interactions subsequently restore s_0 to a physical value.

Nowadays, people are more interested in the QCD corrections than in the zeroth-order parton model. Nevertheless, it is to the zeroth-order model that I shall direct most of my attention; I shall then discuss QCD corrections at the end. This is because the dominant features of the hadronic final state are determined by the zeroth order part of the theory.

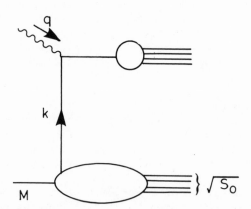

Fig. 1. The parton model impulse approximation to deep inelastic scattering. Not shown are the vital final-state interactions that must be present to ensure quark or colour confinement.

Mainly because of limitations of beam energy, these dominant features have not yet been given adequate experimental study. It is important that they should be studied, because in several respects our theoretical understanding is such that we cannot make well-agreed predictions about the dominant background, and until we know its features we shall find it hard unambiguously to separate from it any effects of QCD corrections.

PARTON TRANSVERSE MOMENTUM

The parton k in **Figure 1** has a transverse momentum k_T before it absorbs the photon or W-boson q. In modern language, this is referred to as "intrinsic or "primordial" transverse momentum, to distinguish it from any further transverse momentum that the parton may acquire subsequently by gluon radiation. The latter represents a QCD correction to the transverse momentum, and I discuss it briefly at the end of the talk. Because of momentum conservation, the transverse momentum of the parton k is equal to the total transverse momentum of the jet of hadrons into which the parton fragments after it absorbs q. Apart from some uncertainty about the effect of the eventual confining final-state interaction, which here is unlikely to be serious, the parton transverse momentum is therefore directly measurable.

In the naive parton model, simple kinematics predict that the intrinsic transverse momentum satisfies[1]

$$\frac{\sigma_L}{\sigma_T} \sim \frac{4}{Q^2} \; [\; < k_T^2 > + m^2 \;] \qquad\qquad (2)$$

where m is the parton mass. This result does not hold in the covariant parton model, and it is also not applicable to the QCD corrections. Strictly speaking, the covariant parton model makes no prediction of the type (2). This is because, according to the Callan-Gross relation, σ_L vanishes in leading order, so that (2) is a relation between non-leading-order effects. The naive parton model assumes that these effects are fully described by the impulse approximation of Figure 1, but in the covariant model the impulse approximation describes only the leading-order terms. To understand why this must be the case, notice that if the parton k is allowed to go off shell, Figure 1 is gauge invariant in leading order only: to respect gauge invariance in non-leading orders one must consider also terms with final-state interactions added in, in this case final-state interactions resulting from the "ordinary" part of the force, and not just from its long-range confining part.

The consequence is that any connection between σ_L and $< k_T >$ is very model dependent. The k_T distribution can be calculated

from Figure 1 alone, but even it is model-dependent[3]. It very
likely changes with the value of theBjorken variable x. It is
likely also that at large k_T the intrinsic transverse momentum
distribution tails off like an inverse power, and not like the
exponential that is often assumed. This can be seen from the
kinematic relation (1): large k_T^2 corresponds to large k^2, but also
$x \to 1$ corresponds to large k^2. Hence if, as is accepted, $F_2(x)$
behaves like a power of $(1 - x)$ near $x = 1$, then the k_T distribution
has to fall off like an inverse power at large k_T. In fact $(1 - x)^3$
corresponds to k_T^{-8}.

As a very simple model, replace the hadronic system s_0 of Fig. 1
by a single "core" of fixed mass M_0, as in Fig. 2 This diagram contains the coupling of the target hadron to parton plus core, and
also the parton propagator. I shall take the product of these factors to be $(k^2 - m^2)^{-2}$; this then gives $F_2(x) \sim (1 - x)^3$ as $x \to 1$.
One then finds that[3]

$$< k_T^2 > = \tfrac{1}{2} \, [x \, M_0^2 + (1-x)m^2 - x(1-x) \, M^2] \qquad (3)$$

That is, $< k_T^2 >$ increases steadily from the value $\tfrac{1}{2}m^2$ at $x = 0$ to
$\tfrac{1}{2}M_0^2$ at $x = 1$. The model also has built into it the k_T^{-8} fall-off of
the parton intrinsic transverse momentum distribution at large k_T.
It is only a model, but I suspect that these two features may be
rather realistic. It seems sensible to take m in the MeV region and
M_0 in the GeV region, so that the variation of $< k_T^2 >$ with x in (3)
is rather large.

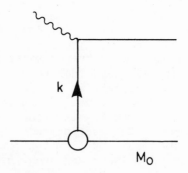

Fig. 2.   Model for Figure 1, in which the hadronic system $\sqrt{s_0}$
          is replaced by a core of fixed mass $\sqrt{s_0} = M_0$.

Likewise, in the fragmenting parton jet in Figure 1, the hadrons have an intrinsic transverse momentum k_T' relative to the jet axis, and $< k_T' >$ is rather likely to vary with the fractional longitudinal momentum z of the jet fragments. Feynman and Field[4] have pointed out that such an effect will surely occur if there is substantial resonance production from the jet. If one measures the transverse momentum p_T of the hadron fragments of the parton jet relative to the initial direction of q, the effects of k_T and k_T' combine according to the formula[5]

$$< p_T^2 > = z^2 < k_T^2(x) > + < k_T'^2(z) > \qquad (4)$$

So one might have quite a complicated variation with x and z, and one would expect $< p_T^2 >$ to increase with increasing z.

RAPIDITY DISTRIBUTION

If I write the kinematic formula (1) a little differently,

$$s_o = \frac{x M_o^2 + k_T^2 + k^2(x-1)}{x} \qquad (5)$$

This tells me that, if I allow x to become small, the invariant mass $\sqrt{s_o}$ of the lower system of hadrons in Figure 1 becomes large. According to familiar ideas of strong-interaction physics, this means that Regge-pole effects occur at small x. In particular, it is responsible for the relation[2]

$$F_2(x) \sim x^{1 - \alpha_o} \qquad \text{as } x \to 0 \qquad (6)$$

where α_o is the intercept of the leading Regge trajectory, the pomeron in the case of the singlet part of F_2 and the A_2' for the non-singlet part. The Regge effects are also reflected in the rapidity distribution of the final state hadrons.

I remind you that in pp inelastic scattering at high energy, Regge effects lead to the production of "pionisation products" in the centre of the rapidity plot. The dynamical mechanism is the multiple pomeron exchange shown in the upper part of Figure 3, and the resulting rapidity distribution is shown in the lower part of the figure. The composition of the central plateau is supposed to be independent of the initial-state hadrons. It is commonly described as a plateau, though experiments at the CERN ISR cast doubt on the extent to which this description is accurate. Certainly its height, which was once believed to be constant, is now known to rise by as much as 40% over the ISR energy range[6]. The consequences of this for deep inelastic lepton scattering should be explored, both theoretically and experimentally.

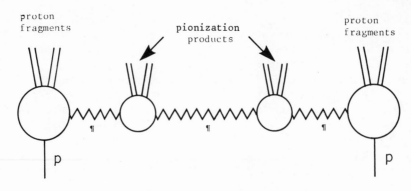

proton
fragments

pionization
products

proton
fragments

p   p

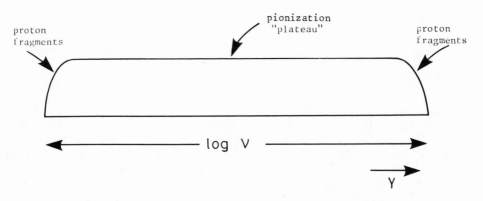

proton
fragments

pionization
"plateau"

proton
fragments

log ν

Y

Fig. 3. Pionization in high-energy pp scattering, with the
         corresponding regions in the rapidity plot. The plot of
         particle density against rapidity has width $Y_{max} - Y_{min}$
         ~ log ν, and the two fragmentation regions each occupy
         about two units of Y.

        The rapidity plot for pp interactions breaks up into three
regions. At small x, we have seen that the variable s_o in Figure 1
is large, so that the rapidity plot corresponding to the lower
part of the diagram similarly breaks up into three regions. These
are the target fragmentation region and a pionization "plateau",
exactly as in the pp case, and then a "hole fragmentation" region
corresponding to the parton k that has been pulled out. See Figure 4.
The parton fragmentation region corresponds to the upper system
of hadrons in Figure 1. The region between it and the other regions
has to be filled in somehow, for if it were isolated from them it
would contain total charge and total baryon number whose values
would be fractional. The filling in is caused by the confinement
force. This is not understood, but it is usually assumed[7] that the

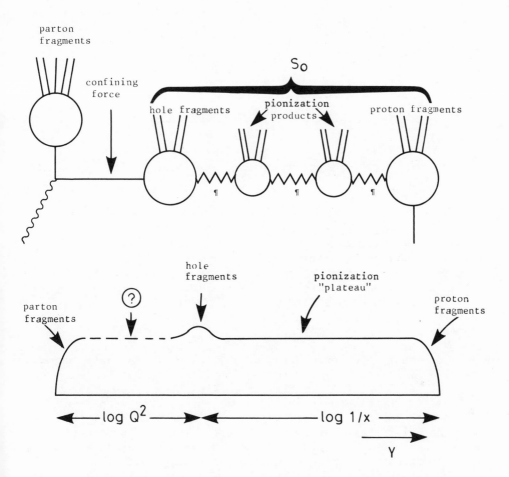

Fig. 4.   The various rapidity regions in deep inelastic lepton
          scattering at small x. The proton fragmentation region
          and the **pionization** plateau are as in pp scattering. It
          is often conjectured that there is a similar plateau
          connecting the parton fragmentation region to the other
          regions, caused by the confinement force.

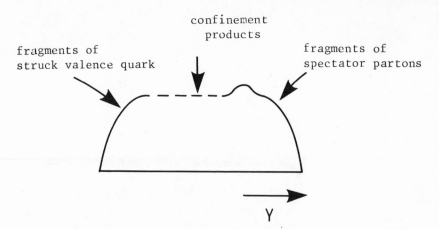

Fig. 5.   The rapidity regions in deep inelastic scattering at non-
          small x.

infilling corresponds to another plateau, similar to the pionization
plateau. However, the theoretical status of this is different from
that of the pionization plateau; as it is produced by the confinement
force, Regge theory in its conventional form has nothing to say about
it unless one makes new assumptions[7]. For this reason, experimental
investigation is of particular interest. For example, do the two
plateaux have the same height?

     Note that experience with ISR physics suggests that, to separate
the five different rapidity regions in Figure 4, one needs to have
available at least two rapidity units for each. Thus one requires
at least 10 units altogether. However, as a first step it will be
good enough to consider not small x, but more moderate values, so
that the hole fragmentation region merges into the proton fragmentation region, squeezing out the pionisation plateau and giving
a single spectator-parton fragmentation region on the right-hand
side of the rapidity plot (Figure 5). This is still separated from
the parton-fragmentation region by the plateau(?) corresponding to
the confinement. Hopefully, the three rapidity regions are well
separated when six rapidity units are available. The corresponding
three regions in e^+e^- annihilation should be distinguishable at
the upper PETRA energies.

QCD CORRECTIONS

     Whether one takes as the basis of the zeroth-order calculation
the naive parton model or the covariant parton model, one predicts
that in μp scattering the semi-inclusive cross-section for hadron
production in the parton-fragmentation region takes the form

$$\frac{d^4\sigma}{dx\ dy\ dz\ d\phi} \sim \frac{\alpha^2}{yQ^2} [1 + (1-y)^2] \sum_{\substack{\text{flavours} \\ j}} e_j^2\ f_j(x)\ d_j(z)$$

$$(7)$$

Here x is, as usual, the Bjorken variable and y is the fraction of the beam energy taken by the virtual photon. The variables z and ϕ refer to the detected hadron: z is the fractional momentum relative to the total momentum of the parton jet and ϕ is the azimuthal angle measured from the lepton scattering plane. The functions f_j and d_j are the distribution within the target proton and the fragmentation function of the parton of flavour j.

The formula (7) is the asymptotic form of the zeroth-order contribution to the cross-section. The following properties are evident for the contribution from each flavour j.

  (i)   It scales
  (ii) It factorises
  (iii) It is independent of ϕ

We have seen also that

  (iv) $< p_T^2 >$ is a function of x and z only, not of Q^2.

All the properties (i) to (iv) are spoilt by the QCD corrections. In first order, these corrections arise from the diagrams of Figure 6; these diagrams have to be added to the zeroth-order diagram of Figure 1. The big problem in studying how these corrections violate the properties (i) to (iv) is that they are violated also in the zeroth-order part of the cross-section; the expression (7) is valid only asymptotically. If one measures a violation of one of the properties, it is difficult to determine how much of this violation should be attributed to the QCD corrections, and how much to the largely unknown sub-asymptotic behaviour of the zeroth-order term. One can say that one expects the effects of the latter to die away with increasing Q^2 or p_T, leaving just the interesting QCD effects at very large Q^2 or p_T, but it is hard to be confident about how large Q^2 or p_T must be before things become reasonable clear-cut.

ϕ DEPENDENCE

When one takes into account the QCD corrections and the sub-asymptotic zeroth-order terms, the semi-inclusive cross-section for hadron production in μp scattering becomes

$$\frac{d^4\sigma}{dx\ dy\ dz\ d\phi} = A_o + A + B \cos \phi + C \cos 2\phi \qquad (8)$$

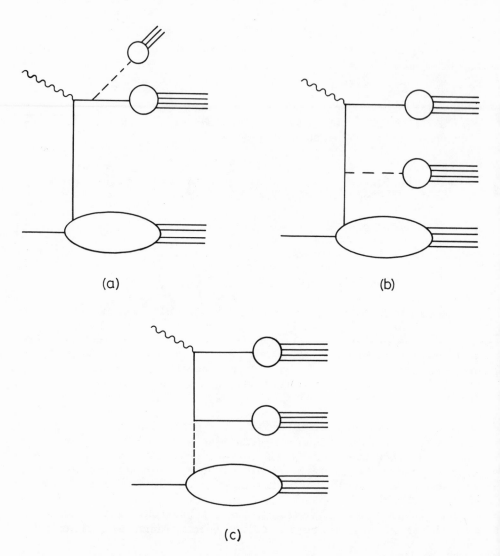

Fig. 6.   First-order QCD corrections which must be added to Figure 1.
The diagram (c) contributes only to the single cross-
section. Each diagram requires additional confinement
final state interactions.

Here A_0 is the leading zeroth-order term (7). In ν and $\bar{\nu}$ scattering, where parity is not conserved, there are additionally terms in $\sin \phi$ and $\sin 2\phi$, but their coefficient functions are likely to be smaller than the functions B and C, by order α_s so far as QCD corrections are concerned.

The dependence on ϕ is connected with the transverse momentum of the active quark in the interaction. The effect arises because a change in the orientation of k_T corresponds to a change in the centre-of-mass energy of the current/quark interaction. In the naive parton model the association between k_T and ϕ dependence is simple[8]:

$$< \cos \phi > \quad \sim \quad < 2k_T > / \sqrt{Q^2}$$

$$< \cos 2\phi > \quad \sim \quad < 4k_T^2 > / Q^2 \qquad\qquad (9)$$

More generally, in the covariant parton model[9], there is no <u>simple</u> relation of the type (10), though one does expect that the sub-asymptotic non-perturbative contributions to $< \cos \phi >$ and $< \cos 2\phi >$ will decrease respectively like $1/\sqrt{Q^2}$ and $1/Q^2$ with unknown multiplying coefficients. The relation (9) also does not apply to the perturbative QCD contributions, but these may be calculated explicitly[10]. In Figure 7 I show the predictions of Mendez et al. for νp scattering at 20 and 2000 GeV/c beam momentum. I have also drawn in the curves $1/Q^2$ and $1/\sqrt{Q^2}$, so as to give a rough idea of the severity of the problem of separating the perturbative and non-perturbative effects.

Transverse momentum

When the active parton radiates before absorbing the current q, as in Figure 6, it can acquire transverse momentum additional to the primordial or intrinsic transverse momentum associated with the target wave function. This perturbative component to the transverse momentum[11], unlike the intrinsic component, increases with Q^2 at fixed x. At given values of Q^2 and x, the perturbative transverse momentum distribution is predicted to have a tail which should be clearly visible above the intrinsic component. Most authors assume that the latter falls off exponentially, but I have already explained that there is reason to doubt this. In Figure 8 I show predictions of Mendez and Weiler[11] for the distribution in the variable

$$\Sigma_T = (\sum_i | p_{\sim T}^i |)^2 \qquad\qquad (10)$$

where i runs over all the final-state hadrons. The estimates of the non-perturbative or intrinsic contribution assume exponential fall-off.

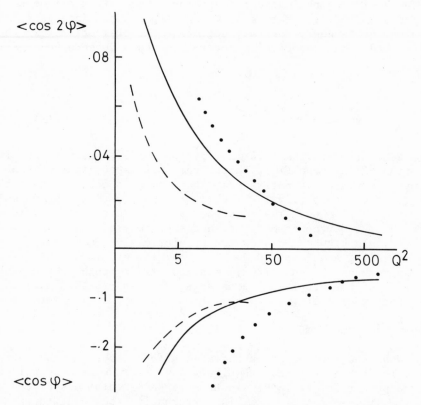

Fig. 7. Predictions of Mendez et al.[10] for perturbative contribution
to $< \cos 2\phi >$ and $< \cos \phi >$ in $\overline{\nu}p$ interactions. The dashed
lines are for 20 GeV/c beam momentum and the solid lines
are for 2000 GeV/c. The dotted lines represent a guess as
to the possible magnitudes of the non-perturbative contributions.

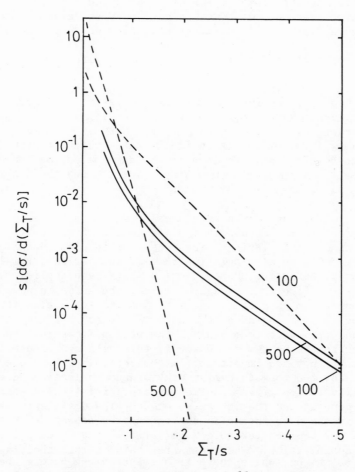

Fig. 8. Predictions of Mendez and Weiler[11] for distributions in
the variable Σ_T, defined in (10), for μp interactions at
100 and 500 GeV/c beam momentum. The solid curves are
the perturbative contributions, and the broken curves are
estimates for the intrinsic, non-perturbative components.

ABSENCE OF FACTORIZATION

Because of the difficulty in understanding how QCD causes
partons to materialize as hadrons, Sterman and Weinberg[12] emphasize
the usefulness in the study of jets of considering just the total
energy flowing into some region of phase space. The hope is that,
by not taking account of how this energy is shared among hadrons,
the unknown features of the materialization process become important.

An analysis of deep inelastic scattering from this approach has
been described by Stevenson.[13] In the zeroth-order parton model
of Figure 1, the final-state hadrons arrange themselves in two jets,
but the QCD perturbative corrections of Figure 6 can evidently lead
to 3-jet events. However, the diagrams of Figure 6 do not yield
three jets in every event. As a matter both of practice and of
principle, if two of the three final state systems in the diagrams
of Figure 6 emerge more or less in the same direction, together they
cannot be distinguished from a single jet. Hence the diagrams of
Figure 6 yield also two jet events.

In the zeroth-order parton model the jet that emerges in the
forward hemisphere is exactly similar to one of the pair of final-
state jets characteristic of the continuum in e^+e^- annihilation.
Stevenson performs a calculation to test whether this remains true
after the perturbative QCD corrections have been applied to both
reactions. Unlike in e^+e^- annihilation, one knows in advance the
direction of the axis of the forward jet in deep inelastic scattering:
in the absence of the non-perturbative effects that generate the
primoridal transverse momentum of the parton it is just the direction
of the current momentum q. However, this difference between the two
reactions is not too important. If one calculates for e^+e^- annihilation and for the forward hemisphere in deep inelastic scattering the
probability that a given fraction of the initial energy emerges in
a given cone around the jet axis, one obtains different answers.
In fact in the deep inelastic scattering the answer depends on the
value of the Bjorken variable x, a variable that does not even appear
in e^+e^- annihilation. The reason for this is that the diagrams of
Figures 6b and 6c, where there is bremsstrahlung before the current
is absorbed by the parton, do not have counterparts in e^+e^- annihilation.

CONCLUSIONS

It will not be easy to identify the perturbative QCD effects
in the hadronic final state in deep inelastic scattering. The
dominant features of the final state are determined by the non-
perturbative contributions of the zeroth-order parton model. In
various respects the latter are not well understood theoretically,
and so their experimental study is interesting for its own sake,
and necessary first if the perturbative effects are eventually
to be separated.

ACKNOWLEDGEMENTS

I am very grateful to Jean-Jacques Aubert and Giuliano Preparata for warm hospitality in Erice.

REFERENCES

1. R. P. Feynman, Photon-Hadron Interactions, (Benjamin).
2. P. V. Landshoff and J. C. Polkinghorne, Physics Reports $\underline{5C}$, 1 (1972).
3. There have been many papers on this subject. Two of the earliest are P. V. Landshoff, Phys. Lett $\underline{66B}$, 452 (1977); J. F. Gunion, Phys. Rev. $\underline{D15}$, 3317 (1977).
4. R. P. Feynman and R. D. Field, Nuclear Physics $\underline{B136}$, 1 (1978).
5. M. Gronau and Y. Zarmi, Phys. Rev. $\underline{D18}$, 2341 (1978).
6. K. Guettler et al., Phys. Lett $\underline{64B}$, 111 (1976).
7. See, for example, the talk by G. Preparata at this Seminar.
8. R. Cahn, Phys. Lett. $\underline{78B}$, 269 (1978).
9. R. L. Kingsley, Phys. Rev. $\underline{D10}$, 1580 (1974).
10. H. Georgi and H. D. Politzer, Phys. Rev. Lett. $\underline{40}$, 3 (1978); A. Mendez, Nuclear Physics $\underline{B126}$, 417 (1977); A. Mendez, Raychaudhuri and Stenger, Nuclear Physics $\underline{B148}$, 499 (1979); J. Cleymans, Phys. Rev. $\underline{D18}$, 954 (1978); G. Kopp, R. Maciejko and P. Zerwas, Nuclear Physics $\underline{B144}$, 123 (1978); P. Mazzanti, R. Odorico and V. Roberts, preprints CERN TH. 2584 and IFUB/78-11.
11. E. Floratos, Nuovo Cimento $\underline{43A}$, 241 (1977); G. Altarelli and G. Martinelli, Phys. Lett. $\underline{76B}$, 89 (1978), H. Georgi and H.D. Politzer, Phys. Rev. Lett. $\underline{76B}$, 89 (1978); A. Mendez and T. Weiler, Oxford preprint 93/78.
12. G. Sterman and S. Weinberg, Phys. Rev. Lett. $\underline{39}$, 1436 (1977).
13. P. M. Stevenson, preprint ICTP/78-79/1.

REVIEW OF ELECTROPRODUCTION EXPERIMENTS AT

CORNELL AND M.I.T. - SLAC

L.S. Osborne

Physics Department
Massachusetts Institute of Technology
Cambridge, MA 02139

I. INTRODUCTION

This paper presents and up-to-date summary of electroproduction experiments by three groups. All of these groups are in the final stages of data analysis and much of this data has already been published. This paper will attempt to summarize what has been learned from all this. There are two groups from Cornell, DECO[1] and LAME[2], and an MIT-SLAC group[3]. The layout of the three experiments is shown in Figure 1.

The two Cornell experiments have the merit that they cover a large solid angle; this has allowed the study of $x_F < 0$ for DECO and full reconstruction of events in the LAME case. One pays a price in background for this. The MIT-SLAC apparatus covered only positive x_F. However, it used D and nuclear targets and also had particle identification over about 1/6 of the solid angle with a threshold Cherenkov counter.

II. INCLUSIVE HADRON PRODUCTION

A. Charge Ratios and Quark Model Parameters

Charged lepton scattering has the power to reveal details of the nucleon quark behavior by making use of its different coupling to quarks (e.g., the u and d quarks). The fullest use of this property is obtained when neutrons as well as protons are used as targets. The extra degree of freedom allows us to investigate separately phenomena due to quark charge asymmetries from majority to minority quark distribution asymmetries.

a) DECO

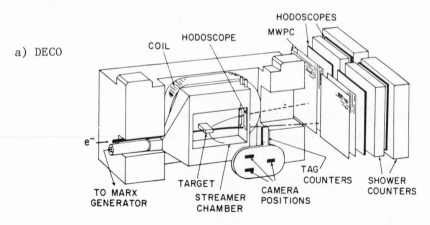

b) LAME

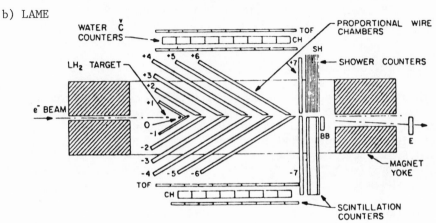

c) MIT-
SLAC

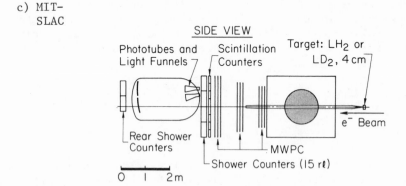

Fig. 1   Layouts of the three experiments described in this paper:
a) DECO[1], b) LAME[2] and c) MIT-SLAC[3].

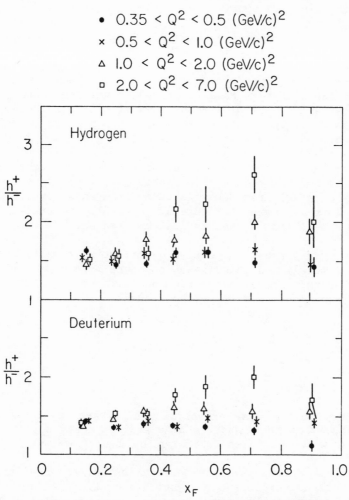

Fig. 2   The ratio of + to – hadrons for various Q^2 bins plotted
          versus x_F.

The basic charge asymmetry built into nucleon structure is
illustrated qualitatively in Figure 2. The +/- ratio for hadrons
is plotted as a function of x_F for various Q^2 bins and separately
for hydrogen and deuterium. We can establish the following results:

1. The +/- ratio is everywhere greater than one. This is not sur-
    prising since we always have a target that is positively charged.

2. The +/- ratio depends on $x_F = P_\ell /P_{\ell max}$ in the center-of-mass
    showing the "leading particle" effect.

3. The +/- ratio depends on x. This is expected from any argument
    which explains this ratio in terms of valence quarks. It defi-
    nitely cannot be explained by a model that attributes charge
    ratios to charge conservation constraints at small multiplicities;
    all data is taken from the same s bins ($15 < s < 36$ GeV^2).

A more dramatic manifestation of nature's charge asymmetry is
obtained by looking at the π^+/π^- ratio (π's were identified by the
Cherenkov Counter) from neutrons taken from an x_F bin, $0.3 < x_F <$
< 0.85. This is shown in Figure 3. A totally neutral initial state
leads to a π^+/π^- ratio that is greater than one.

We may go further and derive quantitative results from this
data. All models and theories which start with the basic photon-
nucleon interaction being a photon-quark interaction will write
the <u>cross-section</u> for meson production as

$$\pi_p^\pm = u_p q_u^2 D_u^\pm + d_p q_d^2 D_d^\pm + \beta$$

$$\pi_n^\pm = u_n q_u^2 D_u^\pm + d_n q_d^2 D_d^\pm + \beta$$

(1

$u_i(d_i) \equiv$ up (down) quark probability distribution; i = p or n,
proton or neutron; q = quark charge and $\beta \equiv$ contribution from the
sea quarks which is taken as charge and source symmetric.

The D functions give the correlation between the struck quark
and the charged particle coming out and distinguish one model from
another. In principle, they could be functions of all variables,
$x = q^2/2M\nu$, s, x_F, and p ; Equation 1 is perfectly general. We
make a further assumption based on isospin invariance

$$u_p = d_n \equiv u_1 \qquad\qquad\qquad u_n = d_p = u_2 \qquad (2$$

$$D_u^+ = D_d^- \equiv D_1 \qquad\qquad\qquad D_u^- = D_d^+ = D_2 \qquad (3$$

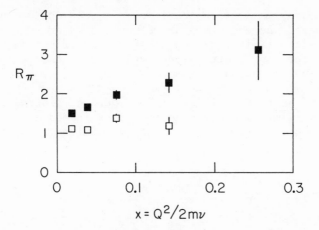

$$x = Q^2/2m\nu$$

Fig. 3   The ratio of positive to negative π mesons from protons ( )
         and neutrons ( ) in the bin $0.3 < x_F < 0.85$. The errors
         on the two lowest x_F points are smaller than the squares.

Equation 2 is solid; however, Equation 3 is applicable only
for π production not K production. We have corrected the data for
K and proton contributions based on our measurement of their production[4,5]. We may now calculate four sums and differences of meson
cross-sections:

$$A'(B) = (\pi_p^+ + \pi_p^-) + (-)(\pi_n^+ + \pi_n^-)$$

$$C(D) = (\pi_p^+ - \pi_p^-) + (-)(\pi_n^+ - \pi_n^-)$$

(4

If we define $A = A' - 4\beta$ and $R = A/A' =$ (valence), (all quarks),
then A through D, when written in terms of u, q, and $D^{\pm}$, are products of sums and differences, e.g.,

$$B = (q_u^2 - q_d^2)(u_1 - u_2)(D_1 + D_2)$$

(5

Combining these as ratios we get

$$r_1' = B\ C/A\ D = (q_u^2 - q_d^2)/(q_u^2 + q_d^2)$$ (6a

$$r_2' = B\ D/A\ C = (u_1 - u_2)^2/(u_1 + u_2)^2$$ (6b

$$r_3' = C\ D/A\ B = (D_1 - D_2)^2/(D_1 + D_2)^2$$ (6c

Thus one may measure the ratio of q's, u's and D's directly from experiment. However, the above equalities are only directly comparable in the limit that the sea quark contribution vanishes; they will differ from their correct value by R, the ratio of valence to all quark contributions.

A plot of $r_1 = (Rr_1)$ is shown in Figure 4, plotted versus x = = $1/\omega$. In the limit x $\gtrsim$ 0.2, the data does indeed agree with 9/25 from Equation 6a. The ratio $R_1 = |q_u/q_d|$ derived from these experi-

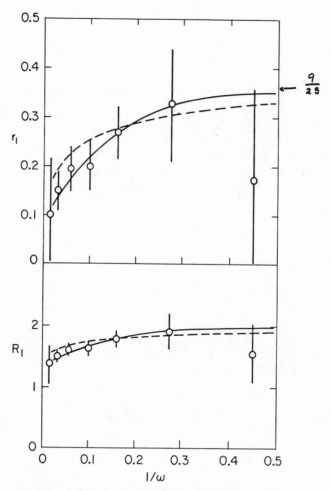

Fig. 4a   The measured value of r_1 vs $1/\omega$.
     4b   The calculation, from r_1, of the ratio R_1.
          In both figures, the solid lines are the expectations
          from Ref. 6, the dashed lines from Ref. 7.

mental numbers is also shown in Figure 4 and agrees with $R_1 = 2$ at high x.

Given that we have confirmed these charge assignments, we can form other ratios

$$r_4 = D/C = (u_1-u_2)/(u_1+u_2) \quad (q_u^2+q_d^2)/(q_u^2-q_d^2) \tag{7a}$$

$$r_5 = D/B = (q_u^2+q_d^2)/(q_u^2-q_d^2) \quad (D_1-D_2)/(D_1+D_2) \tag{7b}$$

$$r_6 = B/C = (u_1-u_2)/(u_1+u_2) \quad (D_1+D_2)/(D_1-D_2) \tag{7c}$$

which do not involve R. A plot of u_p/d_p is shown in Figure 5.

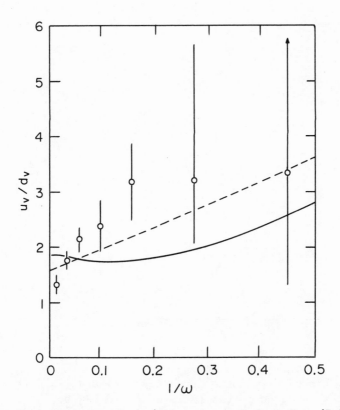

Fig. 5   A plot of u_v/d_v (= u_1/u_2) extracted from Eq. (7a) under the assumption that $(q_u/q_d)^2 = 4$. The solid (dashed) line is the expected behaviour from Refs 6 and 7.

We note that it is never constant versus x at the value of 2, its expected average value. It is consistent with the disappearance of the minority quark at high x which is expected from the ratio of neutron to proton structure function. Note, however, that it approaches 1 at x near 0. This is not expected from the two models[6,7] we use for comparison. Note again that we are comparing only <u>valence</u> quark ratios; the sea contribution has been removed. Using Eq. 7b we also plot in Figure 6

$$\eta = \int_{.3}^{.85} D_1(x_F)dx_F / \int_{.3}^{.85} D_2(x_F)dx_F \tag{7d}$$

A simple quark model would have this independent of x and this appears to be so. In summary, the basic elements of a simple quark model are confirmed though the behavior of the u/d ratio remains unexplained.

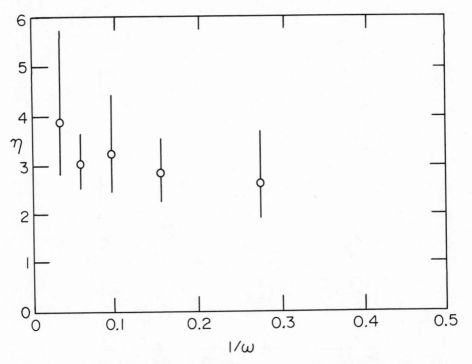

Fig. 6 A plot of η extracted from Eqs (7b) and (7d), assuming $q_u^2/q_d^2 = 4$. η is the ratio of $\bar{D}_1$ to $\bar{D}_2$, to i.e., D and D_2 averaged over x_F from $x_F = 0.3$ to 0.85 and over ϕ^2 and p_T^2.

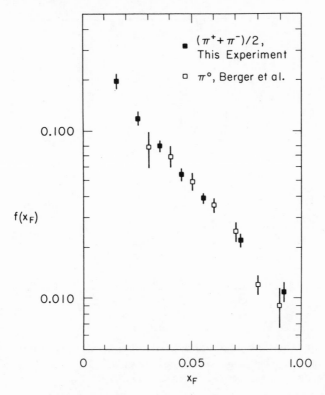

Fig. 7   A comparison of the average charged meson yield to the
π⁰ yield measured in a DESY experiment[8].

B.   General Properties of the Data

Charged to Neutral Ratio.   Figure 7 shows a comparison of the
MIT-SLAC data using $1/2(\pi^+ + \pi^-)$ yields compared to π^0 data from a
DESY experiment at lower energies[8].   It can be observed that the
data points are in agreement which is an expected result from most
quark models.

p_T dependence.   Figure 8 shows the dependence of the average
transverse momentum $\bar{p}_T$ with s and q^2.   We see that $\bar{p}_T$ increases
with s but, for fixed s, shows no change with q^2.   The increase of
$\bar{p}_T$ with s is also observed in hadron-hadron reactions[9].   A plot such
as this one is helpful in determining whether changes in $\bar{p}_T$ are due
to s or to q^2 dependence.   In particular, plots of the q^2 dependence
for an analogous reaction such as neutrino interactions may be mis-
leading.   For a given set of data, which is a function only of x,
an increase of q^2 is also an increase in s and an apparent q^2 dependence may simply be reflecting an s dependence.

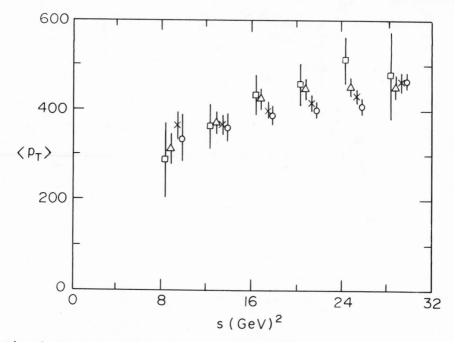

Fig. 8 A plot showing the behaviour of the average transverse
momentum, $\langle p_T \rangle$, as a function of q^2 and s. The q^2
bins have the same labels as in Fig. 2.

Also shown, in Figure 9, is the dependence of $\bar{p}_T$ on x_F. This
is the analogue of the "seagull" effect which is seen in hadron-
hadron collisions and neutrino-hadron collisions[10].

Particle Type Ratios. In hadron-hadron reactions, there appears to be a general rule that the ratio of heavy particle production compared to π-meson production increases with p_T and x_F. This
is well documented for the k meson[11,12] and, with less accuracy,
for the ρ meson[13]. The difference between the ρ⁰ and k behaviour
in this respect is of some interest since the ρ⁰ behaviour would
seem like a pure mass effect whereas change in the k/π ratio may
depend also on strangeness factors. The contribution of electroproduction data to this subject can be seen in Figures 10 and 11.
The k/π ratio is shown, measured in the MIT-SLAC experiment, and
it demonstrates the same behaviour seen in hadron collisions.

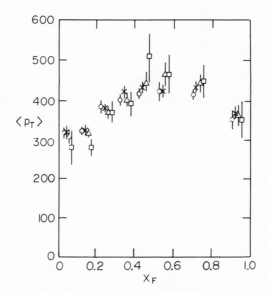

Fig. 9 A plot of $\langle p_T \rangle$ as a function of x_F. The data is separated into q^2 bins as in Fig. 2.

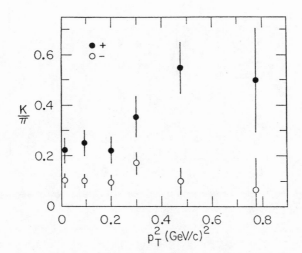

Fig. 10 The K/π ratio from hydrogen and deuterium combined as a function of p_T^2.

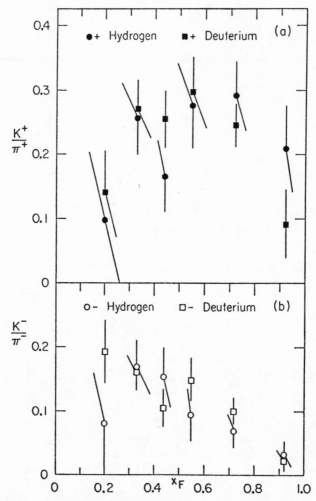

Fig. 11   The K/π ratio as a function of x_F shown separately
for hydrogen and deuterium.

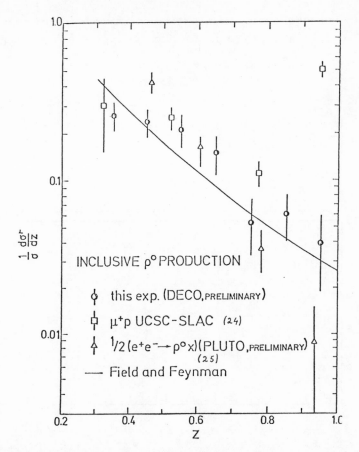

Fig. 12 Inclusive ρ^0 production by electrons and μ mesons.
The solid line is a good fit to average π production.
We see that ρ production is greater than production
for $x_F > 0.4$. The elastic ρ^0 production is not
excluded in the UCSC-SLAC data, thereby giving the
high point at $x_F \simeq 0.95$.

The Cornell-DESY group has measured ρ^0 production and their
ρ^0 to π ratio is shown in Fig 12. This ratio is greater than one.
If we correct for a factor of three because of spin 1, we might
claim that the K and ρ^0 show the same behavior; we would then deduce that the K/π ratio, being less than one, is a kinematic effect
for high mass particle. However, the p_T distribution of the ρ^0's
does not show the rise with respect to π's exhibited by K's.

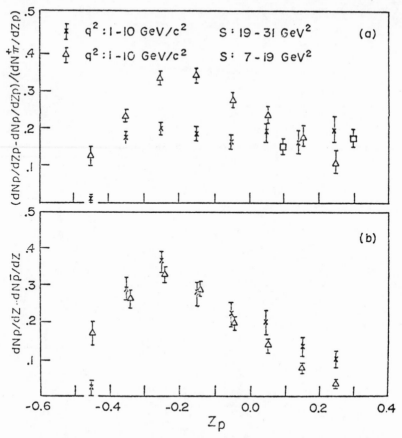

Fig. 13 A plot of the "residual" proton distribution as a
         function of x_F of the protons.

Proton yields have been measured in the MIT–SLAC experiment[14]
at the lower energies by means of time-of-flight. The measurements
were restricted to $x_F > -0.5$. The yield of antiprotons was subtracted from the protons to give the so-called "residual proton" yield;
we obtain the distribution of the center of baryonic charge after
all mesons are emitted (Figure 13). It is interesting to note that
this yield peaks at $x_F = -0.2$; this is not a behavior expected from
"string" type models in which the baryon charge remains in the
target fragmentation region[15]. The total yield under this bump
increases with x leading to a natural (but probably not necessary)
assumption that these protons represent the spent baryon for interactions where a valence quark is struck. The increase in total
yield as a function of x is shown in Figure 14 and increases in the
manner expected but is less than the 1/2 one would get at $x \to 1$ if
half the spent baryons are protons and half are neutrons.

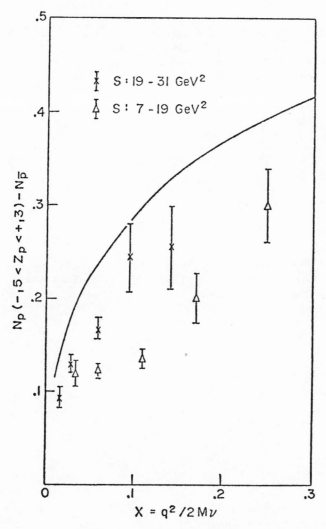

Fig. 14 A plot of the total "residual" proton yield for
$x_F > -0.5$ as a function of x. The curve is the ratio
of valence to total quarks as given by a model (Ref. 7)
but multiplied by 1/2 in the expectation that half
the "residual" baryons are protons.

C.   Comparison With a Model

An exposition of all the data in a large experiment is too cumbersome for a short article. It appears more informative and convenient simply to make a comparison with a model; to the extent that the model matches the data, the model is at least a mnemonic for the data. We pick a "cascade" type model previously proposed by this author[16]. This model has the property that one writes a probability for the deexcitation of an excited hadron by some fractional amount, u, and the general deexcitation probability is written as

$$P(u)du = C\; u^s(1-u)^t du \qquad\qquad C = \frac{\Gamma(s+t+2)}{\Gamma(s+1)\Gamma(t+1)} \qquad (8$$

so that $\int Pdu = 1$. A solution to the integral equation

$$G(u)du = \int_{u'=u}^{u=1} G(u')du'\; P(u/u')du/u' +\; P(u)du \qquad (9$$

can be found where $G(u)du$ is the probability that the hadron was at an excitation state between u and u + du durint the cascade. Solutions to this equation are

$$G(u) = \frac{\alpha}{u} + \text{Polynominal in u}$$
$$\qquad\qquad\qquad\qquad\qquad\qquad\qquad\qquad (10$$
$$\frac{1}{\alpha} = \frac{1}{s+t+1} + \frac{1}{s+t} + \; \cdots \; \frac{1}{s+t}$$

The meson spectrum $F(z)$ is found by

$$F(z) = \int_{u=z}^{u=1} G(u)du\; P(1-z/u)\; dz/u$$

where z is the fractional energy of the mesons related to the initial excitation of the hadron. F is of the form

$$F(z) = \frac{\alpha}{z} + \text{Polynominal in z}$$

when integrated to get the total multiplicity, $\bar{n}$

$$\bar{n} = \text{constant} + \alpha \log (1/z_{min})$$
$$\qquad\qquad\qquad\qquad\qquad\qquad\qquad\qquad (11$$
$$z_{min} = \Delta/W$$

Δ is the ground state $\simeq M_p \simeq 1$ GeV; W is the initial hadron excitation energy.

In our specific model we have t = 0 and s = 2. The model assumes a quasi-plane-wave quark-wave function and massless quarks. Since the nth excitation state has n nodes in its one dimensional wave, if the state is z_n long with excitation q_n

$$n = z_n q_n \tag{12}$$

If we also take the conventional relation between n and q_n given by

$$n \alpha q_n^2 \tag{13}$$

then

$$z_n \alpha q_n \tag{14}$$

The hadronic state "size" increases linearly with energy. The overlap integral squared between two states to get the transition probability is then just the ratio of sizes of u. We get another power of u from the density of final states, Eq. (13), thus leading to s = 2.

From Eq. (11) we get a charged multiplicity

$$\overline{n}_{ch} = \tfrac{2}{3} \overline{n} \ll 2 \log W = \log W^2$$

A more exact calculation gives[16]

$$\overline{n}_{ch} = 1.11 + 1.00 \log W^2 \tag{15}$$

A fit made to neutrino data[10] gives

$$\overline{n}_{ch} = (1.09 \pm 0.36) + (1.09 \pm .03) \log W^2$$

This is good agreement.

We may go further and assume that only one quark is hit and only this quark emits π-mesons changing from u to d as it emits a π^+ meson, etc. We may then write two coupled equations similar to Eq. 9 and obtain the D functions. We get[16]

$$D_u^+ = D_d^- = \frac{1}{z}(1-z)^3(1-\frac{z}{3}) + 2(1-z)^2 \tag{16a}$$

$$D_u^- = D_d^+ = \frac{1}{z}(1-z)^3(1+\frac{z}{3}) \tag{16b}$$

These expressions are just dN/dz for a neutrino experiment. For an electron scattering experiment we will ignore the x dependence of the quark distributions and write for the proton

$$dN^{\pm}/dz = (8/9)D_u^{\pm}+(1/9)D_d^{\pm} \tag{17}$$

When z is not small, we can equate $\varepsilon = p_{\ell}$ and therefore $z = |x_F|$. A comparison of the above with data at $x_F > 0$ fails for both neutrino and electron scattering data; it is too large by a factor of 2. Since the multiplicity was quantitatively explained we must conclude that Eqs. 16 and 17 apply to both positive and negative x_F; we then write

$$dN^{\pm}/d|x_F| = \frac{1}{2}dN^{\pm}/dz \tag{18}$$

We compare the model with our data in Figure 15; we plot $x_F dN/dx_F/(1-x_F)^2$. Note that for all charge states

$$\lim_{|x_F| \to 0} x_F dN/dx_F = \frac{\alpha}{6} = \frac{1}{2} \tag{19}$$

This is, of course, connected with the coefficient of the log term, $\alpha/3$, in Eq. 11. The agreement is satisfactory.

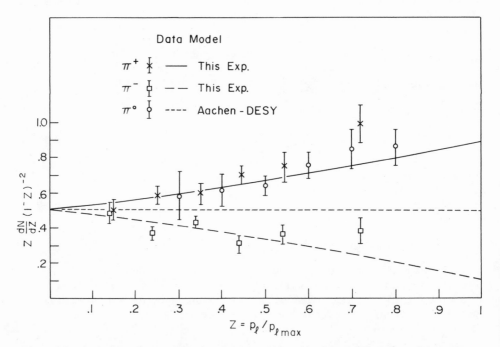

Fig. 15   A comparison of the model (lines) described in the text with the experimental yield of π's of all charges. Note that the data does indeed extrapolate to 1/2 at $x_F = 0$.

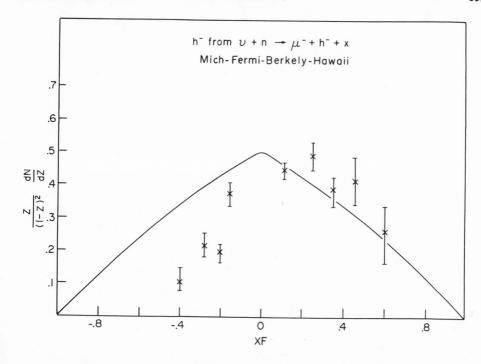

Fig. 16 A plot of the model predictions for neutrino production
         of negative particles (Ref. 10).

In Figure 16, we show the same comparison with neutrino data[10];
here it is possible to observe both plus and minus x_F. We note that
the data is <u>not</u> symmetric about x_F = 0 as was pointed out by the
authors. Both in this data and electroproduction data, the peak in
the yield occurs at slightly positive x_F ($x_F \simeq + .05$) I know of no
explanation for this but it is consistent with the peaking of the
residual proton yield at negative x_F discussed in Section II b just
from considerations of momentum conservation. There it was found
that the peak moves toward x_F = 0 for higher energies; possibly a
shift in the peak x_F downward occurs for mesons.

The rather radical idea that there is some symmetry around
x_F = 0, including "leading" particle effects, would naturally be
explained by our model by simply having the quark "plane wave" states
represent a relativistic oscillator whose period is short compared to
emission times. The data are not in disagreement with this. The
neutrino data show a symmetry around x_F = 0 for the +/- ratio although
one might argue that proton contamination is doing this at negative x_F.
The dilemna is really somewhat model independent; for whatever model
the "leading" particle effect is observed to be too small. Figure 17
shows data from the Cornell-DESY experiment on the average total
charge observed for x_F greater than several values.

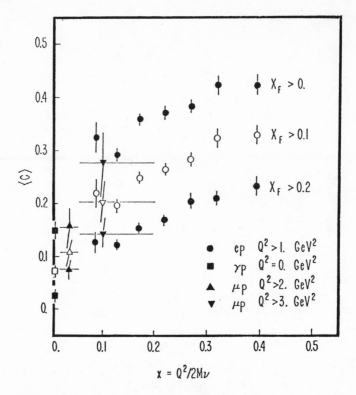

Fig. 17   The net average charge carried forward of a given x_F cut
for photo, electro and muon production.

Even with $x_F = 0$, the charge never reaches the values $+2/3$ that one
might expect from the u quark. Since the quarks start with a charge
state of $+1$, it appears that they are getting close to symmetry,
i.e., $\langle c \rangle = 1/2$.

We go further by looking at charge correlations in electroproduction[17]. We look at events with two or more detected hadrons and
define ratios $R_{++}^{+-}(y_2)$ or $R_{--}^{-+}(y_2)$. The first is the $-/+$ ratio of
the residual hadrons, given that there was a positive leading
hadron at some x_F larger than x_F (cut). The variable y_2 is a new
$P_\ell/P_{\ell max}$ in the centre-of-mass of the residual hadronic state
after the emission of the leading particle.

These ratios are plotted in Figure 18 along with the $+/-$ ratio
from a neutrino experiment[10]. The comparison is to test whether the
single quark emission hypothesis is correct. If a leading π^+ has
been emitted then the remnant excited quark is a d-quark (or possibly

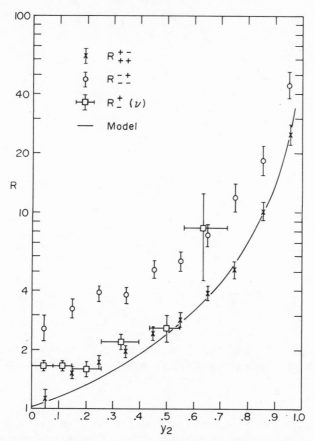

Fig. 18   A plot of charge ratios for electroproduction and neutrino production. $y_2 = p_\ell/p_{\ell max}$ in the hadron system
left by the leading particle.

s).   This is a situation identical to neutrino excitation and the
charge rations should agree.   The prediction of the model is given by
the ratio of Eqs.   16a to 16b.   The agreement is good for R_{++}^{+-} but
off by a factor of two for R_{--}^{++}.

There is some proton contamination at small y_2 to explain part of
this asymmetry.   There is also an asymmetry in K^+K^- which can contribute.   Shown in Figure 19 is the same ratio plotted for various s
bins and y_2 bins against M_3^2, the mass^2 of the residual hadron after
the second emission.   This is done to show that the preference for
unlike charges is not a reflection of charge conservation.   One can
get to a residual M_3 by large s and large y_2 or small s and small y_2;
the ratio is seen to depend on y_2 and not on s or M_3.

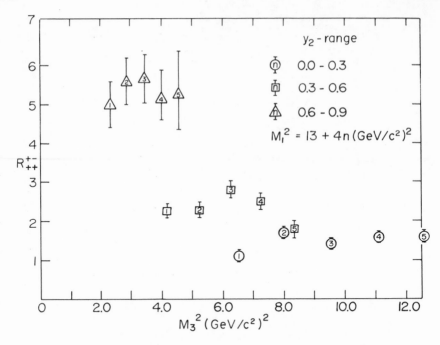

Fig. 19   A plot to show that the charge ratios shown in Fig. 18
are predominantly functions of y_2, not s and M_3^2.

In e^+-e^- hadron production there is no leading particle charge
effect since one is producing quark pairs from the vacuum. However,
the above analysis can be carried out in the same way by taking a
"leading" particle, $x_F > 0.5$, and noting the ratio of unlike to
like charges in the remnant hadronic system and plotting versus x_F^*
in that system. Experimental data was taken from all runs above 7 GeV
in the centre-of-mass at SPEAR. The data was taken by the SLAC
Group C - LBL Group; however, the analysis and conclusions are solely
those of this author and he assumes sole responsibility for the following. The analysis is carried out without benefit of efficiency
corrections; however charge ratios are not very susceptible to this
correction. The plot of like to unlike is shown in Figure 20 together
with the expectations of the model, i.e., the same curve shown in
Figure 18. Positive x_F^* represents the away side with respect to
the leading particle. The line is a fair fit to the data and therefore to the equivalent ratios in lepton scattering but it appears
that the symmetry point is nearer $x_F = +0.1$ than 0.0. Such a shift
was pointed out above with respect to lepton-nucleon scattering.
It should be noted that the symmetry point (if there still is one)
would be shifted to even larger x_F had we used x_F in the laboratory
system.

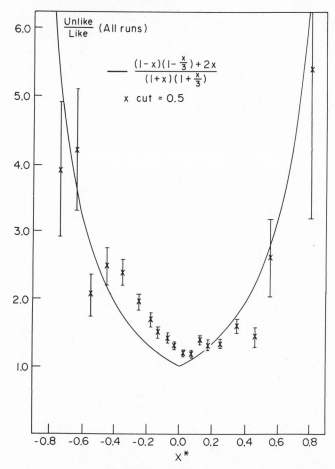

$\dfrac{\text{Unlike}}{\text{Like}}$ (All runs)

$$\dfrac{(1-x)(1-\frac{x}{3})+2x}{(1+x)(1+\frac{x}{3})}$$

x cut = 0.5

X^*

Fig. 20   The charge correlation ratios between leading and
residual hadrons measured in $e^{+}e^{-}$ hadron production.

III. ELECTROPRODUCTION FROM COMPLEX NUCLEI

The MIT-SLAC Group also used nuclear targets as well as deuterium
and hydrogen[18]. This is a **complementary** experiment to those experiments examining hadron-nucleus collision; the latter[19] find that
such a collision gives rise to a multiplicity distribution as a
function of rapidity, y, which is higher at low y, and lower at
high y compared to hadron-proton collisions. There is a simplicity
in electroproduction due to the transparency of the nucleus to the
electron scattering process[20]; the trigger is independent of the
hadronic final state and one may visualize the interaction occurring
homogeneously throughout the nuclear volume.

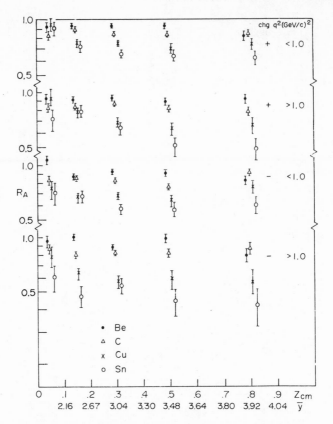

Fig. 21 A plot of yield per event from electroproduction to that
for deuterium. The plot demonstrates the A dependence of
R_A and its insensitivity to hadron charge, q^2 and z_{cm}
except for $z_{cm} < 0.15$. An average rapidity scale is also
shown for those who prefer this variable.

We have plotted our data as a ratio , R_A, of yields of hadrons
in a nuclear event compared to an event in deuterium. In Figure 21
we plot this ratio to test its variation with hadron charge, q^2 and
$z_{cm}(=x_F)$. We have also tested its sensitivity to p_T and s. We find
only a variation with z_{cm} and A; for the former this occurs only at
small z_{cm}. We have parametrized the ratio by

$$R_A = \exp \ - \beta \ (A^{1/3} - 2^{1/3})$$

or $R_A = (A/2)^\alpha$

Plots of β and α are shown in Figure 22 as a function of z_{cm}. For
comparison, $\beta = 0.22$ corresponds to an absorptive cross-section of

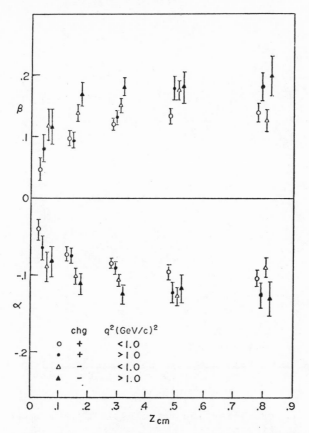

Fig. 22 A plot of the attenuation parameters as a function of z_{cm}.

20 mb, the π-p absorptive cross-section. We also find an increase
in the ratio $r_{p\pi}$, of medium momentum ($\sim$ 1 GeV/c) protons compared to
π's in the same momentum range of .435 $\pm$ .033 for Sn, to .232 $\pm$ .015
for deuterium indicating the increase in secondary collisions.
Our result agrees with a model that would have the excited hadrons
catastrophically absorbed on their way out of a nucleus with a
conventional nuclear absorptions probability but, if out, the state
"decays" as if free. Our result is also compatible with theoretical
models which predict absorption for high secondary particle energies[21].

IV. EXCLUSIVE CHANNELS

Contribution to measurements in the exclusive channels comes
mainly from the LAME Group at Cornell. They have concentrated on
those events which could be completely reconstructed and have looked
for various two body final states. The two most prominent of these
are the $\rho^0 p$, $\pi^- \Delta^{++}$ and ωp [22].

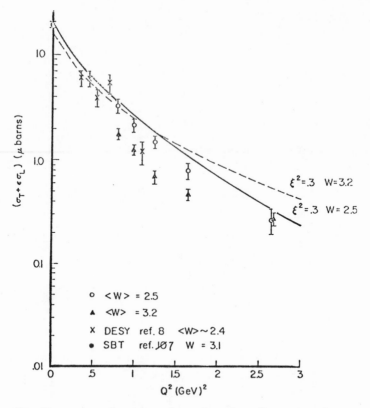

Fig. 23   The cross-section for elastic ρ^0 production as a
function of q^2.

Figure 23 shows their data for the $\rho^0 p$ channel; it exhibits the
classic fall with q^2.   The same behavior is exhibited by the other
two channels.   This group has made an effort to look for other
channels. They conclude that such final states as are seen in photoproduction occur with small or vanishing cross-sections in electroproduction.   The Table illustrates this point.   Considering that $\pi^- \Delta^+$
falls rapidly with cross-sections, it appears that at high energies,
the only "elastic" channels that remain are vector meson (ρ, ω, ϕ)
production.

V.   CONCLUSIONS

An overall impression given by the analysis of hadron electroproduction is that it there is some basic simplicity to it that we
do not yet completely understand.   It certainly reflects the simplest
ideas we have about quarks and we can describe it reasonably
quantitatively with a simple model.   On the other hand, there are

still many bothersome points which might very well be cleared up by
higher energy experiments. What is the charge distribution of mesons
at $x_F < 0$? Will it show itself to be a reflection of the charge
distribution at $x_F > 0$? There appears to be a symmetry point in
yields and charge ratios but it is not exactly at $x_F = 0$; is this
energy dependent? Do the residual protons move to higher x_F with
energy? Is the increase in $<p^2_\perp>$ at higher energies a q^2 or s
effect? Answers could very well be forthcoming at higher energies
but emphasis must be made on particle identification, backward angle
coverage, and high x capability.

Table $\quad Q^2$ Dependence of the Measured Cross-Sections

	Photoproduction[*]	$0.5 < Q^2 < 1.2$ GeV^2	$1.2 < Q^2 < 3.0$ GeV^2
$\gamma p \to \omega p$	2.9 ± 0.4	$0.70 \pm .47$	$0.40 \pm .32$
$\gamma p \to \rho p$	(a)	$0.38 \pm .49$	$0.33 \pm .45$
$\gamma p \to \rho^- \pi^+ p$	0.8 ± 0.5 (b)	$0.14 \pm .22$ (c)	$0.25 \pm .36$ (c)
$\gamma p \to \rho^0 \pi^0 p$	0.5 ± 0.5 (b)	$0.44 \pm .57$ (c)	$0.12 \pm .19$ (c)
$\gamma p \to \rho^+ \pi^- p$	1.8 ± 0.5 (b)	$0.13 \pm .31$ (c)	$0.37 \pm .51$ (c)
$\gamma p \to \rho^- \Delta^{++}$	1.8 ± 0.4	$0.15 \pm .20$	$0.05 \pm .08$
$\gamma p \to \rho^0 \Delta^+$	0.1 ± 0.2	$0.05 \pm .08$	$0.06 \pm .08$
$\gamma p \to \rho^+ \Delta^0$	0.1 ± 0.2	$0.04 \pm .08$	$0.11 \pm .15$
$\gamma p \to \phi p$	$.059 \pm .015$	$.059 \pm .015$	$.017 \pm .005$
$\gamma p \to \pi^- \Delta^{++}$ (d)	8.6 ± 2.0	1.8 ± 0.4	$.53 \pm .13$
$\gamma p \to \rho^0 p$	~ 14		

[*] Photoproduction data from Ref. 23 at $9.7 < s < 11.8$ GeV^2:
(a) Not available; (b) Does not include $\rho\Delta$; (c) Does include
$\rho\Delta$; (d) Falls steeply with energy; this point is an average
$2.8 < w^2 < 11$.

VI.   ACKNOWLEDGEMENTS

The author wishes to thank his colleagues at SLAC (Group E) and
MIT and acknowledge the enormous amount of work they all did to make
the experiment successful; this thanks is particularly extended to
J.F. Martin. He is grateful to the groups at Cornell for sharing
their results. His thanks and gratitude are extended to J.J. Aubert
and G. Preparata for the effort and patience to initiate and carry-out
a successful conference such as this one at Erice. The authors
admiration and appreciation is acknowledged for the staff of the
Institute "Ettore Majorana" and for its Director, A. Zichichi; surely
the setting and style of the Institute is a model for a cultural
center. The imagination and effort that are clearly evident in the
foundation and running of this Institute deserve the greatest
recognition.

REFERENCES

1. I.Cohen, R. Erickson, F. Messing, E. Nordberg, R. Siemann,
       J. Smith-Kintner and P. Stein, Cornell University; G. Drews,
       W. Gevert, P. Joos, A. Ladage, H. Nagel and P. Soding,
       Institut für Experimentalphysik, Universität Hamburg;
       A. Sadoff, Ithaca College.
2. J.T. Linnemann, L.A. Ahrens, K. Berkelman, D.G. Cassel, C.T. Day,
       B.G. Gibbard, D.J. Hadring, D.L. Hartill, J.W. Humphrey,
       T.J. Killian, J.S. Klinger, E.A. Treadwell and D.H. White,
       Laboratory of Nuclear Studies, Cornell University.
3. J.F. Martin, G.J. Feldman, G. Hanson, D.E. Lyon, M.L. Pearl
       and T.P. Pun, SLAC; C. Bolon, R.L. Lanza, D. Luckey,
       L.S. Osborne and D.G. Roth, Laboratory for Nuclear Science,
       MIT; J.T. Dakin, University of Massachusetts.
4. J.F. Martin et al., Phys. Rev. Lett. 40:283 (1978).
5. L.S. Osborne et al., Phys. Rev. Lett. 41:76 (1978).
6. Listed by R. Blankenbecler et al., SLAC Report, SLAC-PUB-1531
       (1975) and attributed to G. Farrar, G. Chu and J. Gunion.
       See also, J.F. Gunion, Phys. Rev. D10:242 (1974) and
       G. Farrar, Nucl. Phys. B77:429 (1974).
7. T. Kuti and V.F. Weisskopf, Phys. Rev. D4:3418 (1971);
       R. McElhaney and S.F. Juan, Phys. Rev. D8:2267 (1973).
8. Some data is compiled in L.S. Osborne, Phys. Rev. Lett.
       34:106 (1975).
9. Berger et al., Phys. Lett. 70B:471 (1977).
10. J. Bell et al., University of Michagan preprint UMBC 78-6 (1978)
       submitted to Phys. Rev.
11. International Conference on High Energy Physics, Tokyo, Japan,
       August 1978.
12. J.R. Johnson et al., Fermilab-Pub 77/98 Exp. 71000.284, submitted
       to Phys. Rev.
13. D. Fong et al., Phys. Lett. 60B:124 (1975).
14. L.S. Osborne et al., Phys. Rev. Lett. 41:76 (1978).
15. R.P. Feynman, "Photon-Hadron Interactions" Benjamin, New York
       (1972);
       S.D. Drell and T.M. Yan, Phys. Rev. Lett. 24:855 (1971);
       J.D. Stack, Phys. Rev. Lett. 28:57 (1972);
       T.P. Cheng and A. Zee, Phys. Rev. D6:885 (1972);
       N.K. Pak, SLAC Report, SLAC-PUB-1744 (unpublished).
16. L.S. Osborne, Phys. Lett. 63B:456 (1976).
17. L.S. Osborne et al., Phys. Rev. Lett. 41:$273 (1978).
18. L.S. Osborne et al., Phys. Rev. Lett. 40:1624 (1978).
19. C. Halliwell et al., Phys. Rev. Lett. 39:1499 (1977).
20. W.R. Ditzler et al., Phys. Lett. 57B:201 (1975);
       W.R. Ditzler, Ph.D. thesis, MIT, January 1972 (unpublished);
       S. Stein et al., Phys. Rev. D12:1884 (1975);
       J. Eickmeyer et al., Cornell University Report CLNS-310 (1975);
       B.D. Dieterle et al., Phys. Rev. Lett. 20:1187 (1969);
       M. May et al., Phys. Rev. Lett. 35:407 (1975).

21. A good summary of various theoretical models and their refe-
     rences will be found in B. Andersson, Proceedings of the
     Seventh International Colloquium on Multiparticle Reactions,
     Tutzing, Germany (1976). For specific models, see, for
     example, N.N. Nikolaev, Phys. Lett. 60B:363 (1976);
     A. Capelia and A. Krzywicki, Phys. Lett. 67B:84 (1977);
     V.V. Asinovich, Yu.M. Shabelsky and V.M. Shekhter, Nucl. Phys.
     B133:477 (1978); K. Gottfried, Phys. Rev. Lett. 32:957 (1974);
     S.J. Brodsky, J.F. Gunion and J.H. Kuhn, Phys. Rev. Lett.
     39:1120 (1977).
22. J.T. Linnemann et al., Phys. Rev. Lett. 41:1266 (1978).
23. Y. Eisenberg et al., Phys. Rev. D5:15 (1972).
24. C. Del Papa et al., Phys. Rev. Lett. 40:90 (1978).
25. K. Wacker, PLUTO Collaboration, private communication.

FUTURE PLANS ON LEPTON PHYSICS

WITH THE ENERGY DOUBLER AT FERMILAB

Thomas B. W. Kirk

Neutrino Department Head
Fermilab
P. O. Box 500
Batavia, IL 60510

INTRODUCTION

We define lepton physics as being experiments using neutrino, muon, and electron beams. Furthermore, at the present time, we expect that the major use of the Fermilab electron beam will be to serve as a source of tagged photons. By reduction, the neutrino and muon beams will be the beams used for lepton physics in the foreseeable future. Neutrino physics will continue the investigation of charged and neutral current physics to Tevatron energies. Muon physics will concentrate on electromagnetic interactions and may be useful for experiments that investigate parity violating effects due to interference between electromagnetic interactions and weak neutral current amplitudes. The practical beam energies are expected to reach 750 to 800 GeV. We discuss the plans for the experimental areas and then comment on the physics that is accessible with these facilities.

NEUTRINO AREA GEOGRAPHY

The present Neutrino Area will remain geographically fixed in the Tevatron era. Existing principal neutrino detectors may be upgraded in appropriate ways, but their locations will be unchanged (Fig. 1). The necessity arises, therefore, for strengthening the absorption capability of the earthen muon absorber shield. We are beginning this upgrade by adding 1.1×10^7 kg of steel in the summer of 1979. Additional steel will be added until the necessary stopping power is achieved. The exact final value will depend upon

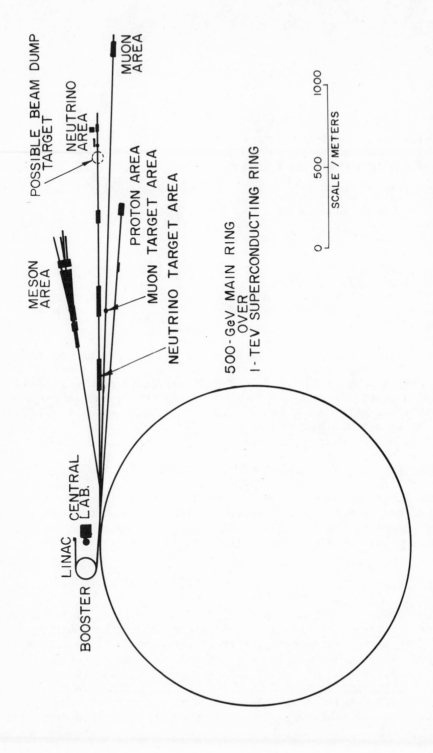

Fig. 1. Plan view of Fermilab site at Tevatron start-up. Lepton physics will take place in neutrino and muon beams.

the physics that is scheduled to be explored and, in consequence, upon the neutrino beams to be used. Adding the first increment of steel will enable the stopping power of the beam to be raised from 500 GeV to 630 GeV.

A second necessary upgrade of the area itself will be the lengthening of the neutrino target tunnel in which the beam forming elements are located. In the summer of 1980 we will extend our present target tube by about 90 meters to accommodate a new stretched out dichromatic neutrino beam focussing system. The new system will provide narrow band neutrinos up to 750 GeV.

All the present Fermilab neutrino beam focussing systems are mounted on narrow gauge railroad cars, making up "trains" which can be rather quickly changed from one focussing system to another. This concept will be retained in the Tevatron era. The length of the vacuum decay space in which pions and kaons decay to form the neutrino beams will not be changed from its present value of 400 meters.

A third proposed change in the area itself, but one not yet adopted as a plan, is to construct a beam dump area just upstream of Lab E in which beams of 1000 GeV protons could be cleanly targeted for the purpose of doing studies of promptly emitted neutrinos, the so-called "beam dump" experiments. This change is shown in dotted lines in Fig. 1 to emphasize that it has not yet been adopted as an approved project.

In addition to the major civil construction projects noted above, there will be a number of changes in the beam lines to allow transmission of higher energy hadron beams for apparatus calibration and testing. The precise scheme hasn't yet been finalized, but the possibility of finding acceptable solutions is assured.

NEUTRINO DETECTOR FACILITIES

Several large neutrino detector facilities already exist at Fermilab or are being constructed at present. These include the 15 Foot Bubble Chamber, The Lab E Spectrometer, and the Lab C Spectrometer. It is expected that these detectors will be useful at Tevatron energies, perhaps with some augmentation or modification. We describe each briefly in turn.

Figure 2 shows the Fermilab 15 Foot Bubble Chamber. It has been instrumented with an External Muon Identifier (EMI) and an Internal Picket Fence (IPF). The EMI has two planes of proportional wire chamber modules with delay line readout. Each module measures x, y and u particle coordinates. The IPF is used for time tagging of hadrons before they enter the absorbers which are used to separate muons from hadrons.

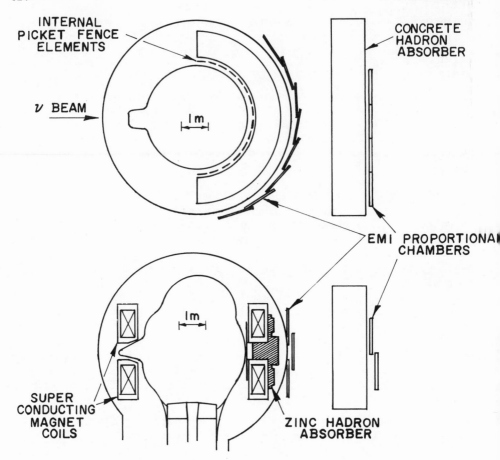

Fig. 2.   Schematic views of 15 Foot Bubble Chamber and associated
          detectors.

The Bubble Chamber holds 2400 kg of liquid hydrogen, 5500 kg
of liquid deuterium or 26,000 kg of neon-hydrogen mixture.   The
usual fiducial volume for neutrino interactions is about 80% of the
total mass.   The sensitive volume is viewed by two sets of three
wide angle cameras to allow full flexibility of chamber use.
Reaction rates for neutrino physics necessitate only a single
expansion per accelerator cycle although the chamber has the
capability of multiple pulsing.   This capability will likely be
utilized by the Tevatron spill sequence.

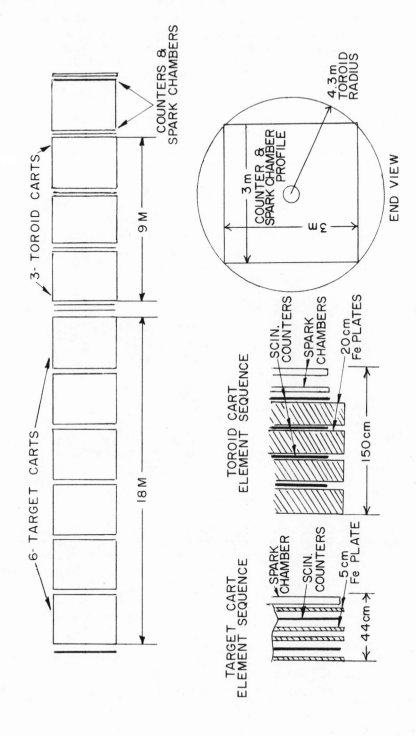

Fig. 3. Large Neutrino Spectrometer in Lab E. The spectrometer parameters are given in Table I.

Studies are underway for the incorporation of additional electronic calorimetry to enhance the chamber capability for the Tevatron. A liquid argon and lead plate calorimeter may be incorporated in part of the liquid volume for measurement of photons. A study is also being made of the measurement resolution which can be obtained on high energy hadron tracks together with the implications for Tevatron regime kinematics.

The Lab E Spectrometer is shown in Fig. 3. Developed by physicists from Fermilab, Cal Tech, and Rockfeller University, it consists of six "carts" in which 5 cm steel plates, liquid scintillation counters and magnetostrictive spark chambers are combined sequentially as a 5×10^5 kg instrumented target calorimeter module. The target carts are followed by four magnetized iron toroids with similar instrumentation and coarser sampling. The toroids are used primarily for muon momentum measurements, but they also serve as target material for the multimuon neutrino physics. A compilation of relevant parameters for the Lab E Spectrometer is given in Table I.

The measurement capabilities of the Lab E Spectrometer are indicated by Figs. 4, 5 and 6. Although the spectrometer has

Table I. Properties of the Lab E Spectrometer

Type	Fe Calorimeter and Toroids
Target Mass	Calorimeter Carts 6.3×10^5 kg (690 tons) Fe Magnetized Toroids 3.6×10^5 kg (400 tons) Fe
Dimensions	Length 18m Calorimeter 9m Toroids Diameter 3.5m O.D. .25m I.D.
Detectors	72 Planes Magnetostrictive Spark Chambers 84 Liquid Scintillation Counter Planes 24 Plastic Scintillation Counter Planes (all are 3.0m square)
Hadron Shower Sampling Distane	Non Magnetic Target, 10 cm Fe Magnetized Toroids, 80 cm Fe
Particle Tracking Sampling Distance	Non Magnetic Target, 20 cm Fe Magnetized Toroids, 20 cm Fe
Toroid Average Magnetic Field	1.7 W/m^2

significant capabilities for neutral current weak interaction
physics, it was primarily designed for precision charged current
measurements in a narrow band dichromatic neutrino beam. The CITFR
Group which developed the spectrometer also has been deeply involved
in the design and implementation of a sophisticated and redundant
monitoring system for the dichromatic neutrino beam. The goal
of the group has been to hold overall systematic uncertainties
in the apparatus to less than 3%. Presently evolving results seem
to indicate that this goal is being met.

There is no aspect of the Lab E Spectrometer that is not
immediately and directly applicable for use with dichromatic
neutrino beams up to 750 GeV. No changes in instrumentation will
be needed when the Tevatron commences. Some developmental work is
going on, nevertheless, with the aim of providing a totally live
target for neutrino physics by means of instrumented liquid argon
modules in which ions are collected after drifting up to 50 cm
The dE/dx is measured with longitudinal and transverse position
resolution in the neighborhood of a few mm in space and energy
deposition resolution that is capable of distinguishing one, two,
three, etc., minimum ionizing particles. If the liquid argon
drift system is successful it will be useful for high statistics
measurements of neutrino electron scattering. A 4,500 kg proto-
type module is currently being developed. If it is successful

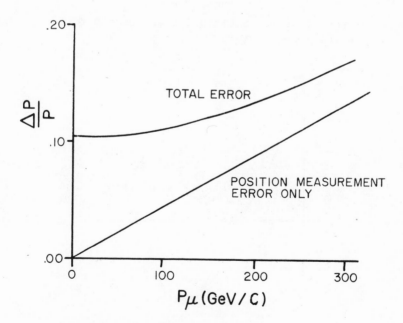

Fig. 4.   Muon momentum measurement capability for the Lab E
          Spectrometer

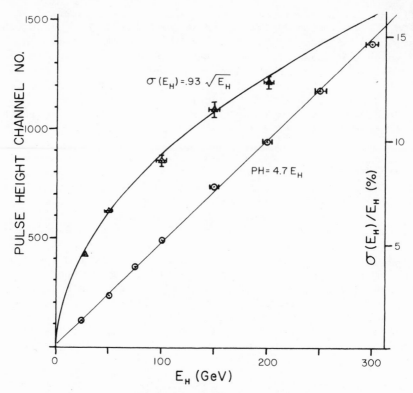

Fig. 5.   Hadron calorimeter linearity and energy resolution for
the Lab E Spectrometer.

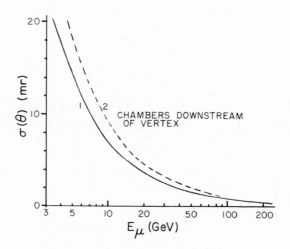

Fig. 6.   Muon angle resolution capability for the Lab E
Spectrometer

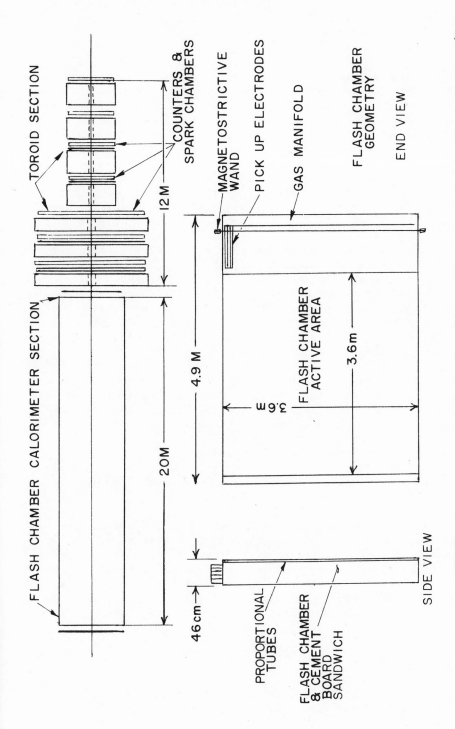

Fig. 7. Large Flash Chamber Neutrino Spectrometer in Lab C. The spectrometer parameters are given in Table II.

a 45,000 kg module will be constructed and tested, probably about the time the Tevatron becomes operational.

The Lab C Spectrometer is currently being constructed by a collaboration of Fermilab, MIT, Michigan State, and Northern Illinois Universities. It is shown schematically in Fig. 7. It differs from the Lab E Spectrometer principally in the much finer granularity of the calorimeter target. This spectrometer, aimed primarily at neutral current physics, consists of 2 cm plates of asbestos loaded cement board (a building material) alternated with planes of neon flash chambers. The flash chambers have parallel rows of flash cells each of 4 mm width arranged at three angles with respect to the beam axis to measure x, y and u coordinates. A layer of multi-wire drift-chamber proportional tubes is located with a 60 cm periodicity in the cement board/flash chamber matrix.

The hadron shower energy and energy flow direction in neutral current events are determined by means of the flash chamber cell pattern and by measuring the charge collected in the proportional tube layers. The energy of the incoming neutrino is known from the setting of the dichromatic beam system and the impact point in the target. Figures 8, 9 and 10 indicate the measurement resolution capabilities of the Lab C Spectrometer. Table II lists the properties of this detector.

Charged current weak interactions can also be reasonably well measured in this apparatus, the muon momentum being measured by the magnetized toroids in the usual way. Again, the Tevatron neutrino physics should be directly accessible with this spectrometer, although there may be some question about the very high energy hadron measurement capability of the flash chamber calorimeter. If the calorimeter shows high energy saturation effects, a calibration beam can be used to measure directly the non-linearities.

In addition to the three major installations noted, there is an expectation that a fourth large neutrino detector will be constructed for use in the Tevatron. No precise plan has been developed but a number of proposals have been discussed. Among these have been fine grained calorimeters based on iron and liquid argon, a large detector consisting of ten liquid argon tanks with bubble chamber optics and drift electronics, and a larger, finer grained version of the CDHS spectrometer at CERN.

TEVATRON NEUTRINO BEAMS

There will probably be two basic neutrino beams for Tevatron physics, a wide band, sign selected beam based on a pulsed horn

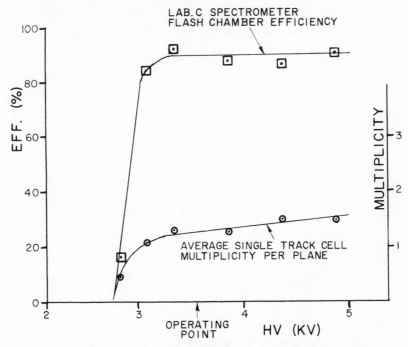

Fig. 8. Efficiency and multiplicity curves for the Lab C
         Spectrometer flash chamber calorimeter planes.

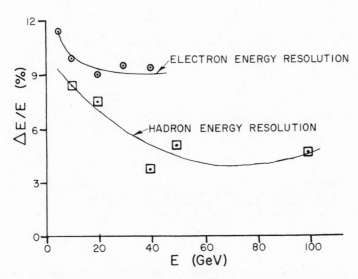

Fig. 9. Energy resolution for electrons and hadrons in the Lab C
         flash tube calorimeter.

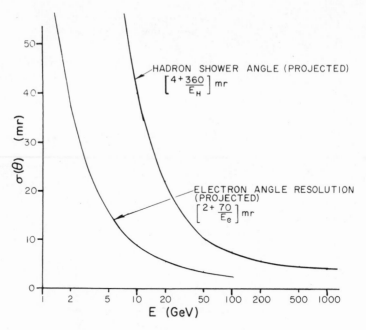

Fig. 10. Angle resolution for electron and hadron showers from the
         Lab C flash tube calorimeter.

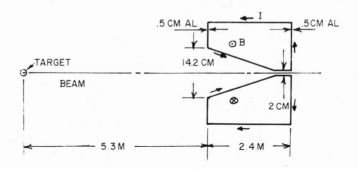

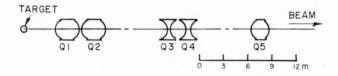

Fig. 11.   Tevatron wide band neutrino beam focussing devices.   The
           single horn system is shown above and the quadrupole
           triplet below.

Table II. Properties of Lab C Spectrometer

Type	Fine Grain Flash Tube Calorimeter + Magnetized Toroids (5.7°, 90° Stereo)
Target Mass	3.6×10^5 kg (400 tons) Cement Board

Dimensions	Length 20m Calorimeter
	          12m Toroids
	
	Diameter 4m x 4m Calorimeter
	         7.3m O.D.
	         .30m I.D. Large Toroids
	
	         3.6m O.D.
	         .30m I.D. Small Toroids

Detectors	600 Planes Multi Cell Flash Chambers 36 Planes Proportional Counter Tubes 7 Planes Wide Gap Spark Chambers 4 Planes Plastic Scintillators 5 Planes Fe-Liquid Scintillator Planes
Hadron Shower Sampling Distance	Fine Grain Calorimeter 2cm Cement Fe-Scin. Calorimeter 20cm Fe
Toroid Average Magnetic Field	$1.7 \ W/m^2$

focussing device and a narrow band dichromatic beam with capability of forming neutrino beams up to 750 GeV. The former is a natural focussing device for charged and neutral current physics in the 15 Foot Bubble Chamber and for low cross section experiments like neutrino electron scattering. The latter is directed toward precision structure function measurements in the very large high resolution spectrometers. Some discussion has taken place on the feasibility of making a short neutral beam for electron neutrino beam physics, but no clear plan for this facility has emerged.

Schematic diagrams for the wide band and dichromatic focussing trains are shown in Fig. 11 and 12. The associated flux curves are shown in Fig. 13. An elaborate set of monitoring devices is used to measure the intensity and optical quality of the dichromatic beam and to measure the kaon to pion particle ratio as a function

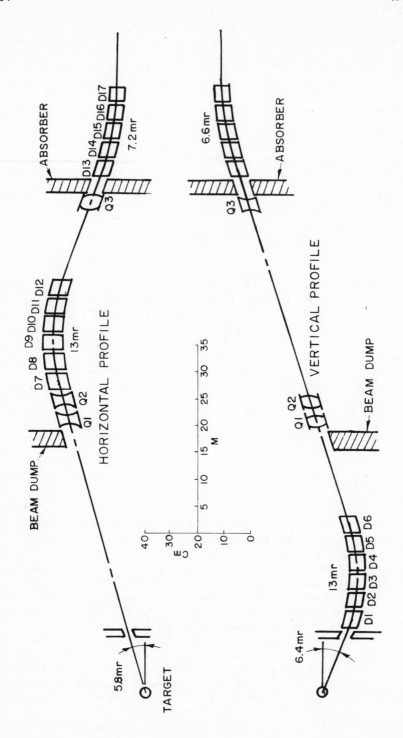

Fig. 12.   Tevatron dichromatic neutrino beam focussing and momentum selection elements.

of beam energy. Devices used include ion chambers, secondary
emission monitors (SEMs), wire chamber profile monitors, magnetic
toroid transformers, radio-frequency cavities and an integrating
Cerenkov light monitor. The combination of these devices is
expected to measure and monitor the narrow band neutrino flux to
about 2%.

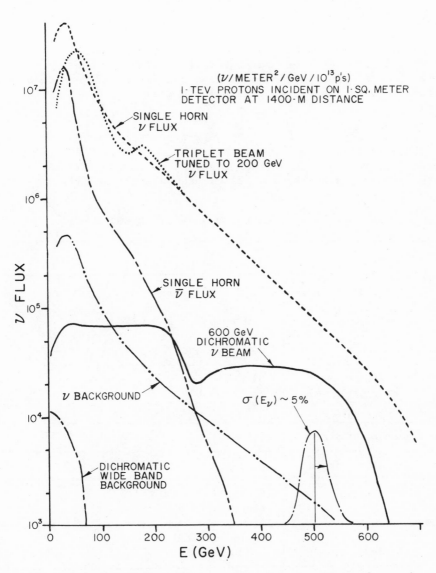

Fig. 13.   Neutrino fluxes from various Tevatron beam focussing
           devices.

The monitoring and flux measurement scheme for the wide band Tevatron beam is just beginning to be considered. It is expected that there will be about six monitoring and flux measuring stations located at intervals within the iron shield. Detectors will be placed in these locations for measurement of the muon flux as a function of position in the shield. From these measurements, the neutrino flux can be unfolded.

MUON AREA GEOGRAPHY AND BEAMS

A completely new muon beam and a new experimental area are planned for the Tevatron era at Fermilab. The new beam is located to the east of the neutrino beam line and extends a considerable distance beyond the existing neutrino experimental facilities (Fig. 1).

The new beam will be of the FODO type and will be similar in concept to the muon beam presently in operation at CERN. It will have useful intensities at muon energies up to 750 GeV. A summary of the properties of the new beam is shown in Table III.

Table III.   Tevatron Muon Beam Parameters

Beam Energies	275 – 750 GeV μ^+, μ^-
μ/p* interacting	1.2×10^{-5} @ 750 GeV μ^+ 9.0×10^{-5} @ 550 GeV μ^+ 30×10^{-5} @ 275 GeV μ^+
Halo/Beam (3m x 3m area)	8% @ 750 GeV 4% @ 550 GeV 7% @ 275 GeV
Beam Spot Size	4.0 cm Horiz. x 2.0 cm Vert. FWHM
Momentum Spread	$\Delta p/p$ = 20% FWHM (tagged to 2%)
Beam Length	920m Decay FODO 430m Muon FODO
Quadrupoles Needed	15 Decay FODO 4 Muon FODO

*calculated from C. L. Wang, Phys. Rev. D10, 3876 (1974).

Principal design goals for the new beam are: maximization of the muon flux per incident proton, minimization of muon flux (halo) not in the beam phase space, and elimination of the close coupling of muon and neutrino experimental programs through the sharing of mutual hadron decay space. In retrospect, combining the muon and neutrino beams, as is done at present at Fermilab, is a bad idea. Several long-term problems are eliminated in the new design. The muon beam fluxes that can be achieved should also be higher by a factor of fifty over the present design.

MUON DETECTOR FACILITIES

A new laboratory will be built for Tevatron muon experiments at the end of the new muon beam line. It will be similar in structure to the existing Fermilab Muon Laboratory, having a "thin" target multiparticle spectrometer placed in front of a "thick" calorimeter apparatus. The principal expected change comes from the fact that the new beam will be about 4 meters below the earth's surface. In consequence, the new muon laboratory will have a concrete pit of size 10m wide by 70m long and with a beam height reaching 1.5m above the floor. The electronics, computers and personnel will be located above the experimental pit in a metal building at ground level. Heavy equipment will be moved in and out of the pit through hatches or via a rear ramp. This approach has been used successfully at Fermilab in the Tagged Photon Facility.

The "thin" target spectrometer will probably look very much like the existing Chicago Cyclotron Magnet Spectrometer (CCMS) which has proved successful in muon experiments in the existing beam line. That magnet itself may be moved to the new site or an equivalent new magnet constructed. Figure 14 shows how a multi-particle spectrometer might be constructed with the existing Chicago magnet. This magnet will be fitted with superconducting coils in the near future, has a very large magnetic aperture, a large magnetic field integral (P_{mag} = 2.2 GeV/c), and would continue to serve as a powerful base for a multiparticle analysis facility.

The CCMS will have a liquid hydrogen (deuterium) target followed by about 20 planes of multiwire proportional chambers extending into the magnet. These "upstream" chambers are useful for measuring low momentum particles by their deflection in the magnet field, and for measuring the upstream segment of high momentum tracks. Several planes of drift chambers will be placed around a particle identifier (type not yet decided) located downstream of the magnet. These chambers measure the recoil muon with high precision as well as forward hadrons. After the drift chambers and particle identifier there will be a liquid argon-lead calorimeter for the detection of photons.

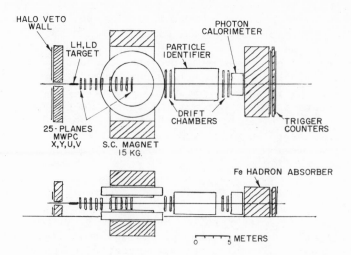

Fig. 14.   A possible thin target muon multiparticle spectrometer
           for use with Tevatron muon beams.

Following the photon calorimeter will be a steel hadron
absorber with counters behind it for detecting scattered muons
and triggering the apparatus. A beam veto counter is located in
the beam to help guard against halo triggers.

This spectrometer will be used for the study of hadrons
recoiling from deeply inelastic muon nucleon scatters. The work
begun at Fermilab and now being continued at CERN on high energy
quark fragmentation will be extended in the Tevatron muon spectrometer.

Following the CCMS will be the existing Multi-Muon Spectrometer (MMS) constructed by a Berkeley-Fermilab-Princeton collaboration in the existing Fermilab muon beam and subsequently used
by them to measure multi-muon production and high Q^2 deeply
inelastic muon nucleon scattering. The MMS will be moved and
re-erected in the new muon laboratory where it will be even more
effective at measuring very rare muon induced processes.

The MMS is a magnetized iron, calorimeter type detector
(Fig. 15), but it has a transformer geometry magnetized iron
element rather than the more common toroidal type. The homogeneity
of the MMS design eliminates some of the difficult acceptance
problems that are endemic to the toroidal geometry. The properties
of the MMS are shown in Table IV. It is seen that the MMS can
offer effective targets of up to 6,000 gm/cm^2 of iron with
excellent acceptance.

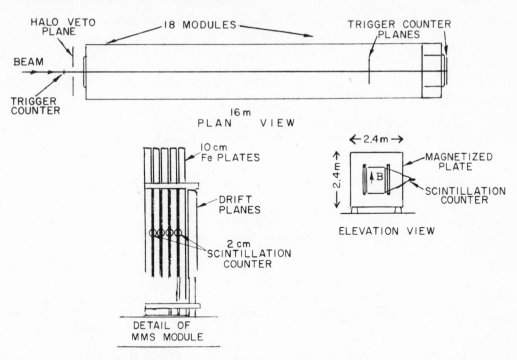

Fig. 15.   The Multi-Muon Spectrometer as it exists and as it would
be used in the Tevatron muon beam.

Table IV.   Properties of Multi-Muon Spectrometer

Target	5900 gm/cm^2 Fe
Dimensions	Length 14m Target Modules 2.7m Meas. Modules Calorimeter 1.0m x 1.0m Square
Detectors	60 Scintillation Counter Planes 18 Drift Chamber Planes
Hadron Shower Sampling Distance	10cm Fe
Particle Tracking Sampling Distance	50cm Fe
Magnetized Fe Field	1.7 W/m^2

Plates of iron 10cm thick are interspersed with drift chambers and scintillation counters in a cyclic pattern to make up the spectrometer. The calorimetry is not as fine as the Lab C neutrino spectrometer, but the known incident beam energy and precise measurement of the scattered muon obviate the necessity of maximum hadronic energy resolution.

No other muon spectrometer installations are being planned at the present time, but there have been informal proposals for alternative measuring approaches. It is also clear that experiments with intense hadron beams could also be carried out with the muon beam without its hadron absorber. No explicit plans are in hand for this type of operation.

PHYSICS ACCESSIBLE TO TEVATRON LEPTON BEAMS

The kinematic range available for lepton physics at various laboratories is shown in Fig. 16. With many-hundred ton neutrino detectors and high intensity neutrino and muon beams, the full range of ν and spacelike Q^2 can be explored with good statistical precision for both electromagnetic and weak interactions. The

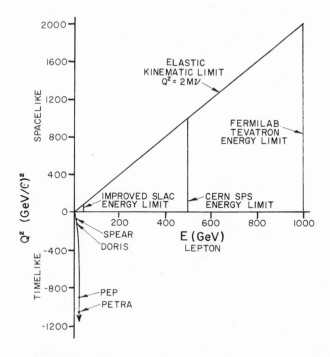

Fig. 16. Kinematically accessible regions for various accelerators based lepton and photon physics.

Fermilab Tevatron, taken together with e^+e^- colliding beam machines, will make it possible to explore both the fundamental elements (quarks and gluons) of particle physics and their dynamic properties (structure functions F_1, F_2, F_3) in the next ten year period. The results of these explorations, in conjunction with guidance from the developing theory of Quantum Chromodynamics (QCD), may finally provide a description of strong interactions based on a true theory rather than upon phenomenological constructions of limited application.

The Tevatron lepton physics program will contribute to the process in two basic ways: i) measurement of nucleon structure functions; ii) measurements of properties of recoil hadrons. We already know that lepton structure functions differ from the naive parton model only by a logarithmic variation (scale violation). It will be necessary to make high statistics measurements over enormous ranges of the kinematic variables to confirm the QCD predictions for scale violations. Furthermore, the measurements will have to be made with both neutrinos and muons to separate out effects due to the boson propagator in weak interactions (the effects which can be confused with scale violations). The muon scattering does not suffer from propagator problems, but it cannot measure XF_3 since the electromagnetic interaction is parity conserving and F_3 is explicitly parity violating.

To gain an idea of the event rates that can be expected in Tevatron detectors, we show in Figs. 17 and 18 the charged current event yields for the Lab E neutrino spectrometer with a 600 ± 10% GeV narrow band neutrino beam and for the relocated Multi-Muon Spectrometer with a beam of 600 GeV muons incident. We see that respectable yields out to the kinematic limits are possible in conventional runs of 1000 hours. In Fig. 19, we show the effects of QCD and of an intermediate boson propagator on the averaged F_2 structure function. Since the slope of the total neutrino cross-section versus energy must be experimentally determined in addition to the propagator effects, and since this slope is also dependent on F_2, it will be essential to take measurements with both neutrino and muon probes to separate the QCD effects from the W propagator effect. Taking the measurements with sufficient statistical precision and with adequate control of systematic errors will be a challenge to experimenters, but the rewards will be great. ·

The measurement of recoil hadron properties from deeply inelastic lepton-nucleon collisions will be done in the 15 Foot B.C. for neutrinos and in the multiparticle muon spectrometer. The easiest measurements to make at very high energies are those which study inclusive hadron spectra and jet behavior. We believe that each hadron jet in a deeply inelastic lepton nucleon collision begins with a single quark suddenly displaced to a distant

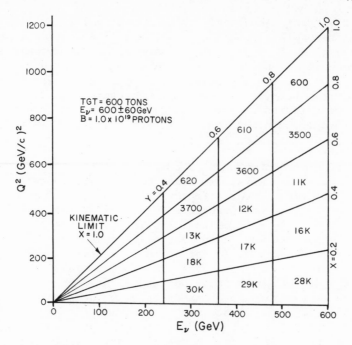

Fig. 17.   Event yields for a Tevatron dichromatic neutrino beam
incident on the Lab E neutrino detector.

part of momentum phase space by absorption of a virtual photon
W(or Z^0) boson. It is still close to its original point in
configuration space, however, and as it tries to move away, the
strong quark gluon forces cause a cascade of quark-antiquark pairs
and gluons to develop which results in the quarks grouping themselves rapidly into quasi-stable configurations as mesons. These
mesons are called "primordial" mesons as they may not even be
stable against strong decays (ρ^0, K*, A_2, etc.), much less against
electromagnetic (η^0, χ, etc.) or weak decays (π, K, etc.). At
any event, this so-called quark fragmentation process is characterized by a severely limited transverse momentum spectrum about
the initial quark direction (i.e., the momentum phase-space
between the struck quark and the remaining nucleon fragment is
filled in by the quark fragmentation process). The process gives
rise to the "jet" of hadrons.

Since the initial struck quark has perhaps 5 GeV/c of transverse momentum with respect to the lepton beam direction, the jet
character of the hadrons in high energy lepton scattering will be
far more clear and dramatic than it can ever be in hadron-hadron
collisions. Both the 15 Foot B.C. and the CCMS will be well
equipped to study these processes. Careful study of the kinematic
properties of the jets and of the quantum number flow in them with

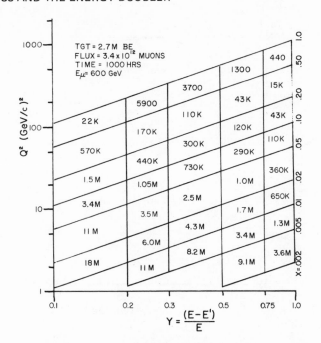

Fig. 18. Event yields for a Tevatron muon beam incident on the
          Fermilab muon spectrometer.

adequate statistics can, one hopes, unravel some of the strong
interaction behavior of quarks and gluons. In particular, there
are QCD predictions which can already be made for the average
hadron transverse momentum in a jet as a function of the Q^2 value
of the lepton scattering. This field is still in its infancy.

     In addition to the QCD related physics just noted, there are
other exploratory experiments that can be done with neutrinos
from neutral hadron beams and from prompt (beam dump) neutrino
sources. The former give electron neutrinos and antineutrinos
as well as muon neutrinos. The latter give tau neutrinos from
the semileptonic decays of F mesons.

     The Tevatron energy boost is very advantageous for $\nu_e(\bar{\nu}_e)$
physics. The event rates in the 15 Foot B.C. benefit from several
sources: i) the basic cross-section rises linearly with energy;
ii) the beam that strikes the fiducial volume is increased by
$(E/E_0)^2$; iii) the production at a fixed energy increases rapidly
with proton beam energy. S. Mori has estimated that the 15 Foot
B.C. with a fill of heavy neon (20 tons) could obtain 100 (50)
events per week for charged current $\nu_e(\bar{\nu}_e)$ interactions. The mean
neutrino energy of these events will be 100 GeV. As a result, a

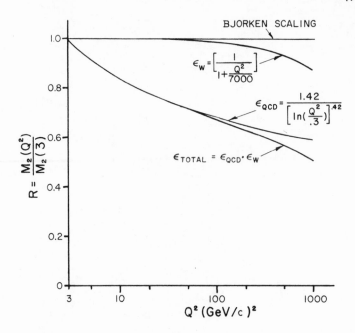

Fig. 19.   Effects of scale breaking on $F_2(x,Q^2)$ in neutrino
scattering arising from QCD and from W boson propagator.
The ordinate is the ratio of the second moment of $F_2(Q^2)$
to its value at $Q^2 = 3$.

test of μ–e universality can be performed with relatively small
systematic errors by comparing events with muons and electrons
in the final state.

The source strength of tau neutrino beams and the separation
of ν_τ induced events is much more speculative. The manner in
which beam dump experiments are done and the physics that will
emerge is not yet clear but more will be known by the time the
Tevatron is ready (from experiments already underway or planned at
lower energy).

A last area of lepton physics which may emerge as very interesting is the study of multi-muon neutrino events from conventional
neutrino beams. These events have already been studied extensively
at present energies with mixed results. The existence of prompt
dimuons and trimuons is clearly established, but the results so
far seem to be from "conventional" sources such as charmed production, leptonic decay or vector mesons and QED processes. It is
tempting to accept the possibility that some trimuons (or even tetra-
muons) could be observed and related to quark "flavor cascades"

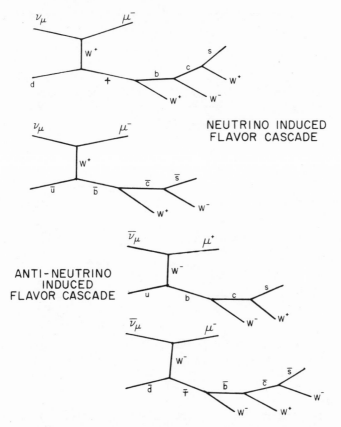

Fig. 20.   Schematic representation of a neutrino or antineutrino
            induced quark flavor cascade. Such processes give rise
            to multimuon neutrino events.

(Figure 20), but at the present time this hypothesis is not convincing. No doubt extensive multimuon work will be carried out at
Tevatron energies and it may yield surprises. Time will tell.

RELATED PROCESSES WITH LEPTONS

      Although it is slightly outside the subject of this paper as
it was defined in the Conference schedule, I can't resist briefly
referring to a process that is not a lepton <u>induced</u> nucleon structure probe, but which involves leptons as a nucleon structure
probe anyway. This process is, of course, the inclusive production of high mass continuum lepton pairs by means of the Drell-Yan
mechanism. In particular, the measurement of dimuons in the mass
region above about 5 GeV (excluding the upsilon resonances) is
thought to arise almost exclusively from the Drell-Yan annihilation

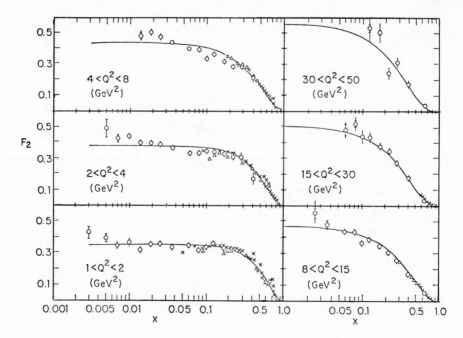

Fig. 21. $F_2(x,Q^2)$ structure function from muon proton inelastic
          scattering at Fermilab. Bjorken scale violation is
          apparent.

of quarks from the projectile and antiquarks from the target (or
vice versa).

Figure 21 shows the $F_2(x,Q^2)$ distribution obtained from
deeply inelastic muon proton scattering at Fermilab. In the
parton picture, F_2 is the momentum distribution of <u>all</u> the quarks
and antiquarks in the proton, weighted by the squares of their
electric charges. The Drell-Yan dimuon distribution is derivable
directly from the product of this structure function and the sea
quark-antiquark distribution which can be extracted from neutrino
experiments (see Figures 22, 23).

Turned around, this dimuon distribution can be used with F_2
from muon scattering to derive the sea quark distribution. The
result of this procedure can then be compared with the sea
distribution obtained independently from neutrino experiments.
Results of this type of analysis have already been carried out by
the Fermilab dimuon group with the result shown in Figure 24. The
neutrino data are still not known over a large kinematic regime,
but the preliminary results are already in striking agreement with
the Drell-Yan/F_2 predictions.

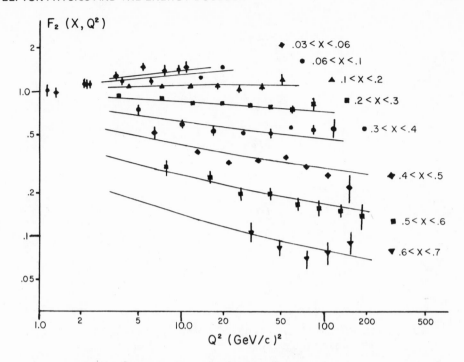

Fig. 22. $xF_3(x,Q^2)$ nucleon structure function as derived from a
neutrino scattering experiment at CERN.

        The same structure function quark distributions seen in muon
scattering and neutrino scattering (Figures 21, 22, 23) are now
appearing in the dimuon production spectrum from pp collisions!
This result makes us feel even more confident that we are at last
on the track of a truly self-consistent picture of lepton structure
(quarks and gluons) and a genuine theory of their interactions
(QCD).

        In consequence, we at Fermilab plan to pursue dimuon conti-
nuum physics and neutrino-muon physics with equal vigor. Even if
no new quark flavors appear as the mass scale is extended, the
study of high mass dimuon pairs and their relationship to the
nucleon quark-gluon distributions will surely prove richly
rewarding as we enter the Tevatron energy regime.

CONCLUSIONS

        I have attempted to outline the lepton physics capability of
which Fermilab will have when the Tevatron arrives. The physics that
can already be predicted will be adequately covered by the planned

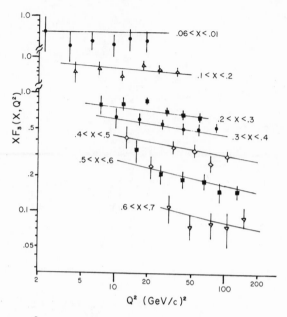

Fig. 23. $F_2(x,Q^2)$ nucleon structure function as derived from a neutrino scattering experiment at CERN.

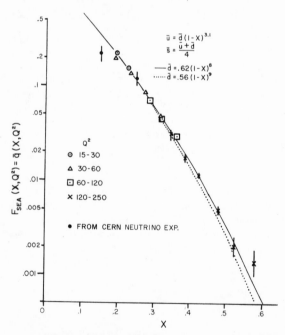

Fig. 24. Sea quark structure function $F_s(x,Q^2 = M^2)$ as derived from the Fermilab dimuon experiment and $F_2(x,Q^2)$. The two neutrino structure function points are derived from the data in Figures 22, 23.

detectors and beams. The program as we see it now will be strongly
influenced by the impact of QCD and tests of this hopeful new theory.
Provision will be made, however, for the study of more speculative
physics as well.

To a large extent, existing detectors will be used for the
first round of Tevatron lepton experiments so that we will not
commit ourselves to new detectors prior to evaluating the first
Tevatron results and determining the most fruitful directions in
which to proceed. We are, however, already designing and building
new muon and neutrino beams to take advantage of the 1000 GeV protons
when they first emerge from the accelerator. We will also have
available a powerful new dimuon spectrometer for continuing the pp in
induced dimuon work.

It will be an exciting time!

ACKNOWLEDGEMENTS

I would like to express my sincere thanks to Dr. Jean Jacques
Aubert and Dr. Giuliano Preparata, the organizers of the Highly
Specialized Seminar on Nucleon Structure, at Erice, for a stimulating
and highly enjoyable session. The size and informal nature of the
sessions were perfect for the communication of new knowledge among
experts, and I have personally not spent a more productive week in
physics anywhere else. Much of the intellectual sparkle and the
very pleasurable social events were clearly the result of careful
planning and hard work on the part of the organizers and the staff
of the Centre for Scientific Culture. They deserve sincere
congratulations and large thank you.

REFERENCES

Cline, D., Mori, S., and Stefanski, R., 1978, Fermilab TM-802,
        "Neutrino Beam Dump Experiment and a Possible Electron
        Neutrino Beam in the Neutrino Area".
de Groot, J. G. H., Submitted to Phys. Letters B, "QCD Analysis
        of Charged Current Structure Functions".
Evans, R., Kirk, T., Malensek, A., Musgrave, B., and Phelan, J. J.,
        1977, Fermilab TM-754, "Design Study for a High Energy
        Muon Beam".
Fermilab TeV Program, 1977, Fermilab (unpublished).
Mori, S., 1978, Fermilab TM-839, "Do We Need a Horn for Tevatron
        Neutrino Physics?".
Mori, S., 1979, Fermilab TM-848, "Estimated τ Neutrino Fluxes from
        a Beam Dump at 400 GeV and 1000 GeV".
Stutte, L., 1979, Fermilab TM-841, "A Possible Dichromatic Neutrino
        Beam for a 1 TeV Accelerator".

THE WA18 ν EXPERIMENT AT THE CERN SPS[1]

F. Niebergall

II. Institut für Experimentalphysik
Hamburg University

The experiment is being performed by the CERN-Hamburg-Amsterdam-Rome-Moscow (CHARM) collaboration. The detector installation was completed in October 1978 though data taking had already begun three months earlier with an incomplete set-up. There are no physics results to be reported yet. Here I will, very briefly, discuss the physics program, especially regarding its implications for the detector design and then will describe the detector and its performance.

I. PHYSICS PROGRAM

Neutral current inclusive interactions $\nu N \rightarrow \nu \chi$: One of the main aims of the experiment is the measurement of the double differential cross-section $d^2\sigma/dxdy$ and the determination, from that, of the structure functions for neutral current inclusive interactions. To determine completely the kinematical variables of this process, one has to know the energy of the incident neutrino and to measure the momentum of the final state hadronic system. In the CERN narrow band neutrino beam (NBB) the energy is a function of the radial distance of the neutrino from the centre of the beam. It is known with an accuracy of about ± 10%, if the interaction point is measured. The beam is, however, dichromatic; that is, the same spot in the detector is hit by neutrinos from two different sources, π decay and K decay which differ in energy.

To perform the experiment one needs a detector that determines the interaction vertex, measures the energy and direction of the hadronic final state and has a good muon identification capability to select neutral current events reliably.

351

Charged current inclusive interactions $\nu N \to \mu X$: The possibility
of measuring down to very low hadronic energy allows investigation of
the kinematical region of low Q^2 and high X, which, up to now, has
been poorly measured. In addition, this reaction is very valuable for
detector calibration and internal normalization.

Polarization of muons produced in ν-interactions: This experiment allows a very direct investigation of the space-time structure
of weak interactions at high energies. V and A interactions pre-
serve the helicity of the incident neutrino whereas S, P and T interactions flip the spin of the lepton. This experiment is performed
together with the CERN-Dortmund-Heidelberg-Saclay collaboration[2].
Some of the muons that originate from normal charged current or dimuon interactions in the CDHS detector, which is situated upstream
in the beam, are stopped in the CHARM detector and the forward-
backward asymmetry of the decay electron is measured.

Neutrino-electron scattering $\nu e \to \nu e$: This reaction is of
fundamental importance for our understanding of weak interactions
and essential for testing unified theories. It has been shown to
exist but little has been discovered about it until now. Ww hope
to be able to perform an experiment with fairly high statistics in
the wide band beam and collect a few events in the narrow band beam.
Because of the kinematical suppression of the cross-section by almost
four orders of magnitude with respect to neutrino-nucleon interactions a very good electron identification is needed.

Inverse muon decay $\nu_\mu e \to \mu \nu_e$: This is another fundamental
interaction which we hope to investigate. It is essential to have
very good muon momentum and angular determination and detect very
low hadronic energy deposition to eliminate the quasi-elastic neutrino interactions.

Beam dump experiment: The main aim of these experiments is
to investigate unconventional sources of prompt neutrinos
(e.g. charm-decay) and possibly to see an effect from τ-neutrinos.
It is again important to discriminate neutral from charged current
events down to very low hadronic energy deposition and we hope to
discriminate ν_e from ν_μ interactions in certain kinematical regions.

II. THE DETECTOR

The basic novel feature of the apparatus is its capability to
measure the direction of hadronic and electromagnetic showers. The
principle adopted for this measurement is to determine the interaction
vertex and the barycentre of the energy deposition in a fine grained
low density calorimeter and to define the angle by the direction

joining these points. Combined with a precise measurement of the total hadronic energy, one obtains a good estimator of the transverse momentum. To avoid biases due to fluctuations in the electromagnetic
component of the shower, a low Z material has to be used in order
to equalize the longitudinal development of the two components.

For muon identification and momentum measurement the low density
calorimeter has to be surrounded by a magnetized iron frame and
followed by a downstream iron magnet. It is important to detect
and correct for energy leakage so the frame and end magnet have to
be instrumented to measure the energy leakage and the muon track.

The calorimeter is also used as a polarimeter for stopping muons.
The material has to consist of even-even nuclei to avoid depolariza-
tion effects in the nuclear magnetic field. A low external magnetic
field is needed to precess the muon spin.

The detector born out of these considerations is composed of
the elements shown in Fig. 1.

The target calorimeter consists of 78 subunits each of which
is composed of three layers oriented perpendicular to the beam:

-    A marble plate of an area of 3 m x 3 m, 8 cm thick, which is
surrounded by an iron frame of 50 cm wide and 8 cm thick plates.
The vertical iron plates are equipped with coils, which can be
powered to create either a 16 kG toroidal field in the frame or a
50 G dipole field in the marble plates. Marble (CaCO$_3$) consists
of even-even nuclei as required in the polarimeter and therefore
also constitutes an isoscalar neutrino target.

-    A hodoscope of 20 scintillation counters each 3 m long, 15 cm
wide and 3 cm thick covers the upstream face of the marble plate.

-    A plane of 128 proportional drift tubes[3] each 4 m long and
with a cross section of 3 cm x 3 cm covers the downstream face of
the marble plate as well as the iron frame.

Scintillation counters and drift tubes are oriented perpendicular to each other and consecutive subunits are oriented at 90° so
that muon tracks and showers are sampled alternatively in two orthogonal directions in both detection systems.

The calorimeter is followed by the end-magnet section which consists of 4 modules all composed of circular iron plates with a diameter of 3.8 m. There are 15 plates, 3 cm thick in the first module with a drift tube plane after every second plate and 5 plates,

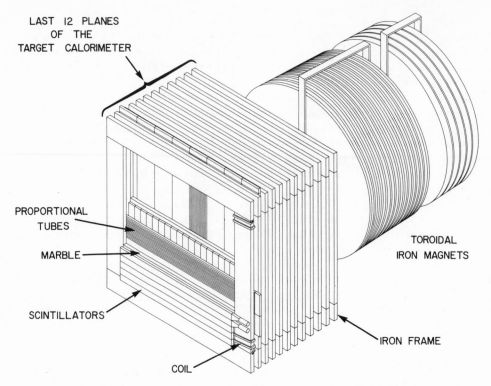

LAST 12 PLANES
OF THE
TARGET CALORIMETER

PROPORTIONAL
TUBES

MARBLE

SCINTILLATORS

COIL

TOROIDAL
IRON MAGNETS

IRON FRAME

Fig. 1   *Schematic view of the elements that constitute the CHARM*
             detector. The rectangular section of the calorimeter is
             followed by the two different types of end magnet
             modules.

15 cm thick in each of the three following modules. Each of the
last three blocks is equipped with two scintillation counter hodoscopes to detect and correct for radiative processes of muons
traversing the magnet. There are also packages of drift tube
planes between the magnet modules and at the end to measure the
muon tracks. A 16 kG toroidal field is created in the iron plates.
The instrumentation of the detector consists of two basic elements:
scintillation counters and proportional drift tubes.

     The scintillation counters are viewed by one photomultiplier
and equipped with a mirror at the opposite side. The signals are
split and sent simultaneously to the electronic logic, where they
are used for triggering the detector, and to two ADC channels of
different sensitivity. Using two analog channels provides a large

dynamic range, a high accuracy for minimum ionizing particles (important for calibration and muon identification) with sufficiently short conversion time. The scintillation counters are used to measure the energy and to determine the barycentre of showers. They also provide some information on the interaction vertex. The drift tubes are operated in proportional mode and each is equipped with a time and an amplitude measuring circuit. The accuracy for track measurement is better than ± 1 mm RMS and the dE/dX resolution for minimum ionizing particles about 60% FWHM. The tubes are used to determine the vertex, to reconstruct muon tracks and to measure the energy leaking out into the iron frame and the end magnet system. They also help in the determination of the shower centre especially for electromagnetic showers.

All active elements are equipped with local buffers so that up to 40 events can be stored during one burst. The system is capable of running with instantaneous event rates of up to 75 kHz.

The most relevant data on the detector and its performance are collected in the Table.

TABLE: Characteristics of the CHARM neutrino detector (E is the energy in units of GeV)

Target mass:	155 to marble, 21 to scintillator
Average density:	1.3 g/cm^3
Sampling step:	1 radiation length .2 collision length
Energy resolution $\sigma(E)/E$:	.53/$\sqrt{E}$ hadrons .35/$\sqrt{E}$ electrons
Angular resolution $\sigma(\theta)$:	(6.5 + 500/E) mrad hadrons (10 + 50/E) mrad electrons
Muon $\sigma(p)/p$:	30% frame magnet 10-20% end magnet
Muon $\sigma(\theta)$:	1 mrad
e-π separation:	1 : 500 at 95% efficiency

The resolutions have been established in a low energy beam (π, p $\leq$ 22 GeV/c and e $\leq$ 8 GeV/c). The extrapolation of the resolution functions to the 100 GeV region has been obtained by Monte Carlo methods and will soon be verified in a high energy beam. The discrimination between electrons and pions is essentially based on the difference in lateral shower development. It has been obtained for low energy test data. Measurements at higher energy are under way.

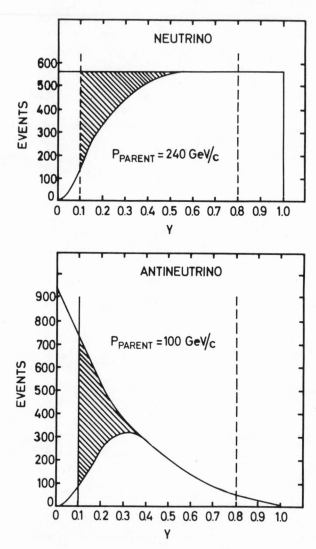

Fig. 2   Characteristic features of the acceptance function for neutral current events.

An important feature of the detector, its acceptance for neutral current semileptonic interactions, is sketched in Fig. 2. The event rates to be expected are compared to the idealized shape of neutrino and antineutrino y distributions. The vertical lines indicate the limits imposed by the intrinsic resolutions. The expression for the error on x is:

$$\frac{\sigma^2(x)}{x^2} = 4 \frac{\sigma^2(\theta_H)}{\theta_H^2} + \frac{1}{(1-y)^2} \frac{\sigma^2(E_H)}{E_H^2}$$

$$+ \frac{y^2}{(1-y)^2} \frac{\sigma^2(E_\nu)}{E_\nu^2}$$

At low y the errors on the hadron angle and energy measurement become big and at high y the denominator causes a divergence of $\sigma(x)$. The shaded area indicates acceptance losses due to the fact that at low hadronic energy the ambiguity between π and K neutrinos can not be resolved and there are two possible values for the energy of the incident neutrino.

Figure 3 shows an on-line display of the two projected views of a charged current neutrino interaction. The beam enters from

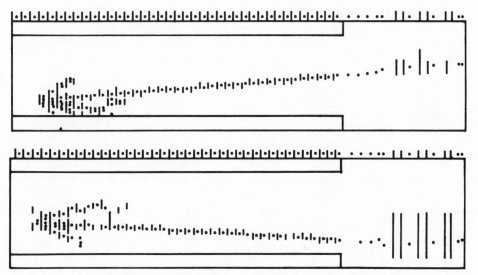

Fig. 3   *Simplified on-line event display of a charged current neutrino interaction.*

the left, the calorimeter section which extends over the range where
the inserted horizontal rectangles indicate the iron frame. The
tag marks on top of the frames indicate the position of detector
planes (dashes for scintillators and dots for drift tubes). Dashes
and dots in the figure show the scintillator and tube elements that
are hit in this interaction. (Some information normally available
on the picture as, for example, the energy deposition in the tube
and scintillator planes has been omitted for clarity.) This figure
gives some impression of the large amount of information available
in this kind of detector.

As mentioned in the introduction, data taking started a few
months ago and we hope that before the end of the year at least
preliminary results will be communicated for all of the subjects
mentioned in the list of the physics program.

REFERENCES

1. M. Jonker, J. Panman, F. Udo, J.V. Allaby, U. Amaldi,
      G. Barbiellini, A. Baroncelli, G. Cocconi, W. Flegel, D.P. Gall,
      W. Kozanecki, E. Longo, K.H. Mess, M. Metcalf, J. Meyer, R.S. Orr
      F. Schneider, A.M. Wetherell, K. Winter, F.W. Büsser, H. Grote,
      P. Heine, B. Kröger, F. Niebergall, K.H. Ranitzsch, P. Stähelin,
      E. Grigoriev, V. Kaftanov, V. Khovansky, A. Rosanov,
      R. Biancastelli, B. Borgia, C. Bosio, A. Capone, F. Ferroni,
      P. Monacelli, F. de Notaristefani, P. Pistilli, C. Santoni,
      V. Valente; to be published in Nucl. Instrum. Methods.
2. M. Holder, J. Knobloch, A. Lacourt, G. Laverrière, J. May,
      H. Paar, P. Palazzi, F. Ranjard, P. Schilly, D. Schlatter,
      J. Steinberger, H. Suter, H. Wahl, E.G.H. Williams, F. Eisele,
      G. Geweniger, K. Kleinknecht, D. Pollmann, G. Spahn,
      H.J. Willutzki, W. Dorth, F. Dydak, V. Hepp, W. Heyde, K. Tittel
      M. Vysočanský, J. Wotschack, P. Bloch, S. Bréhin, B. Devaux,
      M. Grimm, J. Maillard, Y. Malbequi, G. Marel, B. Peyaud,
      J. Rander, A. Savoy-Navarro, G. Tarte, R. Turlay, F.L. Navarria;
      Nucl. Instrum. Methods 148 (1978) 235.
3. C. Bosio, W. Flegel, B. Friend, A. King, K.H. Mess, J. Meyer,
      R.S. Orr, W. Schmidt-Parzefall, F. Schneider, K. Winter,
      F.W. Büsser, D. Gall, H. Grote, P. Heine, B. Kröger,
      F. Niebergall, J. Schütt, P. Stähelin; Nucl. Instrum. Methods
      157 (1978) 35.

STRUCTURE FUNCTIONS FROM NEUTRINO

NUCLEON SCATTERING

Christoph Geweniger

Institut für Hochenergiephysik

Heidelberg, Germany

1. INTRODUCTION

Results of a measurement of the nucleon structure functions
using neutrinos will be presented. The experiment has been carried
out by the CERN-Dortmund-Heidelberg-Saclay collaboration[1] at the
CERN SPS using a counter set-up. The detector was designed for
inclusive studies of neutrino interactions in iron by measuring the
momentum vectors of outgoing muons and the total energy of the produced hadrons. The processes under consideration are

$$\nu N \to \mu^- X \tag{1}$$

$$\bar{\nu} N \to \mu^+ X \tag{2}$$

They are illustrated in Fig. 1. The kinematics can be described
by any three out of the following five variables: E_ν; $Q^2 = -(k - k')^2$;
$\nu = p \cdot (k - k') \sim M E_H$; $x = Q^2/2\nu$ and $y = \nu/p \cdot k$. Here E_ν and E_H

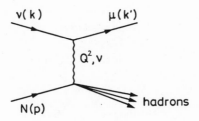

Fig. 1: Diagram for inclusive neutrino nucleon scattering

are the neutrino energy and the total energy of the outgoing hadrons
in the laboratory system, M is the nucleon mass, and the other
variables are defined in Fig. 1.

The data were taken in 1977 in the CERN 200 GeV narrow band
beam. The analysis is based on a final sample of 23,000 events for
reaction (1) and 6,200 events for reaction (2). Details of the
experiment and the results to be discussed here have been published[2]
elsewhere.

This talk is organized as follows: the more important experimental conditions are summarized in Section 2; the determination
of structure functions from measured cross-sections is described in
Section 3. In Section 4 aspects of the data relevant to QCD are
discussed, and conclusions are drawn in Section 5.

2. EXPERIMENTAL CONDITIONS

2.1 Neutrino Flux

The neutrino spectrum is shown in Fig. 2. The spectrum of antineutrinos has qualitatively the same shape. It is continuous from
zero to about 200 GeV with a step around 90 GeV, which separates
neutrinos from pion and kaon decay. The shapes of the two contri-

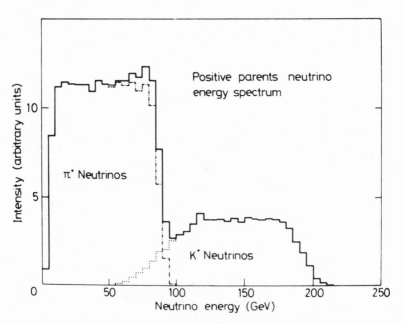

Fig. 2: Neutrino Spectum

butions to the spectrum are well known, whereas the scale of the
upper part relative to the lower part, as given by the K/π ratio
in the parent beam, has an uncertainty of about 10%. The absolute
scale error for pion neutrinos is ± 5%.

2.2 Acceptance

The intrinsic acceptance of the apparatus is limited only by
the trigger threshold at $E_\nu \simeq 3$ GeV and a maximum muon angle of
about 400 mrad. In the analysis, cuts are applied to the kinematic
variables to assure good resolution of the measured quantities (see
Sec. 2.4). In this restricted kinematic region there is practically
no loss of events.

2.3 Calibration and Resolution

The calibration and resolution of the hadron energy measurement
has been determined from special runs in a hadron beam for a test
calorimeter as well as for the detector modules themselves. The
measurement of hadrons and muons can be cross checked taking advantage of the properties of the narrow band beam. The precision is
of the order of a few per cent; at the higher energies it is about
1%.

In order to evaluate cross-sections from the measured distributions, one has to unfold the resolution effects. For this purpose
a Monte Carlo program has been used, taking as input test functions
which are known to reproduce, approximately, the experimental distributions and the known resolution functions. The output consists
of smearing factors which are the ratios of unsmeared to smeared
events in each bin of the kinematic variables. These unsmearing
factors are applied to the data giving corrected distributions. If
these do not agree with the input functions to the Monte Carlo program, the whole procedure is iterated.

2.4 Cuts

Apart from fiducial cuts, the following cuts are applied to
the data:
- a minimum muon momentum of about 7 GeV/c, to assure the reconstruction efficiency for the muon track of (95 ± 2)%.
- $E_\nu > 30$ GeV and $E_H > 5$ GeV to maintain a reasonable resolution.
- $x \gtrsim 0.7$ to avoid smearing corrections of more than 30% (see Sec.
  2.3).
The effect of the cut on the muon momentum is that, events with
$y \geq 0.85$ are rejected.

362   C. GEWENIGER

3.   EVALUATION OF STRUCTURE FUNCTIONS

3.1 Basic relations

The differential cross-section for the process in Fig. 1 is commonly written in terms of three structure functions F_i (x, Q^2):

$$\frac{d^2\sigma}{dxdy} = \sigma_0 \left[y^2 x F_1 + (1-y) F_2 \pm (y-y^2/2) x F_3 \right] \qquad (3)$$

The upper sign refers to neutrinos, the lower to antineutrinos and $\sigma_0 = G^2 M E_\nu / \pi$. The F_i are, in general, different for neutrinos and antineutrinos and depend on the target.

In the quark parton model the linear combinations

$$q\ (x,\ Q^2) = (2xF_1 + xF_3)/2$$
$$\bar{q}\ (x,\ Q^2) = (2xF_1 - xF_3)/2$$

may be interpreted as distributions in the fractional momentum of the nucleon carried by the quarks and antiquarks contributing to the process in Fig. 1. Let us define

$$q = u + d + s + c$$

where u, d, s, c denote the respective distributions for the different quark flavours. Furthermore, let us assume that:
- the target is an isoscalar (u = d);
- strange and charmed quarks are degenerate (s = c);
- the Callan-Gross relation holds ($F_2 = 2xF_1$).

Then we are left with two structure functions F_2 and xF_3, or equivalently, q and $\bar{q}$, and the cross-sections may be written as

$$\frac{d^2\sigma^\nu}{dxdy} = \sigma_0 \left[q + \bar{q}\ (1-y)^2 + c \right]$$
$$\frac{d^2\sigma^{\bar{\nu}}}{dxdy} = \sigma_0 \left[\bar{q} + q\ (1-y)^2 + \bar{c} \right]$$

C and $\bar{C}$ are correction terms to be added in case the above made assumptions do not hold exactly. The structure functions are then obtained simply from the sum and the difference of the cross-sections for neutrinos and antineutrinos, neglecting the corrections for a moment:

$$q \pm \bar{q} = \frac{1}{\sigma_o}\left[\frac{d^2\sigma^\nu}{dxdy} \pm \frac{d^2\sigma^{\bar{\nu}}}{dxdy}\right] / \left[1 \pm (1-y)^2\right] \qquad (4)$$

$q + \bar{q} = F_2$ represents the fractional momentum of all spin 1/2 constituents and $q - \bar{q}$ that of the valence quarks. The x-distribution of the sea quarks may be best determined from the antineutrino cross-section at high y:

$$\bar{q} + (\bar{s}-\bar{c}) = \frac{1}{\sigma_o}\left[\frac{d^2\sigma^{\bar{\nu}}}{dxdy} - \frac{d^2\sigma^\nu}{dxdy}(1-y)^2\right] \qquad (5)$$

This relation is valid to about 1% for $y > 0.5$.

3.2 Corrections

The corrections to Eq. (4) are factors to be applied to the right-hand-side of these equations, where C_2 and C_3 are the corrections to F_2 and xF_3 , respectively:

$$C_2 = C_R \ (1-\delta\eta\ \frac{q - \bar{q}}{q + \bar{q}} \ - 2\eta\ \frac{\bar{s} - \bar{c}}{q + \bar{q}}\ )$$
$$C_3 = (1 + \delta/\eta)^{-1} \qquad (6)$$
$$\eta = \left[1-(1-y)^2\right] / \left[1+(1-y)^2\right]$$

The terms containing δ correct for the neutron excess in iron, causing a small violation of the isoscalar condition u = d. δ is given by (N − Z)/3A = 0.023 and the correction is always smaller than 2.3%.

The term with $(\bar{s} - \bar{c})/(q + \bar{q})$ accounts for the difference of the strange and the charmed sea. We have used $\bar{s} - \bar{c} = 0.2\bar{q}$ in agreement with a measurement of the magnitude of $\bar{s}$ [3], assuming $\bar{c} = 0$. q was determined from Eq. (5). The correction is non-negligible only at small x, where one has a sizeable contribution from sea quarks to F_2. In the limit x = 0 and $Q^2 = Q^2_{max}$ it can be as much as 20%.

A violation of the Callan-Gross relation gives $C_R > 1$. We have assumed the validity of this relation ($C_R = 1$) in agreement with our data. Details will be discussed in the next section.

Radiative corrections have been applied according to Ref. 4.
Fermi motion was not considered, since it does not change the scaling behaviour of the structure functions. However, the structure
functions might be different when measured on different isoscalar
targets.

3.3 The Callan-Gross relation

A violation of the Callan-Gross relation is measured in terms
of the parameter

$$R' = (F_2 - 2xF_1)/2xF_1 \qquad (7)$$

It can be determined from a fit to the y-distribution of the sum
of neutrino and antineutrino cross-sections (Eq. (3)):

$$\frac{d^2\sigma^\nu}{dxdy} + \frac{d^2\sigma^{\bar\nu}}{dxdy} = 2\sigma_o\left[y^2 xF_1 + (1-y)F_2\right]$$

This should be done in bins of x and Q^2. However, because of the
large errors, only the averaged distributions were fitted, yielding
$R' = 0.03 \pm 0.04$ (statistical error only). The result is corrected
for effects from scaling violations and for $s \neq c$. Radiative corrections are included as well. Recently, the data have been reana-
lyzed in a manner which is independent of any variations of the
structure function with Q^2. $R' = 0.02 \pm 0.12$ was obtained[5].

This result is consistent with measurements of $R = \sigma_L/\sigma_T$ in
ep and µp scattering experiments at SLAC and FNAL as presented at
this seminar[6,7]. Remember that:

$$R = \left[(1+Q^2/\nu^2)\ F_2 - 2xF_1\right]/2xF_1$$

Comparing with (7) one finds:

$$R' = (R - Q^2/\nu^2)\ /\ (1+Q^2/\nu^2)$$

R and R' become equal in the limit of vanishing target mass effects:
$Q^2/\nu^2 = 4x^2M^2/Q^2 = 0$. Taking the SLAC result[6] $R = 0.21 \pm 0.10$, one
obtains $R' \approx 0.05 \pm 0.10$ for $Q^2 < 20$ GeV$^2/c^2$ and $x > 0.1$. The over-
all result of $R = 0.52 \pm 0.35$ from the CHIO collaboration[7] comes
from $x < 0.1$ and mostly from $x < 0.02$. Within these kinematical
limits R decreases with increasing x and Q^2, and it is certainly
consistent with zero for $x > 0.05$.

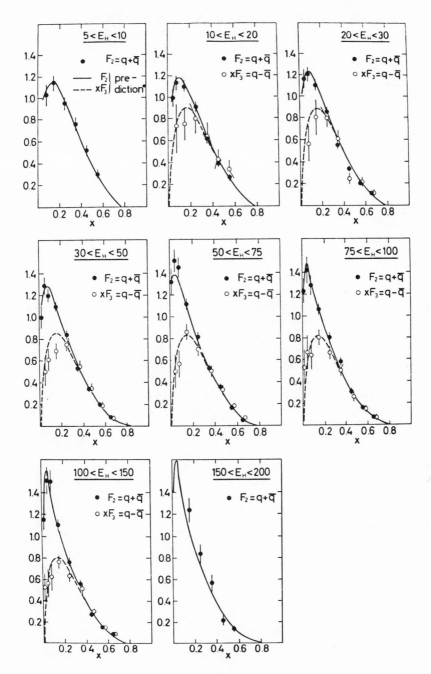

Fig.3: $F_2(x)$ and $xF_3(x)$ for different hadron energies. The curves are explained in the text.

The effect of a parameter $R \neq 0$ on the determination of F_2 is rather small. The correction not yet specified in (6) is

$$C_R = \left[1 - \frac{y^2}{1+(1-y)^2} \cdot \frac{R'}{1+R'} \right]^{-1}$$

where

$$\left\langle \frac{y^2}{1+(1-y)2} \right\rangle \approx \frac{Q^2/x}{500} \quad (Q^2 \text{ in } GeV^2/c^2)$$

To give an example: take the average x and Q^2 from this experiment, $x \stackrel{\sim}{=} 0.2$ and $Q^2 \stackrel{\sim}{=} 20 \, GeV^2/c^2$, and assume $R' = 0.1$; then one gets $C_R - 1 \stackrel{\sim}{=} 2\%$.

3.4 Results

The structure functions are initially analyzed as functions of x and the hadronic energy $E_H \stackrel{\sim}{=} \nu/M$. The reasons for doing this are:

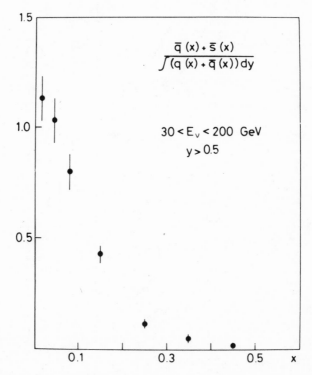

Fig. 4: The antiquark structure function

a) at fixed ν the entire x distribution is measured, and b) its shape is independent of the neutrino spectrum. E_H is then converted to Q^2 using the relation $Q^2 = 2xME_H$, thereby introducing additional errors to the distributions due to the flux uncertainties discussed in Section 2.1.

F_2 and xF_3 are shown in Fig. 3 as functions of x for different E_H bins. The shape of F_2 changes significantly with E_H. Such a behaviour is inconsistent with the assumption of scaling. The curves indicate the test functions used in the Monte Carlo program of Section 2.3. The difference between F_2 and xF_3 is due to sea quarks. Their x distribution, averaged over all energies, is shown in Fig. 4; it is steeply falling with x and vanishes for x > 0.4. In Figs 5 and 6 the data are shown as functions of Q^2. The error bars indicate the statistical errors. The systematical errors are, at most, as large as the statistical errors. The over-all scale errors are ± 6% for F_2 and ± 8% for xF_3. Also shown in Fig. 5 are appropriately scaled ed scattering data from SLAC[8] which are in excellent agreement with our data in the regions of overlap.

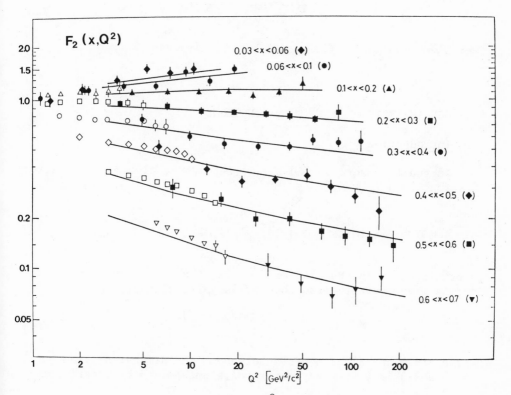

Fig.5: F_2 as a function of Q^2 in different bins of x.

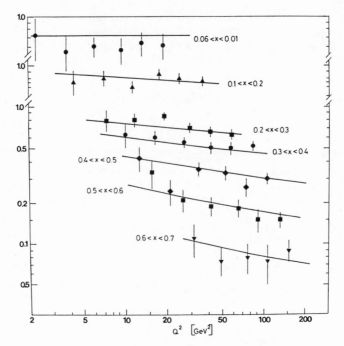

Fig.6: xF_3 as a function of Q^2 in different bins of x.

4. QCD ANALYSIS OF STRUCTURE FUNCTIONS

4.1 Introduction

In the framework of QCD the Q^2 dependence of moments of structure functions

$$M_i(n,Q^2) = \int x^{n-2} F_i(x,Q^2) \, dx \tag{8}$$

is predicted[9]. In the case of moments of xF_3, the prediction, in
leading order perturbation theory, is particularly simple:

$$M_3(n,Q^2) \sim \left[\ln(Q^2/\Lambda^2)\right]^{-d_n} \tag{9}$$

Here d_n is a so-called non-singlet anomalous dimension given by the
theory and Λ is a free parameter to be determined by experiment.

In Section 4.2 Λ is determined from a fit to the structure
functions employing a parametrization that produces the required

Q^2 behaviour of the moments. In Section 4.3 Λ is derived directly from the moments of xF_3. In addition, measured ratios d_n/d_m are compared to the predictions.

4.2 Fit to Structure Functions

Following the procedure proposed by Buras and Gaemers[10], fits to F_2 and xF_3 have been performed. The structure functions are parametrized as follows:

$$xF_3(x,Q^2) = Cx^{a(s)}(1-x)^{b(s)}$$

$$F_2(x,Q^2) = xF_3(x,Q^2) + A(s)(1-x)^{B(s)}$$

$$a(s) = a_0 + a_1 s$$

$$b(s) = b_0 + b_1 s$$

$$s = \ln\left[\frac{\ln(Q^2/\Lambda^2)}{\ln(Q_0^2/\Lambda^2)}\right]$$

In the case of the valence quark distribution xF_3, the normalization C is chosen such that the number of valence quarks is three. The fit has three free parameters: a_0', b_0' and Λ.

In the equation for F_2 the second term represents the sea quarks. $A(s)$ and $B(s)$ are determined from QCD equations for the second and third moments of F_2 which involve two moments of the gluon distribution. One is the total momentum fraction carried by gluons. It is fixed by momentum conservation: $M_G(2,Q^2) = 1 - M_2(2,Q^2)$. The other moment represents the average fractional momentum of gluons $M_G(3,Q^2)$. This is a new piece of information being determined by the fit. Altogether, there are three free parameters in addition to the three parameters for xF_3: $A(0)$, $B(0)$ and $M_G(3,Q_0^2)$.

The data are well described by the fits as can be seen in Figs 5 and 6. Also, in the case of F_2, the curves extrapolate reasonably well to the SLAC ed data[8] which have not been used in the fit. The results for Λ are (0.55 ± 0.15) GeV from xF_3 and (0.47 ± 0.11) GeV from F_2. The systematical error, not yet included here, is ± 0.1 GeV. The shapes of the quark distributions at $Q^2 = 20$ GeV/c^2 are

valence quarks $\sim x^{(0.51\pm0.02)}(1-x)^{3.03\pm0.09}$

sea quarks $\sim (1-x)^{8.0\pm0.7}$

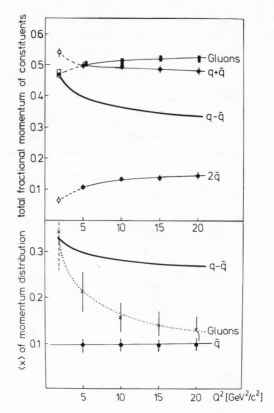

Fig. 7: *The total fractional momenta of all nucleon constituents and the average values of their momentum distributions as functions of Q^2.*

The total and the average fractional momentum of the constituents as a function of Q^2 is shown in Fig. 7. Errors are relative to valence quarks $q-\bar{q}$. However, each point represents the same data such that there are no statistical fluctuations of a moment taken at different Q^2. The points for the total momentum have a common scale error of $\pm$ 6%.

4.3 Moments of xF_3

In order to calculate moments we have supplemented our data at $x \geq 0.4$ with SLAC ed data[8]. It has been demonstrated in Section 3.4 that sea quarks are negligible at these values of x. So we have the simple quark model relation

$$xF_3^{\nu N} = \frac{9}{5} F_2^{ed} \qquad (x \geqq 0.4)$$

Even for the combined data set there are, at fixed Q^2, always regions around x = 0 and x = 1 not covered by data points. In order to compute the integrals, a parametrization, $xF_3 \sim x^\alpha(1 - x)^b$, is chosen to extrapolate the structure function. Only those moments are kept where the extrapolated parts contribute less than 25% to the full integral. This limits the moments to n in the range 2-6 and Q^2 in the range 6.5-75 GeV^2/c^2. Ordinary moments (8), as well as Nachtmann moments[11], are calculated. The latter account for target mass effects of the order of $4x^2M^2/Q^2$.

Two testable predictions of QCD are derived from Eq. (9). The first one states that the logarithm of any two moments should have a linear relationship:

$$\ln M_3(n,Q^2) = (d_n/d_m)\cdot\ln M_3 (m,Q^2) + \text{const.}$$

Moreover, the slope is given by the ratio of the two respective anomolous dimensions d_n/d_m, a unique number in QCD. In Fig. 8 pairs of Nachtmann moments are plotted against each other. The straight lines indicate the predicted slopes; they are in good agreement with the data. All results are summarized in Fig. 9. More specifically, the slopes test the vector nature of the gluons. Scalar gluons are not favoured by the data. The difference between ordinary and Nachtmann moments shows that target mass effects are not negligible, even though the minimum Q^2 is 6.5 GeV^2/c^2.

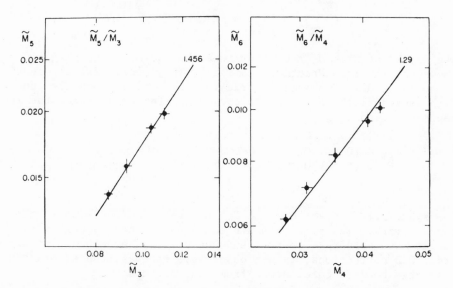

Fig. 8: *Log-log plot for pairs of Nachtmann moments;*
$\widetilde{M}(5,Q^2)$ versus $\widetilde{M}(3,Q^2)$ and $\widetilde{M}(6,Q^2)$ versus $\widetilde{M}(4,Q^2)$.
The straight lines have slopes predicted from QCD.

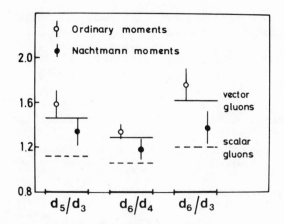

Fig.9: Ratios of anomalous dimensions determined from ordinary and Nachtmann moments. The solid and dashed lines are predictions for vector and scalar gluons, respectivly.

The second prediction states that the inverted moments should follow a straight line when plotted atainst $\ln Q^2$:

$$M_3(n, Q^2)^{-1/d_n} = \text{const.} \ (\ln Q^2 - \ln \Lambda^2)$$

The intercept with $M_3 = 0$ determines the parameter Λ. Fig. 10 shows the inverted Nachtmann moments. They nicely fit straight lines in $\ln Q^2$. The same is true for the ordinary moments, however, with substantially different slopes and intercepts. From the intercepts one finds:

$$\Lambda = \begin{array}{l} (0.60 \pm 0.15) \ \text{GeV (ordinary moments)} \\ (0.33 \pm 0.1o) \ \text{GeV (Nachtmann moments)} \end{array}$$

The systematical error of 0.10 GeV has not been included in the above errors. The results again demonstrate the importance of target mass effects. An n dependence of Λ, as expected from higher order QCD corrections, is not observed. However, the errors are too large to draw conclusions from that circumstance.

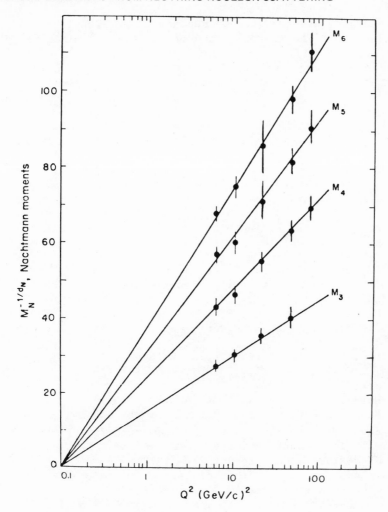

Fig. 10: Inverted Nachtmann moments of xF_3 as functions of Q^2. The straight lines are fits with a common intercept.

5.   CONCLUSIONS

- Nucleon structure functions have been measured up to $Q^2 = 200$ GeV^2. Clear scaling violations are observed in the structure function F_2.

- F_2 and xF_3 are well described by a parametrization suggested from QCD.

- Moments of xF_3 show the Q^2 dependence predicted by leading order QCD. The vector nature of the gluons in confirmed.

- In leading order QCD the value of the parameter Λ is in the range from 0.3 GeV to 0.7 GeV.

REFERENCES

1. The members of the collaboration are: J.G.H. de Groot, T. Hansl
     M. Holder, J. Knobloch, J. May, H.P. Paar, P. Palazzi,
     A. Para, F. Ranjard, D. Schlatter, J. Steinberger, H. Suter,
     W. von Rüden, H. Wahl, S. Whitaker and E.G.H. Williams
     (CERN, Geneva);

     F. Eisele, K. Kleinknecht, H. Lierl, C. Spahn and
     H.J. Willutzki (University of Dortmund);

     W. Dorth, F. Dydak, C. Geweniger, V. Hepp, K. Tittel and
     J. Wotschack (University of Heidelberg);

     P. Bloch, B. Devaux, S. Loucatos, J. Maillard, B. Peyaud,
     J. Rander, A. Savoy-Navarro and R. Turlay (CEN-Saclay);

     F.L. Navarria (University of Bologna).
2. J.G.H. de Groot et al., Z. Physik C1:143 (1979); Phys. Lett.
     82B:292, 456 (1979).
3. M. Holder et al., Phys. Lett. 69B:377 (1977).
4. R. Barlow and S. Wolfram, Oxford report OUNP 24/78 (1978).
5. A. Savoy-Navarro, talk given at "Neutrino 79", Bergen (1979).
6. R.E. Taylor, contribution to this seminar. See also,
     M. Mestayer, SLAC report 214 (1978).
7. H.L. Anderson, these proceedings.
8. E.M. Riordan et al., SLAC Pub. 1634 (1975).
9. D.J. Gross and F.A. Wilczek, Phys. Rev. D8:3633 (1978);
     D9:980 (1974);
     H. Georgi and H.D. Politzer, Phys. Rev. D9:416 (1974).
10. A.J. Buras and K.J.F. Gaemers, Nucl. Phys. B132:249 (1978).
11. O. Nachtmann, Nucl. Phys. B63:237 (1973); B78:455 (1974).

MEASUREMENT OF NEUTRAL TO CHARGED CURRENT RATIO

IN NEUTRINO PROTON INTERACTIONS

Dieter Haidt

Deutsches Elektronen-Synchrotron
DESY
Hamburg, Germany

INTRODUCTION

The measurement of the ratio $R_p = \sigma/\nu p \rightarrow \nu X)/\sigma(\nu P \rightarrow \mu^- X)$ involving neutral current (NC) and charged current (CC) inclusive hadronic states provides information about the neutral weak coupling constants u_L, u_R, d_L, and d_R

$$J_\alpha^{NC} = \bar{\psi}_u \gamma_\alpha \left\{ u_L \ (1+\gamma_5) + u_R \ (1-\gamma_5) \right\} \psi_u +$$

$$\bar{\psi}_d \gamma_\alpha \left\{ d_L \ (1+\gamma_5) + d_R \ (1-\gamma_5) \right\} \psi_d + \ldots$$

using the notation of Seghal[1].

Inclusive measurements on isoscalar targets with neutrinos and antineutrinos[2] distinguish only between right and left handed coupling. The isospin structure of the weak neutral current[3] reveals itself in exclusive channels [e.g., $\nu p \rightarrow \nu \Delta^+$ [4], (see Fig. 1)], in semi-inclusive channels (e.g., $\nu N \rightarrow \nu \pi^\pm X$) or in inclusive scattering off a non-isoscalar target (e.g., $\nu p \rightarrow \nu X$). In the latter case, the target consists of twice as many u quarks than d quarks and thus R_p is roughly related to $2u_L^2 + d_L^2$. Therefore, a comparison of the NC/CC rations obtained from scattering off protons and off isoscalar targets allows $|u_L|$ and $|d_L|$ to be disentangled, provided R_p can be measured to an accuracy of less than 10%. Until now, the accuracy of measurement has been 30%[5].

EXPERIMENTAL CONDITIONS

    The experiment, known as WA 21 (Aachen-Bonn-CERN-München-
Oxford Collaboration),was performed using BEBC filled with hydrogen
and exposed to the CERN SPS wideband neutrino beam derived from
350 GeV/c protons. The bubble chamber was complemented by a two plane
EMI (external muon identifier) in order to ensure a good distinction between events with and without muons.

    The basic sample consists of 2295 charged current candidates
(CC) and 3128 neutral current candidates (NC). The selection criteria are:

    i)   Muon momentum greater than 3 GeV/c
   ii)   Energy of all tracks greater than 5 GeV
  iii)   Only three and more prongs

    An event is then classed CC, if it has at least one identified
negative muon; it is classed NC, if it has no identified negative
muons. . The small number of events without an identified negative muon
but with an identified positive muon are also classed NC candidates.
Figure 2 shows various distributions of the basic sample.

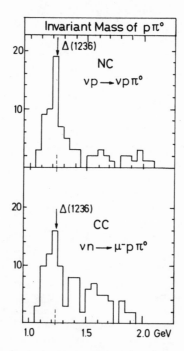

Fig. 1   Invariant mass distribution of the $p\pi^0$ system in the chan-
         nels $\nu p \to \nu p \pi^0$ and $\nu n \to \mu^- p \pi^0$. Already the fact that in
         the NC channel the Δ resonance is excited excludes a purely
         isoscalar weak neutral current.

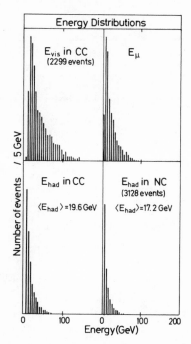

Fig. 2 Various energy distributions of the charged and neutral
current sample.

METHOD

The CC sample is clean and incomplete. The NC sample is
contaminated and incomplete. This is schematically demonstrated in
Table 1. The EMI has a high, although not 100%, efficiency, for
identifying muons. Consequently, some genuine CC events, where the
muon escapes detection, are classed NC.

A high contamination of the NC sample by neutral hadron interactions (produced upstream by neutrino interactions in the shielding) is expected despite the cut in the visible hadronic energy
of 5 GeV. In previous experiments (see P.C. Bosetti et al., in Ref. 2
this background was determined by relating the non associated
neutral hadron interactions to the observable associated neutral
stars[6]. Due to the small number of associated events in hydrogen
(interaction length of about 10 m) this method is not accurate
enough. Therefore, another distinctive feature is used to suppress
the neutral hadron background. In neutrino interactions the transverse momentum of the hadron system, compensating the transverse
momentum of the outgoing lepton, increases with the neutrino
energy E_ν:

$$\langle P_{T,had}^2 \rangle \approx 2ME_\nu \langle xy(1-y) \rangle \sim E_\nu$$

Table 1: Event topologies and their classifications

Topology	Event Type	Classification
	CC μ^- identified	CC
	CC μ^- not identified	NC
	NC no μ^- identified	NC
	Neutral hadron interaction	NC

On the contrary, the hadron shower induced by the incoming neutral hadron has at average the same, and hence small, transverse momentum as the incoming neutral hadron.

In conclusion, cutting off events with low transverse momentum of the hadron system purifies the NC sample. This is demonstrated for various cuts in P_T^{had} in Table 2 and Fig.3. Obviously, both types of background in the NC sample are reduced by the cut off in P_T^{had}. As suggested by Fig. 3, only events satisfyir $P_T^{had} > 1$ GeV/c are retained.

EFFECT OF EMI

Charged current events are only selected if the muon momentum P_μ exceeds 3 GeV/c. For the other events, the muon detection efficiency (see Fig. 4) would be too small. Hence, 2.4% of the CC-events are lost.

The geometrical acceptance of the EMI is shown in Fig. 4 as a function of the muon momentum. Correcting for not identified muons as a function of P_μ leads to a) adding 193 events to the CC sample and b) removing 97 events from the NC sample. The electronic inefficiency causes a loss of $(1 \pm 0.5)\%$ independent of P_μ. Finally, due to bad measurements, about 0.9% of the CC and 0.2% of the NC events are lost.

Table 2: Variation of the basic event sample and
          the raw ratio R_p = NC/CC with the trans-
          verse momentum cut off in the hadronic system

$P_T^{had} >$ (in GeV/c)	CC	NC	R_p (raw)
.0	2299	3128	1.36
1.0	1484	1138	0.77 ± 0.03
1.5	858	554	0.65 ± 0.04
2.0	430	259	0.60 ± 0.05
3.0	107	70	0.65 ± 0.11

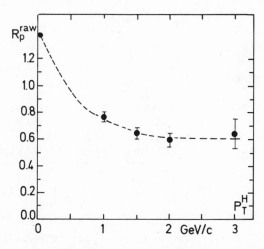

Fig. 3   The raw ratio of neutral to charged current events with
         transverse momentum of their hadronic system exceeding P_T^H
         as a function of P_T^H=

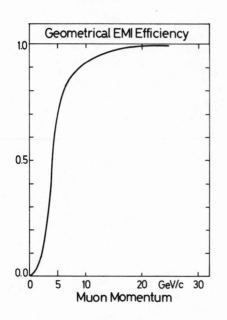

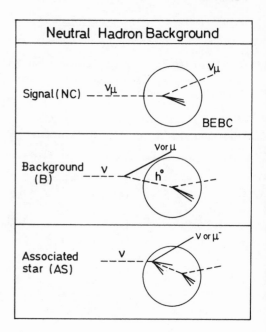

Fig. 4 Variation of the geo- Fig. 5 Sketches of a nutrino inter-
metrical muon detection effic- action with a neutral hadron observ-
iency by the external muon ed in the bubble chamber either iso-
identifier (EMI) with the muon lated (B) or associated (AS).
momentum.

NEUTRAL HADRON BACKGROUND

        In Fig. 5 a neutrino interaction with a secondary neutral
hadron is sketched in its nonassociated (background) and its
associated form. Due to the small number of associated events (AS)
in hydrogen the background (B) is related to the signal (NC) in
three steps:

$$\frac{B}{NC} = \frac{B}{AS} \frac{AS}{\nu} \frac{\nu}{NC} = 13 \times 1\% \times 3 = 0.4 \pm 0.1$$

The first factor B/AS is calculated using Monte Carlo techniques[6].
The second factor is obtained from the more abundant neutrino
stars observed in the bubble chamber BEBC filled with neon (WA 19,
see P.C. Bosetti in Ref. 2. The appropriate changes due to different spectra and interaction lengths are taken into account. The
last factor $\nu/NC = 1 + CC/NC$ is taken to be three. After the application of the cut in p_T^{had} at 1 GeV/c, one obtains B/NC = 0.19 ± 0.06.

OTHER CORRECTIONS AND BACKGROUNDS

Since only **three prong events are selected, the lost NC events with one charged hadron in the final state have to be estimated.** An extrapolation from the multiplicity distribution (see Fig. 6) gives $(5 \pm 3)\%$. A recent Monte Carlo calculation confirms this estimate.

The neutrino beam has a small background of $\bar{\nu}_\mu$ and ν_e leading to events mistaken as NC. Finally, the difference of one unit of charge between CC and NC final hadron states may induce a slightly different energy deposition. This effect is expected to be small. Indeed, Monte Carlo calculations indicate a correction of at most 2%. Within the limited statistics, this is confirmed by comparing the energy depositions of neutrals in NC and CC interactions.

Table 3 summarizes all corrections and background subtractions. The final result is $R_p = 0.50 \pm 0.02 \pm 0.04$, where the statistical and systematic accuracies are 0.02 and 0.04 respectively.

Table 3:   Summary of backgrounds and corrections

Raw sample	1484	1138
EMI geom. acc.	193	−97
EMI electr. ineff. + bad measurements	29	−7
$P_\mu < 3$ GeV/c 2.4%	35	
Neutral hadrons $(19 \pm 6)\%$		−166
NC 1 prong $(5 \pm 3)\%$		+44
$\bar{\nu}_\mu$ 1.3% $\bar{\nu}_\mu/\nu_\mu$		−10
ν_e 2% ν_e/ν_μ		−26
Final sample	1741	876

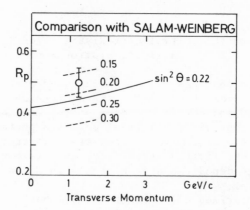

Fig. 6   Frequency of the observed NC and CC events with more than
two charged hadrons.   The dotted line is an eyeball fit.
The extrapolation to one prong NC is indicated.

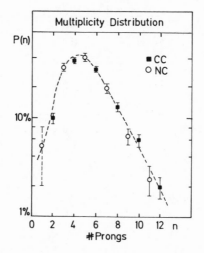

Fig. 7   The measured value R_p is compared with the prediction by
the Salam-Weinberg model.   The theoretical curves are
given for various values of $\sin^2\theta$ as a function of the
transverse momentum of the hadronic system.

INTERPRETATION

Figure 7 shows the corrected ratio R_p for $P_T^{had} > 1$ GeV/c together with the prediction of the Salam-Weinberg model for various values of the parameter $\sin^2\theta$. The measurement of R_p leads to

$$\sin^2\theta = 0.18 \pm 0.04$$

in agreement with Ref. 2.

Within the next two months the WA 21 Collaboration will publish results on R_p based on improved statistics.

ACKNOWLEDGMENT

I would like to thank Drs. Luc Pape and Per-Olaf Hulth for fruitful discussions.

REFERENCES

1. L.M. Seghal, Proc. of the 1977 Int. Symposium of Lepton and Photon Interactions at High Energies, p. 837.
2. P.C. Bosetti et al., Phys. Lett. 76B:505 (1978); M. Holder et al., Phys. Lett. 69B:377 (1977).
3. D. Haidt, Isospin Properties of the Weak Neutral Current, Ben Lee Memorial International Conference on Parity Non-conservation, Weak Neutral Currents and Gauge Theories, 1977, eds. D. Cline and F. Mills, p. 77=
4. W. Krenz et al., Nucl. Phys. B135:45 (1978).
5. F.A. Harris et al., Phys. Rev. Lett. 39:437 (1977).
6. W.F. Fry and D. Haidt, CERN Yellow Report 75/1 (1975).

PROPERTIES OF HADRONS PRODUCED BY NEUTRINOS

IN HYDROGEN

J.C. Vander Velde[*]

University of Michigan
Ann Arbor, Michigan 48109, U.S.A.

INTRODUCTION

I will discuss some results on the properties of single and correlated hadrons made in the hadron "jet" via the reaction $\nu p \rightarrow \mu^- +$ jet. We know the direction of the jet to within $\pm 1^\circ$ in these events and the vector momenta of the individual charged hadrons are measured to better than a few per cent. Hence we have an excellent method for studying the detailed properties of hadron jets.

The elementary process that forms the jets is believed to be the interaction of a virtual W^+ weak boson with a d quark in the proton, changing it to a u quark. In the c.m. of the over-all hadronic system we can picture this as follows:

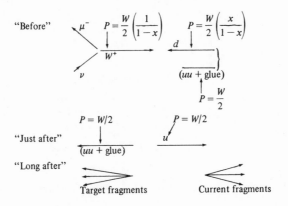

---

[*]Berkeley-Hawaii-Fermilab-Michigan Collaboration(See Ref. 1).

Looked at in this way, each event consists of a pair of back-to-back jets; one formed from the fragmentation of a u quark and the other formed from the spectator remains of the target proton which consists of two u quarks and some gluons, $q\bar{q}$ pairs, etc.. We have indicated on the diagram the momenta of these elemental objects, before and after the collision, expressed in terms of the over-all invariant hadronic mass W and the Bjorken scaling variable x.

BASIC PROPERTIES

We will start by showing a few of the general properties of the jets we observe. Our data indicate that the properties of the "current jet" and the "target jet" are quite similar. In Fig. 1 we compare the rates of hadron production versus rapidity in the two regions and find that they are equal within statistics. (The cuts are made in order to separate the two regions.) This may indicate that in purely hadronic processes the jet formed by a single quark emanating from a hard scattering will be similar to the di-quark formed by the remains of the parent nucleon.

The angular confinement of the total hadronic system, if we think of it as a single "jet", is illustrated in Fig. 2. The larger the jet momentum, the smaller the fraction of its energy that appears outside a given lab angle. Figure 2 should give the hadron jet experimentalist a rough idea of what fraction of the jet's energy can be captured in a fixed-angle apparatus.

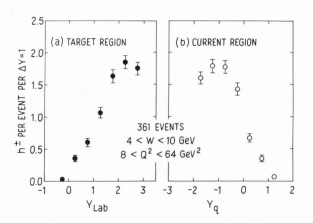

Fig. 1. Charged hadron rate for high-Q^2 events plotted vs. rapidity Y_{lab} and quark-frame rapidity $Y_q = Y_{lab} - \ln (W^2/1 \text{ GeV}^2)$. The approach to and height of the two "plateaux" are very similar.

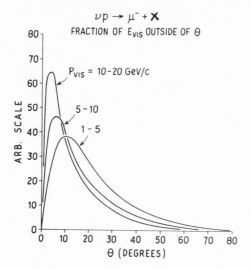

Fig. 2.   Fraction of visible energy E_{VIS} contained inside lab angle θ with respect to the entire hadronic system direction.

The fraction of longitudinal momentum carried by the first and second highest momentum particles is shown in Fig. 3. (Figures 2 and 3 are smooth curves drawn through the data.) The data of Fig. 3 also scale in the sense that the fraction of the remaining momentum carried by the "second" (ignoring the "first") is distributed (not shown) just like the fraction of the total that the "first" carries.

We now turn to some more details of these events looked at in the context of the quark-parton model. The W distribution for 1670 events (which we believe to be $\sim$ 95% pure νp charged current) is shown in Fig. 4. The cuts shown in the figure are made to purify the charged current (CC) sample. They are described in detail in Ref. 1. The sharp peak near 1 GeV represents the exclusive channel $\nu p \to \mu^- + (\Delta^{++} \to p\pi^+)$.

According to Feynman's idea[2] a given quark should develop into a jet of hadrons ("fragment") in a way which is independent of the kinematic variables (W and Q^2) associated with its formation. In Fig. 5 we see that the average number of positive and negative high Z hadrons per event changes with W for low W values but appears to approach a constant plateau for $W \gtrsim 4$ GeV. Hence one of the basic assumptions of the quark-parton model -- that quark fragmentation is independent of W and Q^2 -- is confirmed by Figs. 5 and 6, at least for $W \gtrsim 4$ GeV and $2 < Q^2 < 64$ GeV. (Z is the longitudinal fraction of total hadronic momentum carried by a given hadron.)

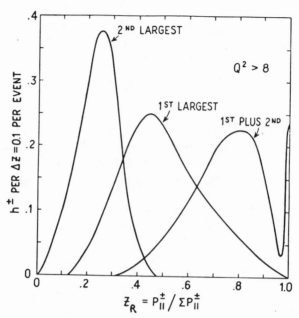

Fig. 3. Distribution of the fraction Z_R of visible
longitudinal momentum carried by the 1st, 2nd,
and (1st plus 2nd) highest momentum charged
hadrons.

The solid curves on Fig. 6 are from a Monte Carlo calculation
described in detail in Ref. 1.   The calculation is meant to account
for the "kinematic" effects of longitudinal phase space and charge
conservation, and so it generates hadrons with a flat distribution
in (the W rest frame) rapidity and an exponential in the transverse
mass but no charge-neutral or charge-charge correlations (e.g.,
clusters or resonances) are put in. We refer to the model calculations as "uncorrelated Monte Carlo" (UMC). The generated events
are passed through the same even reconstruction procedures and cuts
as the real events and thus allow us also to evaluate our biases.

Encouraged by the Q^2 independence and Feynman scaling properties
of the events (Figs. 5,6) we proceed to examine further details of
these back-to-back quark-diquark systems. The first thing we note
is that they have a definite jet-like character, as illustrated in
Fig. 7. We plot the average value of p_T^2 vs W for particles in two
different Feynman x (x_F) regions. In both regions $\langle p_T^2 \rangle$ rapidly
attains a maximum value which is much less than its kinematically
allowed value of $W^2/4$. This means that we are indeed observing
"jets" with a severely limited transverse momentum distribution not
unlike that of the exponentially cut-off UMC model in which $d^2\sigma/dY$
$dp_T^2 = e^{-bm_T}$ (with m_T = transverse mass and $b = 6$ GeV^{-1}). We will
say more about the Q^2 and x_F dependence of $\langle p_T^2 \rangle$ later.

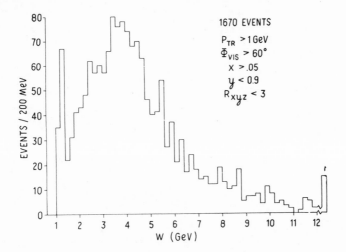

Fig. 4.   Distribution of invariant mass of the hadronic
system for a sample of neutrino charged-current
events.   Typical mass resolution is ±15% for
the bulk of the events, which have missing neutrals.   The peak at the left is due to the final
state $\mu^-\Delta^{++}$.

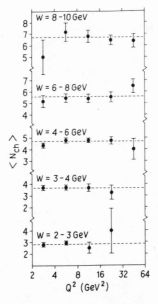

Fig. 5.   Average number of charged particles in the
hadronic system vs. Q^2 for various W slices taken
from Fig. 4.   The dashed lines represent the over-
all average for each W slice.

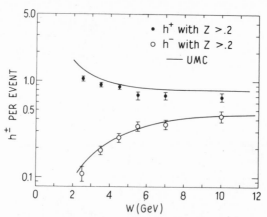

Fig. 6. The rate of production of positive and negative
         hadrons with Z>0.2 as a function of W. We take
         the Feynman scaling (W-independent) region as
         W> 4 GeV. The curves are the result of the
         Monte Carlo model described in the text.

LONGITUDINAL PROPERTIES

     The single particle inclusive properties of the quark-diquark
jet can be examined in terms of longitudinal, transverse, and azimuthal variables. In order to deal only with events in the scaling
region of Fig. 6, we make a "high W" selection of W > 4 GeV, in
addition to the selections shown in Fig. 4. The resulting distributions for h^- and h^+ hadrons are shown plotted vs the longitudinal
x_F variable in Figs. 8 and 9. We see that both distributions, and
hence their ratio, are fairly well represented by the UMC model.
This indicates that a quark fragments into hadrons in a manner which
approximates longitudinal phase space in its single particle inclusive aspects. Figures 8 and 9 represent the so-called D functions
of Feynman[2] integrated over transverse momentum and using x_F instead
of Z as the longitudinal variable.

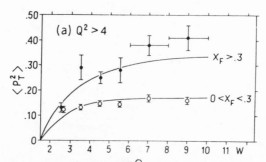

Fig. 7. Average value of P_T^2 vs. W for two X_F regions.

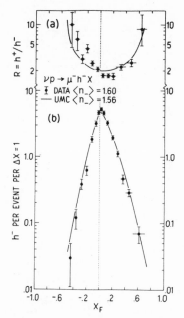

Fig. 8. (a) The (positive/negative)hadron ratio vs. X_F.
        (b) The D_u^+ function vs. X_F. The average number
            of negative tracks (integral of the graph)
            is indicated.

CHARGE DISTRIBUTION

        Field and Feynman suggest (FF2)[3] that when a high-energy quark
fragments, its properties will be most strongly correlated to those
of the highest momentum ("fastest") hadron observed. They have made
predictions for the distribution in Z_R of the fastest positive and
fastest negative hadron in each event, where Z_R is the fraction of
the total charged hadron momentum carried by the fastest charged
hadron. These predictions for a fragmenting u quark are shown in
Fig. 10, and are in good agreement with the data. We also show
the UMC calculation. It is somewhat surprising that the UMC model
gives such a good description, since it has no input charge correlations other than those arising from charge conservation. The
difference between the fastest positive and the fastest negative
distributions in the UMC arises solely from the fact that there are
more positives than negatives generated in each event, and hence
the fastest one is more likely to be positive. This difference between positives and negatives will diminish at higher W values but
the predicted change with W is slow due to the logarithmic increase
in charged particle multiplicity put into the UMC model. In any
case, at our present energies, Fig. 10 does not give evidence for
the charge of a fragmenting u quark, but simply reflects over-all
charge conservation in the events.

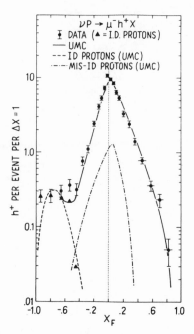

Fig. 9.   The D_u^+ function vs. Feynman-X in the hadronic rest
          frame. Identified (ID) and the estimated mis-
          identified protons are indicated. The norma-
          lization of the UMC model is absolute and not
          adjusted to fit the data.

TRANSVERSE MOMENTUM PROPERTIES

    We now examine the transverse momentum (p_T) properties of the
produced hadrons relative to the over-all hadronic (virtual W^+)
direction in the laboratory. This is of interest in looking for
effects of gluon bremsstrahlung from the struck quark and/or "primordial" parton momentum in the struck proton. Such effects are
conjectured to cause $<p_T>$ to be an increasing function of Q^2 and
Feynman x_F. We see in Fig. 11a that $<p_T>$ is indeed a steep function
of x_F. However, this is primarily due to the kinematic "seagull"
effect" and disappears if we choose another longitudinal variable
such as Y_q (Fig. 11b). The curves of Fig. 11 are from UMC model
calculation and illustrate that the p_T properties of the bulk of
the produced particles can be nicely described by longitudinal phase
space (LPS). If we are to look for any anomalous effects of gluons,
we must first understand and account for the "seagull effect". Not
doing this has often been a source of some confusion in this subject.

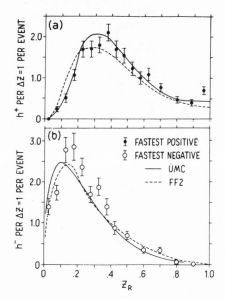

Fig. 10. (a) and (b). The distribution in $Z_R = P_\parallel / P_{visible}$ of the "fastest" positive and nega-tive tracks in each event. The dashed curves are the predictions of Field and Feynman. (a) includes (+) tracks from events where the fast-est track is negative and vice-versa for (b).

The "seagull effect" can be easily understood from Fig. 12 which shows lines of constant c.m. rapidity (Y) on a z_T vs X plot (on the right) and lines of constant X on a z_T vs Y plot (on the left). Imagine that particles populate the plot on the left with a density which is independent of Y (up to the kinematic boundary) but decreasing with the transverse variable $z_T = 2\sqrt{m^2 + p_T^2}/W$ in accor-dance with LPS. If we calculate $\langle p_T \rangle$ vs Y we simply get a flat function which tails off at large Y due to kinematic boundary. However, if we plot $\langle p_T \rangle$ vs X for the same data we get a "seagull". This is because, looking at the two adjacent horizontal bars on Fig. 12, at a given p_T (or z_T) the production rate is proportional to ΔY, and ΔY is a decreasing function of X for a fixed ΔX ($\Delta Y = \Delta X/\sqrt{X^2 + z_T^2}$). That is, the inner bar (X = 0 to 0.1) contains more particles than the one adjacent to it (X = 0.1 to 0.2) and therefore, low p_T is more heavily weighted in the low X interval. Hence, the "seagull effect" arises simply from LPS and the transformation of variables giving $dY = dX/\sqrt{X^2 + z_T^2}$. It even produces the surprising result that $\langle p_T \rangle$ (X = 0) is less than $\langle p_T \rangle$ (Y = 0) along the same line X = Y = 0!

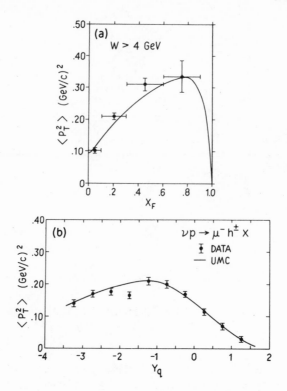

Fig. 11 (a) Average transverse momentum squared vs.
Feynman-X for charged particles in the $X_F>0$
(current fragmentation) hemisphere. The seagull effect, seen here is discussed in detail
in the text. (b) $\langle P_T^2 \rangle$ vs. rapidity in the quark
frame, $Y_q = Y_{lab} - \ln (W^2/1 \text{ GeV}^2)$. $X_F>0$ corresponds roughly to $Y_q > -2$.

Turning back to Fig. 11a, we see that the data show no significant X dependent increase of $\langle p_T \rangle$ above what is expected from
LPS (solid line). To look further for effects of gluon emission
we take events in which the highest momentum (h_1) track is positive
and therefore most closely related kinematically to the parent u
quark. If we plot $\langle p_T \rangle$ for these particles vs Q^2, we do see a rise,
but it is practically all accounted for by a similar rise in the
UMC calculations (Fig. 13). The QCD expected linear rise is only
vaguely hinted at with a coefficient $d\langle p_T^2 \rangle/dQ^2 \approx 0.003 \pm 0.003$.
Still another place where gluon radiation or primordial parton
momentum should show up is in the fragmentation region of the target
proton which has been stripped of one of its quarks. This kinematic
region is best defined in the lab frame where the target is initially
at rest. We again focus our attention on positive particles since

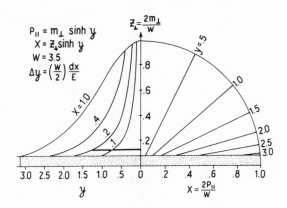

Fig. 12.   Kinematic plot to illustrate the seagull
effect (see text).

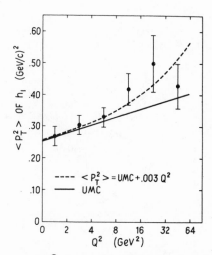

Fig. 13. $\langle P_T^2 \rangle$ vs. Q^2 for h_1 tracks with $Z > 0.3$ in
events where h_1 is $(+)$. There are 497 events
(and 497 tracks) in the plot with average Z
value $\langle Z \rangle = 0.45$. Events with $3\,\text{GeV} < W < 12\,\text{GeV}$
are used. The dashed curve is a fit to a
linear function of Q^2 which has been added to
the baseline curve (solid) calculated with the
UMC model.

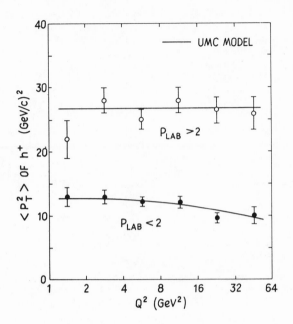

Fig. 14.   Average transverse momentum-squared of positive
hadrons in two P_{lab} regions.   Events with W
between 4 and 12 GeV are used.

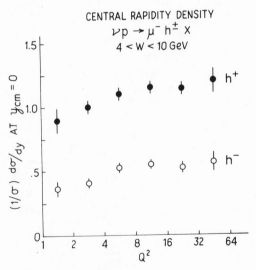

Fig. 15.   Particle density near $Y = 0$ as a function of Q^2.

we are looking for fragments of a uu diquark, and restrict the particle momentum to $P_{lab} < 2$ GEV/c. The $<p_T^2>$ for these particles (and also those with $P_{lab} > 2$ GeV/c) is plotted vs Q^2 in Fig. 14. Again we find no Q^2 dependent effects up to $Q^2 = 64$ GeV^2; the data follow the UMC model extremely well. (There are 337 particles from 180 events in the $16 < Q^2 < 64$ region of the $P_{lab} < 2$ GeV/c data.) If, indeed, the proton contains primordial parton transverse momentum of order $0.8 - 0.9$ GeV/c, as some hadronic experiments have indicated, it does not appear to manifest itself in these neutrino data. We may be just on the verge of having high enough Q^2 to see these effects, if indeed they are there. Hopefully, other deep inelastic and e^+e^- annihilation experiments will settle these questions in the near future.

FRAGMENTATION OF GLUONS?

Virtual gluons, if they "exist", are supposed to produce hadron jets just as virtual quarks do. If the struck quark in our case emits a gluon one might expect two separate jets to develop in the current fragmentation region. These two separate jets would be difficult to observe at our energies, but one might expect that the density of particles in phase space would increase significantly in such events, and that their frequency would increase with Q^2. Hence, we show, in Fig. 15, the density of hadrons near $Y = 0$ as a function of Q^2. After an initial rise there seems to be very little Q^2 dependence. Again we must conclude that either the effects of gluon radiation are quite small, or we need higher Q^2 values to see them.

ACKNOWLEDGEMENTS

I am much indebted to my colleagues of the Berkeley-Hawaii-Fermilab-Michigan E45 collaboration for a lot of hard work in extracting and analyzing these data. In addition, I have benefited particularly from conversations with J.D. Bjorken, R.N. Cahn, S.D. Ellis, R.D. Field, G.L. Kane, and H.D. Politzer.

I am most appreciative of the kind hospitality of the "Ettore Majorana" Center and of the efforts of the seminar organizers, J.J. Aubert and G. Preparata. Anyone must come away from Erice with a beautiful lasting impression.

REFERENCES

1. J. Bell, J.P. Berge, D.V. Bogert, R.J. Cence, C.T. Coffin,
        R.N. Diamond, F.A. DiBiance, R. Endorf, H.T. French, R. Hanft
        F.A. Harris, M. Jones, C. Kochowski, W.C. Louis, G.R. Lynch,
        J.A. Malko, J.P. Marriner, G.I. Moffatt, F.A. Nezrick,
        M.W. Peters, V.Z. Peterson, B.P. Roe, R.T. Ross, W.G. Scott,
        A.A. Seidl, W. Smart, V.J. Stenger, M.L. Stevenson,
        J.C. Vander Velde and E. Wang, Experimental Study of Hadrons
        Produced in High Energy Charged Current Neutrino-Proton
        Interactions, Phys. Rev. D19:1-19 (1979)
2. R.P. Feynman, Photon-Hadron Interactions, "Frontiers in Physics"
        series, David Pines, ed., W.A. Benjamin, Reading, Mass. (1972)
3. R.D. Field and R.P. Feynman, Nucl. Phys. B136:1 (1978).

PROBING NEUTRAL CURRENTS WITH LEPTONS

L.M. Sehgal

III. Physikalisches Institut, Technische Hochschule

Aachen, Germany

1. INTRODUCTION

There is now a considerable consensus[1-3] that the Weinberg-
Salam (W-S) model describes rather closely the properties of the
neutral current interaction revealed by a large number of neutrino
experiments as well as the recent measurements with polarized
electron beams. This development enhances the credibility of the
gauge theory principles underlying this model, and influences
profoundly our thinking about the properties of the quanta that
mediate the weak interactions, the behaviour of weak amplitudes
at high energies, and the possible synthesis of the weak, electromagnetic and strong forces. Because of these implications, it is
important that we examine the evidence carefully, not only to
locate areas of weakness or uncertainty, but also to detect small
but potentially significant departures from the theoretical
expectations. Equally important, we must ask[4] whether the present
successes of the standard theory can be rationalized by arguments
that are not based on the usual principles of weak electromagnetic
unification and spontaneous symmetry breaking.

In this talk, I shall review briefly the status of our
knowledge about the neutral current couplings among neutrinos,
electrons and quarks, and the degree to which the evidence
supports the W-S model. I shall then outline some ideas (due to
Bjorken[4] and Hung and Sakurai[5]) that show how an effective neutral
current interaction of precisely the W-S form can be motivated
by alternative arguments outside the gauge theory framework.
Finally, I shall discuss the prospects of using electron-proton
colliding beams to determine the propagator of the weak boson(s)
mediating the neutral current interaction, thus revealing the
range of the force responsible for neutral current phenomena.

2. EVIDENCE FOR THE W-S MODEL

According to the W-S model, neutral current phenomena are
governed by a single neutral boson Z with the interaction

$$\mathcal{L} = \frac{g}{\cos\Theta} \left[J_\alpha^3 - \sin^2\Theta \, J_\alpha^{em} \right] Z^\alpha \tag{1}$$

where g, the coupling constant associated with weak isospin, is
related to the electric charge by

$$g = e/\sin\Theta . \tag{2}$$

The mass spectrum of intermediate bosons is

$$M_W = 37.3 \text{ GeV}/\sin\Theta \quad , \quad M_Z = 37.3 \text{ GeV}/\sin\Theta\cos\Theta . \tag{3}$$

In addition, a neutral spin 0 particle (the Higgs boson) is
predicted, but with an unknown mass and couplings that play a
negligible role in most low energy phenomena.

Existing measurements have probed only the interaction between fermions in situations where Q^2/M_Z^2 is negligible. The effective Lagrangian resulting from Eqs. (1), (2) and (3) is

$$\mathcal{L}_{eff} = -4 \frac{G}{\sqrt{2}} \left[J_\alpha^3 - \sin^2\Theta \, J_\alpha^{em} \right]^2 \qquad (4)$$

and determines, in particular, the mutual couplings between neutrinos, electrons and quarks. A convenient parametrization of these couplings is as follows.

$$\mathcal{L}_{eff} (\nu q) = -\frac{G}{\sqrt{2}} \, \bar{\nu}\gamma_\alpha(1+\gamma_5)\nu \left[\bar{u}\gamma_\alpha\{u_L(1+\gamma_5) + u_R(1-\gamma_5)\}u \right.$$
$$\left. + \ \ldots \right] \qquad (5)$$

$$\mathcal{L}_{eff} (\nu e) = -\frac{G}{\sqrt{2}} \, \bar{\nu}\gamma_\alpha(1+\gamma_5)\nu \left[\bar{e}\gamma_\alpha\{c_L(1+\gamma_5) + c_R(1-\gamma_5)\}e \right] \qquad (6)$$

$$\mathcal{L}_{eff} (eq) = -\frac{G}{\sqrt{2}} \, \bar{e}\gamma_\alpha(1+\gamma_5)e \left[\bar{u}\gamma_\alpha\{\epsilon_{LL}(u) \ (1+\gamma_5) \right.$$
$$\left. +\epsilon_{LR}(u) \ (1-\gamma_5)\}u + \ldots \right]$$
$$-\frac{G}{\sqrt{2}} \, \bar{e}\gamma_\alpha(1-\gamma_5)e \left[\bar{u}\gamma_\alpha\{\epsilon_{RL}(u) \ (1+\gamma_5) \right.$$
$$\left. +\epsilon_{RR}(u) \ (1-\gamma_5)\}u + \ldots \right] \qquad (7)$$

(We specify the couplings of the u quark only, the dots (.....) representing analogous couplings for other quarks). It is a measure of the predictive strength of the W-S model that all of the above couplings are determined in terms of just one unknown constant, $\sin^2\Theta$. The neutrino couplings are ($x \equiv \sin^2\Theta$)

$$u_L = \frac{1}{2} - \frac{2}{3}x \quad , \quad u_R = -\frac{2}{3}x$$

$$d_L = -\frac{1}{2} + \frac{1}{3}x \quad , \quad d_R = \frac{1}{3}x \tag{8}$$

$$c_L = -\frac{1}{2} + x \quad , \quad c_R = x$$

The electron-quark couplings are related to the neutrino couplings by "factorization" relationships:

$$\varepsilon_{LL}(u) = 2\,c_L u_L \quad , \quad \varepsilon_{LR}(u) = 2\,c_L u_R$$

$$\varepsilon_{RL}(u) = 2\,c_R u_L \quad , \quad \varepsilon_{RR}(u) = 2\,c_R u_R \tag{9}$$

etc.

Our empirical knowledge of the various couplings is summed up in the following section.

a. Neutrino-Quark Sector:

Measurements of the inclusive cross-sections for $\nu(\bar{\nu})\,N \to \nu(\bar{\nu})\,X$, in comparison to those of the charged current reactions $\nu(\bar{\nu})N \to \mu^-(\mu^+)\,X$, using isoscalar targets, have determined the combinations[1]

$$u_L^2 + d_L^2 = 0.29 \pm 0.02 \quad , \quad u_R^2 + d_R^2 = 0.03 \pm 0.01 . \tag{10}$$

Separation of the u- and d-coupling strengths has been effected by examining the π^+/π^- ratio in leading hadrons[6], by comparing the cross-section on protons with that on neutrons[7], and by measuring the NC/CC ratio in hydrogen[8]. Averaging such measurements, one finds

$$u_L^2 \;=\; 0.11 \pm 0.03 \qquad\qquad d_L^2 \;=\; 0.18 \pm 0.03$$

$$u_R^2 \;=\; 0.03 \pm 0.015 \qquad\qquad d_R^2 \;=\; 0 \quad\;\; \pm 0.015 \qquad (11)$$

The relative signs of the couplings have been established by the study of exclusive channels. Thus, the observation of a strong Δ-signal in $\nu p \to \nu\Delta$[9] implies a dominantly isovector current, and in particular

$$u_L d_L \;=\; \text{negative} \;. \qquad (12)$$

Similarly, an analysis of the elastic channel $\nu p \to \nu p$[10], while confirming the pattern (11), reveals that

$$u_L u_R \;=\; \text{negative} \;. \qquad (13)$$

The total picture produced by these measurements is shown in Fig. 1*, and is in excellent accord with the W-S model, for $\sin^2\theta = 0.24 \pm 0.02$.

Impressive as this agreement is, there is room for further investigation. As is evident from Eq. (11) and Fig. 1, the right-handed couplings are poorly determined, and to that extent, the theory is not incisively tested. Also, an important selection rule, namely the absence of an isoscalar axial current ($u_L + d_L - u_R - d_R = 0$)[9] is not tested in a direct way. Finally, there exists a measurement of the relative strength of the one-pion channels $\nu p \to \nu p \pi^0$, $\nu n \to \nu n \pi^0$, $\nu p \to \nu n \pi^+$ and $\nu n \to \nu p \pi^-$ which awaits explanation. The issue is important because the observed difference between $p\pi^0$ and $n\pi^0$ (and between $p\pi^-$ and $n\pi^+$) is a direct measure of the

---

*The over-all sign, unmeasurable in neutrino experiments, has been chosen so that $u_L > 0$ in agreement with the W-S model.

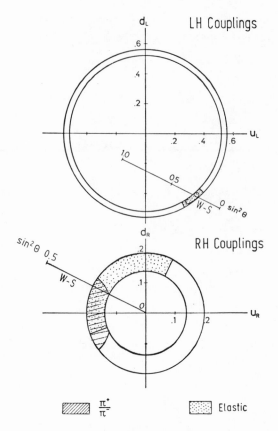

Fig. 1: Allowed domain of neutrino-quark neutral current couplings.

interference between the isoscalar and isovector parts of the neutral current, and one ought to know if at least the sign of the observed asymmetry [e.g., $\sigma(p\pi^{o}) > \sigma(n\pi^{o})$] is consistent with expectations.

These remarks notwithstanding, one must acknowledge that the degree of agreement between the observed νq couplings and the W-S model is highly non-trivial. It should be recalled that variations of the model in which the right-handed u and/or d

quarks are placed in doublets are excluded. Finally, if interaction
(5) is modified by a multiplicative factor ρ (as would be the case
if $M_W^2/M_Z^2 \cos^2\theta = \rho$ instead of unity), the data yield[1]

$$\rho = 0.98 \pm 0.05$$

$$\sin^2\theta = 0.23 \pm 0.03$$

(14)

From the point of view of gauge theories, the closeness of ρ to
unity is a proof of the $I = 1/2$ nature of the Higgs boson in the
W-S model.

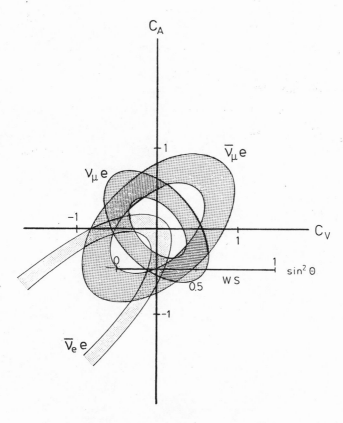

Fig. 2: Allowed domain of neutrino-electron couplings.

b. Neutrino-Electron Sector

It is gratifying that the cross sections for $\nu_\mu e$ and $\bar{\nu}_\mu e$ scattering, measured by several groups now converge to consistent values[11], and constrain the parameters C_V ($= c_L + c_R$) and C_A ($= c_L - c_R$) to the intersection of the ellipses shown in Fig. 2. Together with the reactor measurement of $\bar{\nu}_e e$ [11], the data yield two solutions

$$C_V = 0.0 \pm 0.1 \,, \qquad C_A = -0.5 \pm 0.1 \qquad \text{(a)}$$

and $\qquad\qquad\qquad\qquad\qquad\qquad\qquad\qquad\qquad\qquad\qquad$ (15)

$$C_V = -0.5 \pm 0.1 \,, \qquad C_A = 0.0 \pm 0.1 \qquad \text{(b)}$$

of which the first agrees with the W–S model for $\sin^2\theta = .25 \pm .05$. Curiously, the second solution is compatible with a variant in which $\bar{e}_R$ is placed in a doublet $(E_R^o, \bar{e}_R)$. Despite the limited statistics of these measurements the data are quite restrictive. For example, variations of the W–S model in which the Higgs particle has an exotic isospin $I \geq 1$, or in which $|I_3(e_R)| \geq 1$ can be excluded[12].

c. Electron-Quark Sector

The recent experiment at SLAC[13] has determined an asymmetry between the scattering of left and right-handed electrons on a deuterium target. This asymmetry is interpreted as an interference between the one-photon and Z-exchange amplitudes, and is given by

$$A^{ed} = \frac{\sigma_R - \sigma_L}{\sigma_R + \sigma_L} = \frac{4 G Q^2}{\sqrt{2}\, e^2} \frac{9}{5} \left[(\tfrac{2}{3} u_V - \tfrac{1}{3} d_V) C_A \right.$$

$$\left. + (\tfrac{2}{3} u_A - \tfrac{1}{3} d_A) C_V \frac{q - \bar{q}}{q + \bar{q}} f(y) \right]$$

$\qquad\qquad\qquad\qquad\qquad\qquad\qquad\qquad\qquad\qquad\qquad\qquad\qquad\qquad$ (16)

where q, $\bar{q}$ are the quark and antiquark densities, and
$f(y) = \left[1 - (1-y)^2\right]/\left[1 + (1-y)^2\right]$. If we insert the values of
$u_{V,A}$ and $d_{V,A}$ determined from neutrino data (Fig. 1), the observed
asymmetry A^{ed} translate into an allowed domain in the C_V-C_A
plane. As seen in Fig. 3, the data prefer the solution (a) of the
neutrino-electron sector, and hence the minimal version of the
W-S model. More directly, the observed asymmetry, when compared
with the prediction of the W-S model, gives

$$\sin^2\theta = 0.20 \pm 0.03 \qquad (17)$$

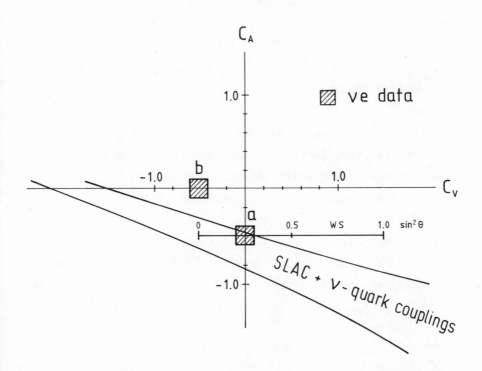

Fig. 3: Constraint on C_V and C_A imposed by parity-violating
        asymmetry in ed scattering together with neutrino-quark
        couplings. (Factorization relations Eq. (9) are assumed.)

[At this meeting, Taylor[14] reported a more precise result based
on measurements of A^{ed} at several values of y:

$$\sin^2\theta = 0.224 \pm 0.012 \quad (\pm 0.02 \text{ systematic})] \ .$$

To summarize, the effective neutral current interaction predicted by the W-S model has received confirmation from several independent
directions, the data pointing to a value of $\sin^2\theta \simeq 0.23 \pm 0.02$.
Clearly, the next challenge is to hunt for the weak quanta $W^{\pm}$ and
Z, predicted to occur with masses $M_W = 78 \pm 3$ GeV, $M_Z = 89 \pm 3$ GeV.
In the next section, however, we discuss a viewpoint according to
which the low energy properties of the neutral current interaction
can be reconciled with a significantly different pattern of masses
for the weak intermediate bosons and for the behaviour of weak
amplitudes at high energies.

3. ALTERNATIVES TO THE W-S MODEL

Bjorken[4] has stressed that the vital tests of gauge theories
(and of weak-electromagnetic unification) lie in the domain of
intermediate bosons and Higgs particles, a domain that remains to
be discovered and explored. It is therefore pertinent to ask
whether the existing evidence on neutral currents admits an
alternative explanation that, while not rooted in a renormalizable
framework, remains open to the possibility of unexpected behaviour
at high energies. As an example of an alternative to the W-S model,
Bjorken considers an effective weak interaction of the form[15]

$$\mathcal{L}^{\text{weak}}_{\text{eff}} = -4 \frac{G}{\sqrt{2}} \vec{J}_\alpha \cdot \vec{J}_\alpha \tag{18}$$

where $\vec{J}_\alpha$ is a weak isospin current

$$\vec{J}_\alpha = \sum_i \bar{\psi}_i \gamma_\alpha \frac{1 + \gamma_5}{2} \frac{\vec{\tau}}{2} \psi_i$$

$$\psi_i = \begin{pmatrix} \nu_e \\ e^- \end{pmatrix}, \begin{pmatrix} \nu_\mu \\ \mu^- \end{pmatrix}, \begin{pmatrix} u \\ d' \end{pmatrix}, \begin{pmatrix} s \\ c' \end{pmatrix} \quad \text{etc.} \tag{19}$$

Observed neutral current phenomena are attributed to the $J_3 J_3$ part of the above interaction, together with the induced effects of single-photon exchange between the interacting fermions. For neutrino interactions, in particular, the effective neutral current Lagrangian takes the form

$$\mathcal{L}_{eff}(\nu) = -\frac{G}{\sqrt{2}} \bar{\nu}\gamma_\alpha (1+\gamma_5)\nu \left[2 J_\alpha^3 - \frac{e^2}{\sqrt{2}\,\Lambda^2 G} J_\alpha^{em} \right] \tag{20}$$

where we define the neutrino-photon vertex as $(\bar{\nu}\gamma_\alpha \frac{1+\gamma_5}{2} \nu)$ eq^2/Λ^2. This has precisely the structure and strength of the W-S Lagrangian, provided one identifies

$$\sin^2\theta \equiv \frac{e^2}{\sqrt{2}\,\Lambda^2 G} \tag{21}$$

Thus all of the successes of the standard theory are duplicated in this simple phenomenological scheme.

The same idea can be illustrated in an intermediate boson model proposed by Hung and Sakurai[5] (following Bjorken[4]). We imagine again the weak and electromagnetic interactions to be distinct (non-unified), and to be described by the interaction Lagrangian

$$\mathcal{L} = g \vec{J}_\alpha \cdot \vec{W}_\alpha + e J_\alpha^{em} A_\alpha , \tag{22}$$

where $\vec{W}$ is a triplet of weak bosons (of mass M_W) and A is the massless photon. Let us add to this system a mixing term of the form

$$\mathcal{L}_{mix} = -\frac{1}{4} \lambda \left[F_{\mu\nu} W^{(3)\mu\nu} + W_{\mu\nu}^{(3)} F^{\mu\nu} \right] \tag{23}$$

where $F_{\mu\nu}$ is the electromagnetic field tensor $\partial_\mu A_\nu - \partial_\nu A_\mu$, and $W_{\mu\nu}^{(3)}$ the corresponding tensor for the neutral W. As a consequence of this mixing, the mass spectrum of the neutral bosons will be perturbed. The inverse propagator matrix of the W^o–γ system is

$$\Delta^{-1}(q) = \begin{bmatrix} q^2 & \lambda q^2 \\ \lambda q^2 & q^2 - M_W^2 \end{bmatrix} , \tag{24}$$

and the perturbed masses are given by the roots of the equation $\det \left[\Delta^{-1} \right] = 0$. One thus obtains

$$M_1^2 \equiv M_\gamma^2 = 0$$

$$M_2^2 \equiv M_Z^2 = M_W^2/(1-\lambda^2) . \tag{25}$$

The fact that the photon stays massless is guaranteed by the gauge invariance of Eq. (23) with respect to the electromagnetic field. On the other hand, the massive neutral boson Z is shifted up in mass with respect to its charged partners $W^\pm$.

We can now examine the effective interaction due to the propagation of γ and Z. This is conveniently written as

$$\mathcal{L}_{eff} = \frac{1}{2} J^+ \Delta J \tag{26}$$

where $J = \begin{pmatrix} e J_\lambda^{em} \\ g J_\lambda^3 \end{pmatrix}$. In full detail,

$$\mathcal{L}_{eff} = \frac{1}{2} \left\{ e^2 J^{em}_\lambda \frac{1}{q^2} J^{em}_\lambda + \frac{g^2}{1-\lambda^2} (J^3_\lambda - \frac{\lambda e}{g} J^{em}_\lambda) \frac{1}{q^2 - M_Z^2} (J^3_\lambda - \frac{\lambda e}{g} J^{em}_\lambda) \right\}$$

(27)

The first term is recognized as the standard electromagnetic interaction, while the second is the neutral current term. In the local limit $q^2 \to 0$, the latter becomes <u>identical</u> to the W-S Lagrangian given in Eq. (4), provided one defines

$$\frac{\lambda e}{g} \equiv \sin^2\Theta$$

(28)

At the same time, the mass spectrum of this model is <u>different</u> from that of the standard theory: the masses of W and Z are not separately fixed by the value of the parameter $\sin^2\Theta$ appearing in the structure of the neutral current. Instead, we have the more general mass relation

$$(\frac{37.3 \text{ GeV}}{M_W})^2 - (\frac{37.3 \text{ GeV}}{M_Z})^2 = \sin^4\Theta \quad .$$

(29)

Curiously, it is possible to recover the mass relations of the W-S model by a postulate of "asymptotic SU(2) symmetry"[4,5] on the effective Lagrangian [eq. (27)]. This is the requirement that as $q^2 \to \infty$, $\mathcal{L}_{eff}$ be free of any cross term $J^3 J^Y$ (where $J^{em} = J^3 + J^Y$). Such a condition is satified provided λ is chosen to be

$$\lambda = e/g \quad .$$

(30)

In this situation, we obtain both the "unification condition"

$$e = g \sin\Theta$$

(31)

and the W-S mass spectrum

$$M_W = 37.3/\sin\Theta \quad , \qquad M_Z = 37.3/\sin\Theta \cos\Theta$$

(32)

While the present model makes no pretense of being a renormalizable
theory, it is certainly intriguing that two of the most important
features of the W-S model (the structure of the NC interaction and
the mass spectrum of W and Z) are reproduced without any reference
to the concept of Higgs particles and spontaneous symmetry breaking.

Bjorken[4] has also considered generalizations of the above
models to cases where the weak quanta contain not only discrete
states, but a continuum as well, including the possibility that
these quanta may be a strongly interacting system or composites
of new types of "quarks". Even within so general a framework, it
is possible to derive the effective NC interaction of the standard
model. At the same time, such an approach allows for unusual and
complicated phenomena beyond the threshold for producing the weak
quanta.

4. SEARCH FOR PROPAGATOR EFFECTS: ELECTRON-PROTON COLLIDING BEAMS

It is evident that the next major test of our ideas about
the electro-weak interactions will take place in the arena of the
intermediate bosons. Hence the utmost importance of attempts to
produce these particles in $\bar{p}p$, pp and e^+e^- collisions. There is
interest, too, in detecting the propagator effects of these quanta
by studying lepton-hadron interactions at very high Q^2. In this
connection, considerable promise is offered by electron-proton
colliding beams of the type currently under discussion[16]. A
typical project envisages 20 GeV electrons colliding against
250 GeV protons to produce c.m. energies s = 20,000 GeV^2. This
opens up the prospect of studying weak-electromagnetic interference
in ep → eX at momentum transfers of several thousand GeV^2, in
the manner of the SLAC parity-violation experiment. The interference

effect is nominally of order

$$\frac{G\ Q^2}{\sqrt{2}\ e^2} \simeq 9 \times 10^{-5}\ Q^2\ \text{GeV}^{-2} \tag{33}$$

which is $\gtrsim 10\%$ for $Q^2 \gtrsim 10^3$ GeV^2. Thus large asymmetries may be
expected between left and right-handed electrons (and between
electrons and positrons). In this circumstance it may be possible
to discern the presence of the Z-propagator, which will modify
the asymmetry by the factor

$$\eta(Q^2) = \frac{M_Z^2}{M_Z^2 + Q^2} . \tag{34}$$

Let us recall briefly the theory of $Z - \gamma$ interference in
$ep \rightarrow eX$ at high energies. The inclusive cross-section may be
written as

$$\frac{d\sigma}{dxdy} = \frac{2\pi\alpha^2}{Q^4}\ s\ \left[F^{em} + F^{int} + F^{weak} \right] \tag{35}$$

where F^{em}, F^{int} and F^{weak} are the electromagnetic, interference
and weak terms. Of these, F^{int} and F^{weak} depend on the helicity
and charge of the incident lepton and on the structure of the NC
interaction. Denoting by e_i the charge of quark i and by $f_i(x)$
the corresponding parton distribution (x times the parton density),
the three terms take the form

$$F^{em} = \sum_i e_i^2 \left[f_i(x) + \bar{f}_i(x) \right] \left[1 + (1-y)^2 \right]$$

$$F^{int} = -\frac{8GQ^2}{\sqrt{2}\ e^2}\ \eta \sum_i e_i \left[A_i \left\{ f_i(x) + \bar{f}_i(x)\ (1-y)^2 \right\} \right.$$

$$\left. + B_i \left\{ \bar{f}_i(x) + f_i(x)\ (1-y)^2 \right\} \right]$$

$$F^{weak} = \left(\frac{4GQ^2}{\sqrt{2} \, e^2} \, \eta\right)^2 \sum_i \left[A_i^2 \left\{ f_i(x) + \bar{f}_i(x) \, (1-y)^2 \right\} \right.$$
$$\left. + B_i^2 \left\{ \bar{f}_i(x) + f_i(x) \, (1-y)^2 \right\} \right] \qquad (36)$$

The coefficients A_i, B_i are given by the coupling constants of the electron-quark interaction defined in Eq. (7). Their values, for the four possible states of the incident lepton, are given below.

	A_i	B_i
e_L^-	$\varepsilon_{LL}(i)$	$\varepsilon_{LR}(i)$
e_R^-	$\varepsilon_{RR}(i)$	$\varepsilon_{RL}(i)$
e_L^+	$\varepsilon_{RL}(i)$	$\varepsilon_{RR}(i)$
e_R^+	$\varepsilon_{LR}(i)$	$\varepsilon_{LL}(i)$

In Fig. 4a, we show the event rates that may be expected in two typical ep colliders, as a function of Q^2. (These have been evaluated using Q^2-dependent parton densities of the type predicted by QCD [17].) Even for the machine with s = 20,000 GeV^2, one can penetrate to momentum transfers as high as $Q^2 \approx$ 5000 GeV^2, with reasonable rates. Figure 4b illustrates how significant the weak electromagnetic and purely weak contributions are, relative to the one-photon exchange cross-section. The large interference effect manifests itself in widely different cross sections for e_L^-, e_R^-, e_L^+ and e_R^+, as shown in Fig. 5a for the case of the W-S model. The resulting left-right asymmetry is displayed in Fig. 5b. Note

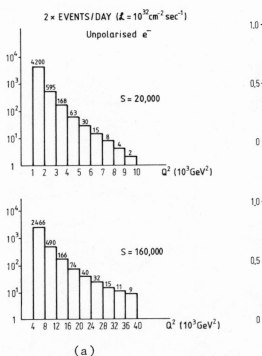

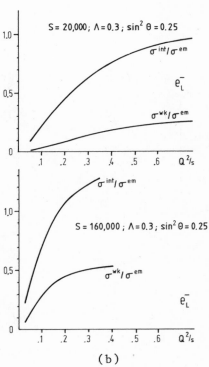

(a)   (b)

Fig. 4: (a) Event rates for ep → eX expected in two typical e p
         colliders: 20 e × 250 p (s = 20,000 GeV^2) and
         100 e × 400 p (s = 160,000 GeV^2).

       (b) Magnitude of weak-electromagnetic interference and
           purely weak contribution, normalised to the one-
           photon exchange cross-section.

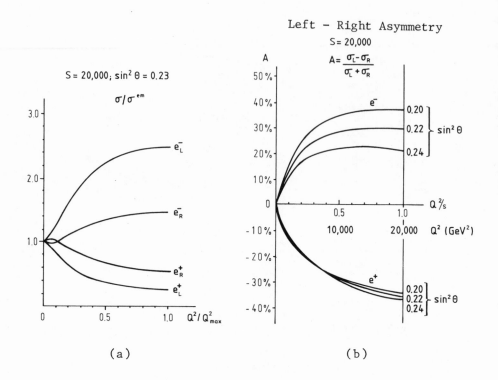

Fig. 5: (a) Cross sections, normalized to one-photon exchange, for
the four types of beam e_L^-, e_R^-, e_L^+ and e_R^+, assuming
$\sin^2\Theta = 0.23$, and the parton distributions of Buras
and Gaemers.[17]

(b) Asymmetry between cross-section of left- and right-
handed leptons, as a function of Q^2, in the W-S model.

Left – Right Asymmetry

$\sin^2\theta = .23$; S = 20,000

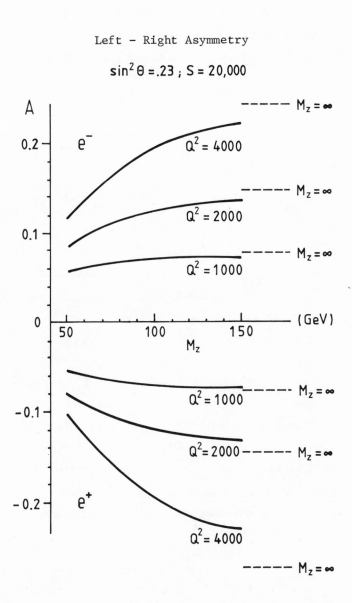

Fig.6: Dependence of asymmetry on mass of Z boson, assuming neutral
current couplings to be those of the W-S model.

particularly that the asymmetry is linear in Q^2 for low momentum
transfers, but flattens out as a result of the Z-propagator when
Q^2 becomes comparable to M_Z^2. It is precisely this departure from
linear behaviour that can serve as a measure of M_Z.

To see the effectiveness of such measurements in determining
M_Z, we have plotted in Fig. 6 the asymmetry as a function of M_Z
for some typical values of Q^2. The neutral current couplings have
been assumed to be those of the W-S model (with $\sin^2\theta = 0.23$)
but the mass M_Z is assumed unknown (as for instance in the Hung-
Sakurai alternative). One can conclude from this figure that an
effective boson mass M_Z as high as 150 GeV is detectable, and a
mass in the neighbourhood of 90 GeV could probably be determined
with an accuracy of 5 - 10 GeV. In any final analysis one must,
of course, take into consideration the uncertainties introduced
by the assumptions on the parton densities etc.. These, however,
appear to be controllable.[16]

To conclude, it would seem that a meaningful measurement of
the Z-propagator, and hence the range of the force mediating
neutral current interactions, is possible with ep colliding beams.
The correlation of such measurements with the information likely
to be delivered by $\bar{p}p$, pp and e^+e^- machines on the properties of
the weak quanta, will undoubtedly bring our understanding of the
weak interactions to a new and more profound level.

References

1. L.M. Sehgal, Proc. of the "Neutrinos 78" Conference, Purdue,
   Ed. E.C. Fowler (Purdue University, 1978) p. 253.

2. J. Sakurai, Proc. of the Oxford Conference on Neutrino Physics
   (Rutherford Lab., 1978); UCLA /78/TEP/18.

3. C. Baltay, Proc. of the Tokyo Conference on High Energy Physics, Ed. S. Homma et al. (Physical Society of Japan 1979), p. 882.

4. J. Bjorken, SLAC-PUB-2062 (1977) and SLAC-PUB-2133 (1978).

5. P. Hung and J. Sakurai, UCLA-78-TEP-8 (1978).

6. L.M. Sehgal, Phys. Letters 71B, 99 (1977);
   L. Abbot and R.M. Barnett, SLAC-PUB-2136 (1978).

7. G. Rajasekaran and K.V.L. Sarma, Tata Institute preprint TIFR/TH/78-60;
   L. Abbot and R.M. Barnett, SLAC-PUB-2227 (1978).

8. D. Haidt, these Proceedings.

9. W. Krenz et al., Nucl. Phys. B135, 45 (1978);
   D. Erriques et al., Phys. Lett. 73B, 350 (1978).

10. M. Claudson, E.A. Paschos and L.R. Sulak, report presented at the Tokyo Conference (1978);
    P.Q. Hung and J. Sakurai, Phys. Lett. 72B, 208 (1977);
    G. Ecker, Phys. Lett. 72B, 450 (1978).

11. H. Faissner et al., Phys. Rev. Lett. 41, 213 (1978);
    J. Blietschau et al., Phys. Lett. 73B, 232 (1978);
    C. Baltay et al., Phys. Rev. Lett. 41, 357 (1978);
    N. Armenise et al., CERN/EP/79-38 (1979);
    F. Reines et al., Phys. Rev. Lett. 37, 315 (1976).

12. H. Faissner, Proc. B.W. Lee Memorial Conf. 1977, ed. D. Cline (FNAL, Batavia, 1978) p. 85.

13. C. Prescott et al., Phys. Lett. 77B, 347 (1978).

14. R. Taylor, these Proceedings.

15. S. Bludman, Nuovo Cimento 9, 443 (1958)

16. C.H. Llewellyn Smith and B.H. Wiik, DESY 77/38 (1977);
    CHEEP, CERN Yellow Report 78-02 (1978), Eds. J. Ellis et al.;
    contributions to the ECFA-DESY meeting on Study of an ep
    facility for Europe, April 1979.

17. A. Buras and K. Gaemers, Nucl. Phys. B132, 249 (1978).

JETS IN LEPTOPRODUCTION FROM QCD

P. Binétruy and G. Girardi

LAPP, Annecy-le-Vieux, France

INTRODUCTION

If quarks and gluons are confined, the best way to "see" them is to study the jets of particles. However, while the domain of perturbative QCD has greatly expanded for the last few years, the precision of its predictions has certainly suffered from our lack of methods to handle the non-perturbative features of the theory. In the following, we shall study what perturbative QCD can predict, and also what it cannot. So far, the rule of the game is to find tricks which eliminate non-perturbative effects as much as possible.

Perturbative QCD is a theory of quarks and gluons. One immediately faces two major problems:
- gluons are massless; how do we deal with the infra-red problems?
- physical particles are bound states of quarks and gluons; how do we make the passage from the parton world to hadrons?

The infra-red problem is very similar to the one already extensively studied in electrodynamics. When one computes cross-sections the masslessness of gluons gives rise to divergences of two different kinds. The first one is the infra-red divergence that comes from the emission of soft gluons by a quark. Infinities of this kind are cancelled order by order by the corresponding virtual corrections. For example, the contribution of diagrams shown in Fig. 1a becomes infinite when the mass of the gluon goes to zero but the total contribution of Fig. 1 is finite.

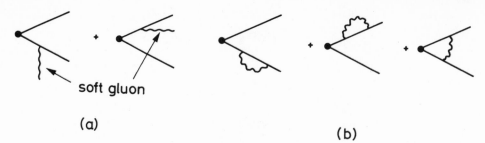

(a)

(b)

Fig. 1   First order QCD corrections to the emission of two quarks
         with (a) one real or (b) virtual radiated soft gluon.

The second kind of infinity is the so-called mass or collinear
singularity; it corresponds to the emission of collinear gluons
(as shown in Fig. 2). The physical meaning of such a singularity
is straightforward: one cannot distinguish between one zero mass
particle of momentum p (i.e., a quark) and two zero mass particles
of momenta xp and (I - x)p, respectively (i.e., a quark and a gluon).
Similarly, the way to get rid of these singularities is obvious:
these infinities cancel in the distributions of sufficiently inclusive variables; that is sufficiently inclusive in order not to
distinguish between the two preceding processes[1].

This required feature provides as a bonus an attack on the
second problem faced by perturbative QCD: the hadronization of
quarks and gluons. For a quantity to be independent of this hadronization process, it must only depend on the properties of a
jet as a whole and this condition has been explained in the preceding paragraphs.

We are now in a position to discuss the suitability of different variables. First of all, sphericity is a good example of
the limitations of perturbative QCD. It is defined as follows[2]

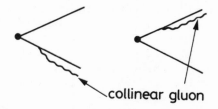

collinear gluon

Fig. 2   First order QCD corrections to the emission of two quarks
         with one gluon radiated collinearly.

$$\hat{S} = \frac{3}{2} \min \frac{\sum\limits_{i} \left| \vec{p}_\perp^{\,i} \right|^2}{\sum\limits_{i} \left| \vec{p} i \right|^2} \tag{1}$$

where the sum is over all particles and the $\vec{p}_\perp^{\,i}$ are transverse to a jet axis chosen to minimize $\hat{S}$. Two massless quanta of respective momentum xp and $(I - x)p$ give a contribution $\left[x^2 + (I - x)^2 \right] \vec{p}_T^{\,2}$ to the numerator whereas a single particle of momentum p gives $\vec{p}_T^{\,2}$. The two contributions are obviously different. This means that:

a) sphericity is strongly dependent on the hadronization process;

b) even when computing cross-sections at the parton level, one encounters, in the course of the calculation, large logarithms of masses that just reflect the fact that sphericity is not an infra-red safe quantity. Sphericity is therefore not computable in perturbative QCD.

A QCD ersatz for sphericity is spherocity[3] whose definition closely follows the preceding:

$$S = \left(\frac{4}{\pi} \right)^2 \min \left(\frac{\sum\limits_{i} \left| \vec{p}_\perp^{\,i} \right|}{\sum\limits_{i} \left| \vec{p}^{\,i} \right|} \right)^2 \tag{2}$$

Since $\left[x + (I - x) \right]^2 \vec{p}_T^{\,2} = \vec{p}_T^{\,2}$, S is an infra-red safe quantity (see above). S = 0 for a two-jet event; S = 1 for a spherical event.

Whereas the transverse momentum broadening of the jet is characterized by spherocity, one defines the thrust[4] to study the features of the longitudinal momentum:

$$T = \max \frac{\overset{\sim}{\sum\limits_{i}} \left| \vec{p}_{\shortparallel}^{\,i} \right|}{\sum\limits_{i} \left| \vec{p}^{\,i} \right|} \tag{3}$$

where Σ runs over all observed particles, $\overset{\sim}{\Sigma}$ over all particles emitted in a hemisphere chosen to maximize T, and the $p_{\shortparallel}$'s are transverse to the plane defining the hemisphere (or parallel to its normal, the jet axis). T = 1 for a two-jet event and T = $\frac{1}{2}$ for a spherical event.

Having recalled the definition of general variables for the study of jets, we will now restrict ourselves to the case of leptoproduction.

KINEMATICS AND THE SPIRIT OF FIRST ORDER QCD CALCULATIONS

The diagrams contributing in the zeroth and first order to leptoproduction are given in Fig. 3 and 4. In the zeroth order, (Fig. 3) one has just the usual parton model with a two jet structure; the struck parton jet and the proton fragmentation jet. First order QCD corrections give rise to a third jet either coming from a gluon radiated by the quark,(Fig. 4a and b) or from new origin: a gluon in the target proton wave function materializes into a quark-antiquark pair (Fig. 4 c).

We first have to choose the frame in which to study those jets. Whereas the angular correlations that we will discuss in a following section are frame-independent (provided those frames are obtained from one another by a boost along the intermediate vector boson momentum), such a choice is important in thrust and spherocity distributions. The most fashionable frames are:

a) the lab: $\vec{p} = \vec{0}$. Certainly, it is the easiest to obtain experimentally, but unfortunately all distributions are collimated around the vector boson momentum.

b) The Breit frame: $\vec{k}_1 + \vec{k}_2 = \vec{0}$. This frame allows us to get rid of the nucleon remnants since they always go backwards. Streng, Walsh and Zerwas[5] have discussed some nice features of this frame and studied the distribution of a quantity similar to thrust.

c) The final hadronic state rest frame: $\vec{p} + \vec{q} = \vec{0}$. From now on, we shall stick to this frame. It certainly is the frame where the jet structure is the most "open" and where the final state is the most similar to the one encountered in e^+e^- annihilation, but for the fact that the recoil jet is not simply made of one single parton. The three outgoing partons lie in a plane, which we shall call the hadron plane.

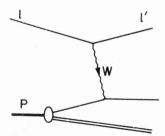

Fig. 3   The zeroth order contribution to two-jet production.

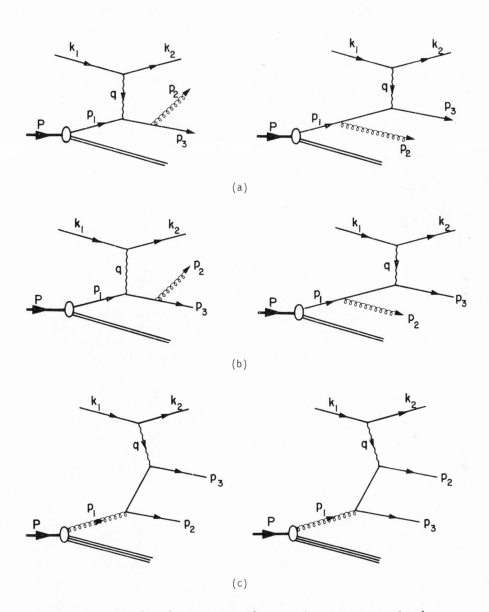

Fig. 4 Contribution of the first order QCD perturbation
theory to three-jet events.
(a) Incident quark;
(b) Incident antiquark;
(c) Incident gluon.

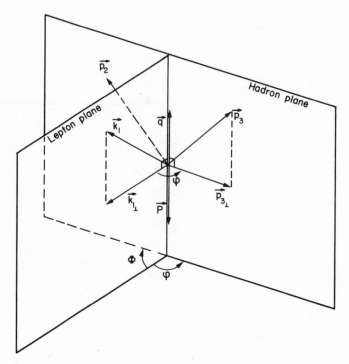

Fig. 5   Kinematic configuration of lepton-hadron
         scattering in the hadronic final state
         rest frame. $\vec{P}, \vec{q}, \vec{p}_1, \vec{p}_2, \vec{p}_3, \vec{k}$ are defined
         in Fig. 4.

In Fig. 5, we give the kinematic configuration and some definitions.
We also define the total hadronic energy $W^2 = (P + q)^2$; $y = (p_1 \cdot q)/$
$/(p_1 \cdot k_1)$;   and $x_i = 2E_i/W$ where E_i is the energy of the ith outgoing quantum.

    We now come to thrust and spherocity. The definitions given
above were originally proposed for the e^+e^- annihilation case.
The main difference between this case and leptoproduction is that
we now have an axis defined by $\vec{q}$. One could then try to define
new variables, taking this fact into account[6]. Unfortunately, it
seems that experimental uncertainties in the determination of $\vec{q}$
undermine the usefulness of such variables. So we keep the original
definitions of thrust and spherocity.

These quantities have been computed by De Rújula et al.,[7] for a three-quantum final state. Their results still hold:

$$T = 2 \max \frac{\sum |\vec{P}_i''|}{\sum |\vec{P}_i|} = \max (x_1, x_2, x_3) \geqslant \frac{2}{3} \qquad (4)$$

$$S = (\frac{4}{\pi})^2 \min \left(\frac{\sum |\vec{P}_\perp^i|}{\sum |\vec{P}^i|} \right)^2 = \frac{64}{\pi^2} (1-x_1)(1-x_2)(1-x_3)/T^2 \qquad (5)$$

STERMAN-WEINBERG JET FORMULA

Mainly for illustrative purposes, we begin by describing the Sterman-Weinberg approach as applied to leptoproduction. Sterman and Weinberg[8] define and compute the fraction of two-jet events $f(\varepsilon,\delta)$ for e^+e^- annihilation in which a fraction of the total hadronic energy smaller than ε is emitted outside two opposite cones of aperture angle 2δ. We shall keep this definition for leptoproduction (for another point of view, see Ref. 6). This quantity is obviously infra-red finite. Requiring that a fraction ε of the energy W is emitted outside the cone eliminates the emission of soft gluons that one cannot handle and giving a finite aperture 2δ to the cone integrates the collinear singularities already discussed. Of course, when one puts ε or $\delta = 0$, the infra-red divergences creep in again. This is easily seen[9] in Fig. 6 which shows $1 - f(\varepsilon,\delta)$ as a function of δ, at fixed ε. For small enough ε or δ, $1 - f$ becomes large and the calculation in first order in x_s is no longer reliable.

If we want the results of Fig. 6 to be meaningful, we have to compare them with a two jet model with a finite transverse momentum spread. We choose the simplest parametrization of $f(\varepsilon,\delta)$ for such a model, namely, at fixed ε:

$$f_{NP}(\varepsilon,\delta) = 1 - e^{-\delta^2/(\Delta\delta)^2} \text{ with } \Delta\delta = n(W) \frac{<p_\perp>}{\varepsilon W} \qquad (6)$$

For numerical results, we take $<p_T>$ to be 300 MeV and[10]

$$n(W) = \frac{3}{2} (\frac{2}{3} + 1.28 \ln W^2)$$

The motivations for Eq. (6) are the following: since δ is strongly related to the transverse momentum, we take a gaussian, remembering the behaviour in $e^{-4p_T^2}$ observed experimentally. Moreover, when $\delta \to 0$ or $\varepsilon \to 0$, $f(\varepsilon,\delta) \to 0$, as one could naively expect. The dashed curves of Fig. 6 show the behaviour of $f_{NP}(\varepsilon,\delta)$. One can see at once that, at presently available energies, the perturbative QCD result is completely lost in the non-perturbative background. The hope is that, at higher energies, as $\Delta\delta$ $\alpha 1/W$ decreases and $f_{NP}(\varepsilon,\delta)$ shrinks around $\varepsilon = 0$, the perturbative curves will show up.

We will refine these gross results in the following section but the conclucions will not change dramatically.

THRUST AND SPHEROCITY DISTRIBUTION

After calculating diagrams of Fig. 4 and then changing to the suitable variables, it is straightforward to get[10,11]

$$\left| \frac{1}{\sigma}\frac{d\sigma}{dT} \right| (Q^2,W^2) = \left| \frac{d\sigma^{(\circ)}}{dQ^2\ dW^2} \right|^{-1} \frac{d\sigma}{dQ^2\ dW^2}$$

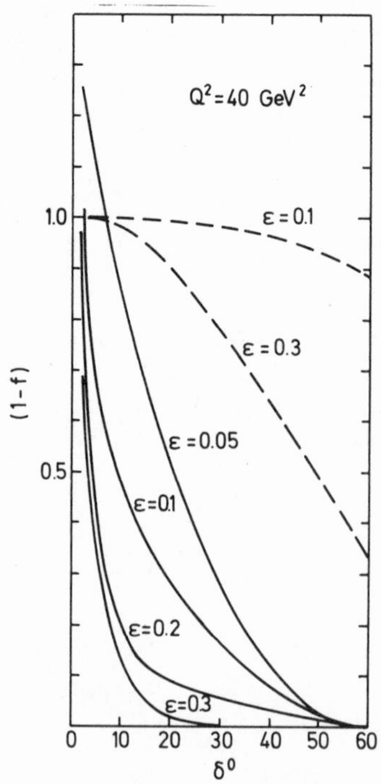

Fig. 6 $1 - f$ as a function of δ at fixed ε and Q^2.

where $\sigma^{(o)}$ is the two-jet cross-section.

The first question of importance is to study Q^2 and W^2 dependence. There are two arguments which tend to opposite conclusions.

First, $(1/\sigma)(d\sigma/dT)$ is of order $\alpha_S(Q^2)$ where $\alpha_S(Q^2)$ is the running coupling constant

$$\alpha_S(Q^2) = \frac{4\pi}{9} \frac{1}{\ln(Q^2/\Lambda^2)} , \qquad \Lambda = 500 \text{ MeV} \qquad (7)$$

chosen to depend on Q^2 (and not W^2) because Q^2 is the only relevant variable for parton diagrams of the type given in Fig. 4. Second, W is the natural variable for the hadronic final state, in particular for the three jet distribution. Figure 7 shows the actual

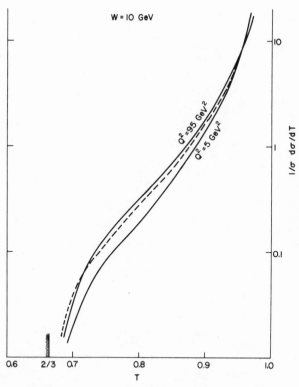

Fig. 7 Q^2 dependence of $(1/\sigma)$ $d\sigma/dT$ at $W^2 = 100$ GeV^2. The dashed curve is the Q^2 integrated version of this distribution.

Q^2 dependence for the QCD distribution. It suggests that the Q^2 dependence in α_s is damped and that W is the natural variable.

The reason for this fortunate behaviour is, of course, kinematic. Let us write the relation between the fraction of momentum ξ carried by the struck parton and the fraction x of energy carried by the nucleon remnant jet

$$1 - \xi = \frac{W^2}{W^2+Q^2} X_1 \tag{8}$$

If the gluon is collinear to the emitted quark, one has $x_1 = T$ and

$$T = (1 - \xi) (1 + \frac{Q^2}{W^2}) = \frac{1 - \xi}{1 - x} \tag{9}$$

where $x = Q^2/2M\nu$ is Bjorken x. This gives a thrust distribution determined by the ξ distribution and, if one neglects dynamics, a thrust distribution approximately Q^2 independent and peaking at

$$T_{max} = \frac{2}{3} (1 + \frac{Q^2}{W^2}) = \frac{2}{3} \frac{1}{1-x} \tag{10}$$

Of course, nothing like this can be seen in Fig. 7 because of the infra-red divergence at $T = 1$. Actually, the expression that one gets for $(1/\sigma)(d\sigma/dT)$ includes terms in $\alpha_s(Q^2) \ln(1 - T)/(1 - T) \sim 1$ that are of the order $[\alpha_s(Q^2)]^0$ and give a contribution to the two-jet distribution. But one can imagine that if one could handle the infra-red divergence, such behaviour would show up.

Now, as Q^2 increases, T_{max} approaches 1 and the tail of the distribution becomes broader. At the same time this tail is of order $\alpha_s(Q^2)$ which tends to decrease. The two effects compensate.

We can now integrate over Q and obtain $[(1/\sigma)/(d\sigma/dT)]$ (W^2). Of course, such a result is of little pratical interest since it deals with quarks and gluons, and we have to "treat" it in order to compare it with experiments.

The way to do it has been suggested by De Rújula et al.,[7] and from now on we closely follow their procedure. It consists in smearing the above results over bins of thrust of width ΔT_{np} to take into account the hadronization of quarks and gluons. ΔT_{np} is taken to be

$$\Delta T_{np} = \frac{1}{2} n(W) \frac{<p_\perp>}{W} \tag{11}$$

where n(W) and $<p_T>$ have been defined in formula (6). On the same basis, the zeroth order (two jets) distribution $\delta(1 - T)$ at the parton level gets broadened by the non-perturbative effects. We choose a simple parametrization similar to (6)

$$\frac{1}{\sigma_0} \left(\frac{d\sigma}{dT}\right)_{NP} = \frac{2}{(\Delta T_{np})^2} \; (1-T) \; e^{- (1-T)^2 / (\Delta T_{np})^2} \tag{12}$$

Here the factor $(1 - T)$ accounts for the fact that two massive particles cannot be collinear [one can also easily check that Eq. (12) is compatible with Eq. (6)].

Results are shown in Figs 8 and 9. At presently available energies (Fig. 8) the situation seems hopeless. But for higher W's (Fig. 9), the QCD tail starts emerging from the non-perturbative background.

It is certainly quite instructive to compare our results with those in e^+e^- annihilation[7]: the QCD tail is smaller in leptoproduction than in e^+e^- by roughly a factor of two. On the other hand, one can easily check from the formulas of Ref. 10, that the leading term in thrust distribution (i.e., $\ln(1 - T)/(1 - T)$ is the same for the two processes. This looks encouraging because it implies a kind of universality between the two processes, which is not surprising when one compares the QCD diagrams) and now experimentalists are starting to find such similarities[12]. We are tempted to explain the apparent discrepancy between the two results by the same kinematic argument as before: quark distributions favour thrust around $T_{max} \sim 1$ and therefore disfavour the QCD tail.

From the thrust distribution, we can infer mean values for thrust and spherocity. We give $<S>$ in Fig. 10 and compare it with a two-jet non-perturbative value that we choose to be:

$$<S>_{NP} = \left(\frac{4}{\pi}\right)^2 \frac{<P_\perp>^2}{W^2} \; <n(W)>^2 \tag{13}$$

where all quantities have already been defined. The left side of the figure gives the situation for νp scattering at variable energies. As one could expect, the perturbative three-jet results are completely lost in the two-jet background. The right part gives $<S>$ for ep scattering at $E_{lab} = 20\ 000$ GeV (as one could find in a next generation ep colliding machine). Of course, the situation improves and the QCD tail shows up.

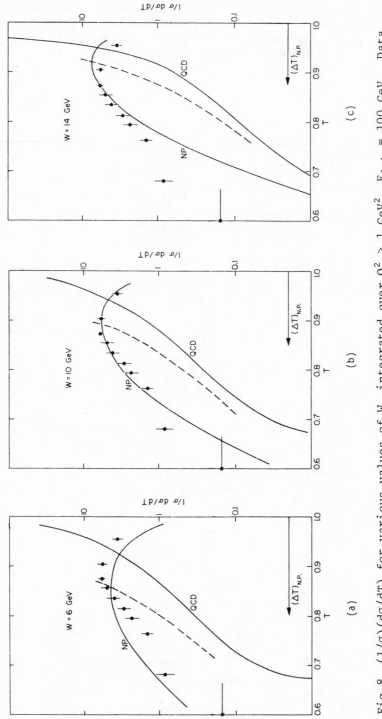

Fig.8 $(1/\sigma)(d\sigma/dT)$ for various values of W, integrated over $Q^2 > 1$ GeV^2, $E_{lab} = 100$ GeV. Data from Ref. 13 correspond to a distribution integrated over $6 < W < 10$ GeV and should therefore be applied only to $W = 10$ GeV. N.P. stands for the non-perturbative distribution given in Eq. (12). The dashed curves correspond to the QCD contribution smeared over bins of with ΔT_{np} indicated in the figures.

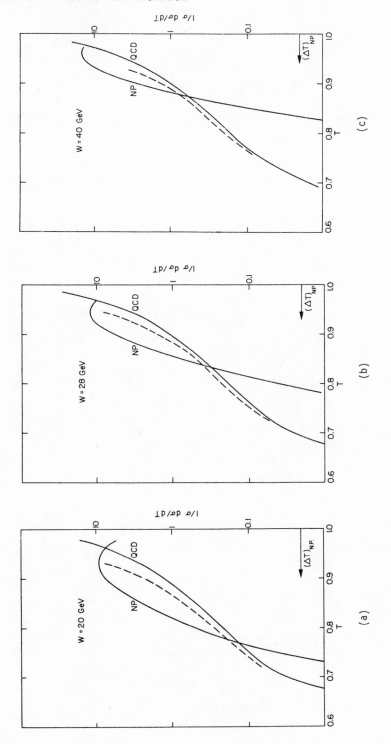

Fig. 9 Analogous to Fig. 8 for larger values of W.

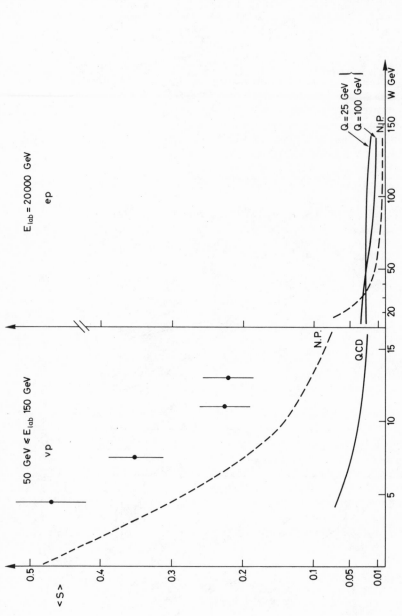

Fig. 10   <S> as a function of W for νp scattering (W < 15 GeV, 50 GeV < E_lab < 150 GeV) at ep scattering (15 GeV < W < 150 GeV). The N.P. dashed lines give the values of <S>_NP, the solid lines give the QCD perturbative result (and experimental points are from Ref. 13. Buras-Gaemers parametrization for structure functions has been used[14].

We end this section by discussing the pointing vector. Roughly speaking, the pointing vector measures the energy flow in the hadron plane as a function of angle. More precisely, it is defined as[7]

$$P\ (Q,W,T,\theta)\ =\ \frac{1}{\sigma_{(\circ)}}\ |p|\ \frac{d\sigma}{dTd\cos^2\theta} \tag{14}$$

For a careful definition of this quantity, see M.K. Gaillard's talk; for the experimental difficulties in measuring it, see H. Meyer's contribution to these proceedings.

$\sigma^{(o)}$ The only difference with the e^+e^- case is that we divide by $\sigma^{(o)}$ in order to eliminate the Q and W dependence of the zeroth order.

It is, of course, of no use to give our estimate of the pointing vector at present energies where the background is so important. So, Fig. 11 gives the predictions for energies available in a high-energy ep colliding machine.

ANGULAR ASYMMETRIES

The distribution in the azimuthal angle ϕ, defined earlier, introduces angular asymmetries whose importance was first stressed by Georgi and Politzer[15]. Cahn[16] has clarified their origin: the initial transverse momentum of the struck quark. Let us sketch his procedure.

First, if $y = 1$, [that is if the incident lepton (neutrino) is collinear with $\vec{q}$], there is no preferred direction in the plane transverse to $\vec{q}$ and therefore no asymmetry. On the other hand, if $y \neq 1$, there is a preferred direction along $\vec{k}_\perp$ (see Fig. 5). One can easily show, for example, on dimensional grounds, that the squared amplitude is of order S , where S is the energy available in the center-of-mass of the incident particles (quark neutrino). If we fix all momentum components parallel to $\vec{q}$, the situation where the struck quark is antiparallel to the neutrino in the transverse plane is favoured, since it maximizes S. Since $q_T = 0$, the emitted quark has the same transverse momentum as the struck quark.

The origin of this initial transverse momentum is twofold: either it is the radiation of a gluon by the quark, before its interaction with the current, or a primordial p_T of the quark in the nucleon.

The asymmetries between the lepton plane and an outgoing hadron have been extensively studied[15-20] (see also P. Landshoff's

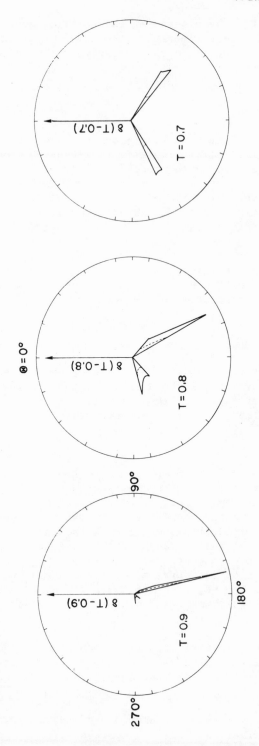

Fig. 11. Pointing vector for different values of thrust at E_{lab} = 20 000 GeV
Solid lines: Q^2 = W^2 = 1 280 GeV; dashed lines $Q^2 \simeq$ 128 GeV^2, W^2 = 8000 GeV^2.
Different scales have been used for the different values of thrust.

talk in these Proceedings). They have three disadvantages: first, one loses most of the information by representing the final state by a single hadron; second, one has to use poorly known fragmentation functions; and third, it is hard, at present energies to disentangle the two origins of transverse momentum.

The asymmetries between the lepton plane and an outgoing jet would be worth studying if one could distinguish between a quark and a gluon jet. Let us only note that, since the quark is preferentially emitted opposite to the incident lepton, the gluon is preferentially radiated on the same side (see Fig. 5).

Finally, we define the angle Φ between the lepton plane and the hadron plane and study corresponding asymmetries[10]. Whereas the definition of a parton plane is obvious (see Fig. 5), we define the hadron plane by minimizing the momentum of all hadrons out of this plane. Since Φ is an angle between planes, we restrict it to the range $-\pi/2 \leq \Phi \leq +\pi/2$. It is straightforward to see that Φ suffers infra-red divergences: when the gluon becomes collinear to a quark, it is no longer possible to define a plane. We therefore, have to make cuts in thrust in order to prevent such a situation. Φ is then an infra-red safe quantity since a quark or a quark plus an extra collinear gluon will give the same contribution.

Results are shown in Figs 12-14 for various cuts in thrust ($T < 1 - \Delta T$). One can see that the effects are sizeable even for large cuts in T.

If we make the following assumption on non-perturbative effects (hadronic broading of jets are rotationally invariant with respect to the jet axis) then those effects are expected to occur at a much lower level than in the previous cases. Two jet events will be cylindrical and will therefore give "no contribution" to the statistical determination of a plane. Moreover, a primordial transverse momentum will shift only slightly the thrust distribution and, thanks to the cuts in T, give no important contribution to the Φ asymmetry.

CONCLUSION

Leptoproduction processes are much richer processes than e^+e^- annihilation since they involve more variables (W,ϕ), a new axis ($\vec{q}$) and a choice between different reference frames. Unfortunately, the situation is not as clearcut as in e^+e^- because we have to deal with uncertaintities in the structure functions, experimental difficulties in determining the q vector (at least, in neutrino processes), and our ignorance of the behaviour of nucleon remnants.

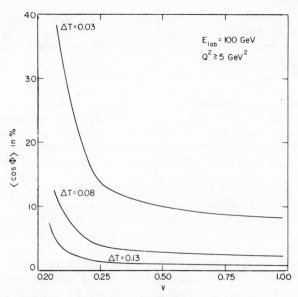

Fig. 12 Angular asymmetry between the hadron and lepton
planes <cos Φ> integrated over Q^2,S,T (2/3 < T <
< 1 - ΔT) as a function of y.

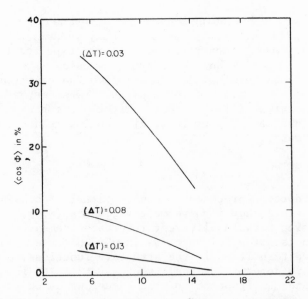

Fig. 13 Same as Fig. 12, <cos Φ> as a function of W.

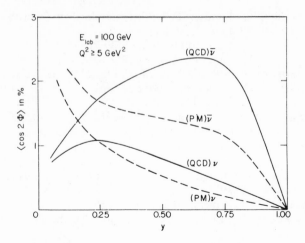

Fig. 14. <cos2Φ> as a function of y.

It is therefore difficult to find clean tests for QCD from jet studies. We believe that the study of the Φ̂ asymmetries may be an exception.

ACKNOWLEDGEMENTS

One of us (P.B.) thanks J.J. Aubert and G. Preparata for giving him the opportunity to participate in the Erice workshop.

REFERENCES

1. T. Kinoshita, J. Math Phys. 3: 650 (1962);
    T.D. Lee and M. Nauenberg, Phys. Rev. 133:B 1549 (1964).
2. J.D. Bjorken and S.J. Brodsky, Phys. Rev. D1: 1416 (1970).
3. H. Georgi and M. Machacek, Phys. Rev. Letters 39: 1237 (1977).
4. E. Farhi, Phys. Rev. Letters 39: 1587 (1977).
5. K.H. Streng, T.F. Walsh and P.M. Zerwas, DESY 79/10 (february 1979).
6. P.M. Stevenson, Nucl. Phys. B150: 357 (1979) and preprint ICTP/78-79/16 (April 1979).
7. A. de Rújula, J. Ellis, E.G. Floratos and M.K. Gaillard, Nucl. Phys. B138: 387 (1978).
8. G. Sterman and S. Weinberg, Phys. Rev. Letters 39: 1436 (1977).
9. P. Binétruy and G. Girardi, Phys. Letters 83B: 382 (1979).
10. P. Binétruy and G. Girardi, Nucl. Phys. B155:150 (1979).
11. Ranft and Ranft, Leipzig University preprint 78-15 (1978).

12. Aachen-Bonn-CERN-London I.C.-Oxford-Saclay Collaboration
    CERN/EP 79-39 Rev. (1979);
    M. Derrick et al., Preprint ANL-HEP-PR-79-17 (1979).
13. A. Vayaki et al., Aachen-Bonn-CERN-London-Oxford-Saclay
    Collaboration, CERN/EP/PHYS 78-28 (1978). Note that in
    Fig. 5c of this paper, the correct scale is $(1/\sigma)(d\sigma/dT) \times 10$
    and not $(1/\sigma)(d\sigma/dT) \times 10^2$.
14. A.J. Buras and K.J. F. Gaemers, Nucl. Phys. B132: 249 (1978).
15. H. Georgi and H. D. Politzer, Phys. Rev. Letters 40: 3 (1978).
16. R.N. Cahn, Phys. Lett. 78B: 269 (1978).
17. A. Méndez, Nucl. Phys. B145: 199 (1978);
    A. Méndez, A. Raychaudhuri and V.J. Stenger, Nucl. Phys.
    B148: 499 (1979).
18. J. Cleymans, Phys. Rev. D18: 954 (1978).
19. G. Köpp, R. Maciejko and P.M. Zerwas, Nucl. Phys. B144: 123
    (1978).
20. P. Mazzanti, R. Odorico and V. Roberto, Preprint IFUB/78-11.

DEEP INELASTIC STRUCTURE FUNCTIONS:

DECISIVE TESTS OF QCD

E. Reya

Deutsches Elektronen-Synchrotron DESY

2000 Hamburg 52

1. INTRODUCTION

The purpose of this talk is to discuss the present status of
scaling violations in deep inelastic lepton-nucleon scattering
processes. Specifically, I will discuss in great detail which experimentally measured structure functions (or moments thereof)
have to be used in order to discriminate between various field
theories of the strong interactions such as (non-abelian) QCD,
abelian vector-gluon $(\bar{\Psi}\gamma_\mu \Psi A^\mu)$, non-abelian scalar-gluon
$(\bar{\Psi}\lambda_a \Psi \phi_a)$, and abelian scalar-gluon $(\bar{\Psi}\Psi\phi)$ theories. Moreover I will delineate the sensitivity of certain structure func-
tions to the fundamental couplings of the theory, i.e., to the quark-
gluon coupling and/or to the gluon self-couplings (triple gluon
vertex) - a measurement of the latter would provide us with a
direct and unambiguous test of the Yang-Mills structure of QCD
which is so very essential for asymptotic freedom. Furthermore,
we will consider three-jet events with special emphasis on deep
inelastic heavy quark (c,b,...) production via virtual Bethe-
Heitler processes. Finally, I shall briefly discuss the relevance
of non-leading 2-loop contributions to anomalous dimensions and
the effects of "finite terms" in Wilson coefficients and their importance for the analysis of deep inelastic reactions such as the
determination of moments of "sea" distributions from measured
neutrino cross sections. The relevance of these "finite terms" for
Drell-Yan dimuon production cross sections will also briefly be
mentioned.

For a detailed discussion of semi-inclusive deep inelastic
reactions, such as hadronic final states and QCD predictions for

gluon jets in lepton-hadron interactions I refer you to the talks
of P. Landshoff and P. Binétruy.

   In Section 2 I will briefly recapitulate the structure of QCD
and other asymptotically non-free (finite fixed point) field
theories and those parts of their predictions for scaling violations in structure functions which will be essential for our present
analysis. Here we shall use the classical language of the light
cone expansion and the renormalization group. Section 3 will be
devoted to a detailed comparison of these predictions with experiment using (of increasing complexity) (A) moments of non-singlet
structure functions ($F_2^{ep} - F_2^{en}$, $F_3^{\nu,\bar{\nu}N}$), (B) moments of flavor singlet
structure functions ($F_2^{\nu N}$, etc.) and (C) the explicit x- and Q^2-
dependence of structure functions. Here, QCD will prove to be the
only renormalizable strong interaction field theory compatible
with experiment! However, as we shall see, present measurements of
scaling violations in $F_2(x,Q^2)$ are rather insensitive to the gluon
content of the hadron. In Section 4 we proceed to discuss (very
fundamental but equally difficult) measurements which are particularly sensitive to the gluon self-couplings of a (locally gauge invariant) Yang-Mills theory, and which would provide us with a direct
and sensitive test of asymptotic freedom. In Section 5 we give predictions for deep inelastic heavy quark (c,b,...) production via
virtual Bethe-Heitler processes and, finally, in Section 6 we briefly discuss non-leading 2-loop corrections to anomalous dimensions
and "finite terms" in Wilson coefficients and their relevance for
comparing QCD and the parton model with experiment; we conclude
with a brief discussion of the effects of "finite terms" in Drell-
Yan processes.

2. FIELD THEORIES AND SCALING VIOLATIONS

   The quark-gluon interactions in QCD are described by the
Lagrangian

$$\mathscr{L}_F = i\,\bar{\Psi}\,\gamma_\mu \left(\partial^\mu - ig\frac{\lambda_a}{2}\,A_a^\mu\right)\Psi \tag{1}$$

where λ_a are the Gell-Mann matrices of color SU(3)$_c$ acting on the
color triplets $\Psi = (q_R,\ q_Y,\ q_B)$ with q = u,d,s,c,... . Renormalizability requires the theory to be locally gauge invariant which,
because of the non-abelian interaction in Eq. (1), implies self-
couplings of the gluon fields A_a^μ via the Yang-Mills Lagrangian
$\mathscr{L}_{YM} = -\frac{1}{4}\,F_a^{\mu\nu}\,F_{a\mu\nu}$ with

$$F_a^{\mu\nu} = \partial^\mu A_a^\nu - \partial^\nu A_a^\mu + g f_{abc} A_b^\mu A_c^\nu \tag{2}$$

where $[\frac{1}{2}\lambda_a, \frac{1}{2}\lambda_b] = i f_{abc} \frac{1}{2}\lambda_c$. It is well known[1] that the
self-interactions of the vector fields A_a^μ implied by Eq. (2) are
responsible for asymptotic freedom, i. e., the effective coupling

constant $\alpha_s \equiv g^2/4\pi$ decreases for increasing momentum transfer squared Q^2 (or decreasing distances)

$$\alpha_s(Q^2) = \frac{12\pi}{(33-2N_f)\ln Q^2/\Lambda^2} \tag{3}$$

with N_f being the number of flavors and Λ the only free parameter of QCD which has to be fixed by experiment ($\Lambda \approx 0.5$ GeV). Equation (3) is the solution of the renormalization group equation $Q\, d\bar{g}/dQ = \beta(\bar{g})$ for the effective coupling $\bar{g}(g,Q^2)$, where the Callan-Symanzik function $\beta[g(\mu)] \equiv \mu\,\partial g(\mu)/\partial\mu$ describes the variation of the renormalized coupling $g(\mu)$ if one changes the arbitrary renormalization mass parameter μ. The reason for asymptotic freedom is that the gluon self-couplings in Eq. (2) allow for a color charge transfer from the quark fields ψ to the gluon fields A_a^μ and thus create an <u>anti</u>-shielding around the bare charge g_0 which dominates the color charge shielding (opposite sign charges) around g_0 stemming from the usual vacuum polarization effects, due to the coupling in Eq. (2), which do not allow for a color charge exchange with the field. The net effect is that $\beta(g) = -b\, g^3 < 0$ as shown in Fig. 1 which implies that perturbation theory becomes better (α_s small) the larger Q^2 ($\gtrsim 2$ GeV^2).

The situation is very different for "conventional" field theories such as abelian vector gluon theories where, as in QED, there is no group structure in the quark-gluon coupling in Eq. (1), i.e., the vertex is described by $g\bar{\psi}\gamma_\mu\psi A^\mu$, and consequently there is no gluon self-coupling term in Eq. (2) and the field strength tensor is simply $F^{\mu\nu} = \partial^\mu A^\nu - \partial^\nu A^\mu$. Similarly, the interaction in Eq. (1) takes the form $g\bar{\psi}\lambda_a\psi\phi_a$ for non-abelian scalar gluon theories, and $g\bar{\psi}\psi\phi$ for abelian scalar gluon (Yukawa) theories. In all these cases we have $\beta(g) = +b\, g^3 > 0$ and thus these theories are not UV stable near the origin $g = 0$, i.e., the coupling will become larger the larger Q^2 since now the only

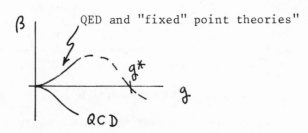

Fig. 1. The β function for QCD and conventional field theories which are assumed to develop a finite UV fixed point g^*.

quantum effects which contribute to charge renormalization are due
to vacuum polarization (shielding, but no anti-shielding exists)
because no self-couplings of the gluon fields (A^{μ} or ϕ_{α}, ϕ) exist.
In order to proceed perturbatively at large energies (small distances)
we have to make the (so far unproven) assumption that there exists a
finite fixed point coupling g^{*} as $Q^2 \to \infty$, i.e. $\beta(g^{*}) = 0$, such
that the effective coupling $\alpha^{*}/4\pi \ll 1$ - a necessary requirement
in order not to reject a priori conventional field theories as
possible candidates for explaining (perturbatively) the experimentally
observed <u>small</u> scaling violations. This situation is shown in Fig. 1
and we shall call hereafter these conventional field theories simply "fixed point theories". The fixed point coupling g^{*} is then the
only free parameter of the theory to be determined experimentally
(as is the case for Λ in QCD).

The importance of deep inelastic lepton-nucleon scattering
processes is that there exists an operator product expansion (OPE)
on the light cone,[3] a direct generalization of Wilson's short
distance expansion of the product of two currents, which can be used,
together with renormalization group techniques, to calculate scaling
violations in field theory. Symbolically we may write

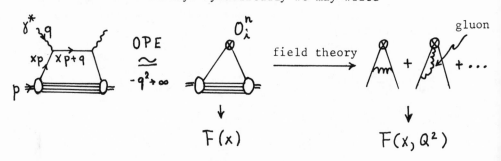

Fig. 2

where O_i^n are the various light-cone (or Wilson) operators of different
type i and spin n, and $x = Q^2/2p \cdot q$, $Q^2 \equiv -q^2 \geqslant 0$. Note that in any
interacting field theory the structure function F will depend on
the <u>two</u> variables x and Q^2, say, in contrast to the naïve parton
model where in the Bjorken limit $F = F(x)$. This additional Q^2 dependence is usually referred to as scaling violation. The OPE together with field theory makes predictions[4] only for (Mellin)
<u>moments</u> of deep inelastic structure functions defined by

$$\langle F(Q^2) \rangle_n \equiv \int_0^1 dx \, x^{n-2} \, F(x, Q^2) \tag{4}$$

where $F = xF_1$, F_2 or xF_3. In terms of these moments the above OPE
can be formally written as

$$\langle F(Q^2) \rangle_n = \sum_i C_i^n(Q^2/\mu^2, \alpha_s(\mu^2)) \langle p | O_i^n | p \rangle \tag{5}$$

where the Q^2-independent matrix elements $\langle p| O_i^n |p\rangle$ which describe
the bound state (wave function) of the nucleon cannot, at the present
state of art, be calculated perturbatively and will be related to the
parton distributions in the nucleon at a fixed value of $Q^2 = Q_o^2$,
which in turn will be determined from experiment. On the other hand,
the Q^2 behavior of the Wilson coefficients C_i^n - the expansion coefficients of the OPE - can be calculated perturbatively which will
uniquely determine the Q^2 evolution of quark and gluon distributions,
provided Q^2 is large enough. To this end one writes a renormalization group (RG) equation for the Wilson coefficients (which ex-
presses the invariance of any measurable physical quantity with respect to changes of the arbitrary renormalization point μ) in order
to sum the leading log contributions to all orders α_s, i.e., one improves simple order by order perturbation theory. For non-singlet
(NS) structure functions there is only one leading fermionic Wilson
operator[4] O_{NS}^n which contributes to the sum in Eq. (5), for which
case the RG equation reads

$$\left(\mu \frac{\partial}{\partial \mu} + \beta \frac{\partial}{\partial g} - \gamma_{NS}^n\right) C_{NS}^n(Q^2/\mu^2, \alpha_s(\mu^2)) = 0 \qquad (6)$$

where the "anomalous dimension" γ_{NS}^n of O_{NS}^n, which governs the Q^2
dependence of structure functions, is the coefficient of the logarithmic contribution[4] of the one-gluon-loops in Fig. 2. The solution
of Eq. (6) is well known to be

$$C_{NS}^n(Q^2/\mu^2, \alpha_s(\mu^2)) = C_{NS}^n(1, \alpha_s(Q^2)) \exp\left[-\int_\mu^Q \frac{dQ'}{Q'} \gamma_{NS}^n(\alpha_s(Q'^2))\right] \qquad (7)$$

where $\qquad C_{NS}^n(1, \alpha_s(Q^2)) \simeq 1 + O(\alpha_s) \qquad\qquad$ results from

and the non-leading $O(\alpha_s)$ contributions are the so called "finite
terms", i.e., all non-logarithmic contributions from $\gamma^* q \rightarrow g q$.
As long as we work in the leading 1-loop approximation of γ_{NS}^n, we
have to keep only the leading naive parton model contribution ($\simeq 1$)
and disregard the $O(\alpha_s)$ terms which we will do for the time being.
Thus the final result for the Q^2 evolution of the moments of a non-
singlet structure function can be written as[4-6] (I follow closely
the notation of Refs. 5 and 6)

$$\langle F_{NS}(Q^2)\rangle_n = \langle F_{NS}(Q_o^2)\rangle_n \, e^{-S \, a_{NS}(n)} \qquad (8)$$

where $Q_o^2 (\simeq 2\text{-}4 \text{ GeV}^2)$ is the input reference momentum at which the

structure function has to be determined experimentally, and for QCD
the renormalization group exponents are given by

$$a_i = \frac{\gamma_i}{8\pi\alpha_s b} \quad , \quad s = \ln\frac{\ln(Q^2/\Lambda^2)}{\ln(Q_0^2/\Lambda^2)} \tag{9}$$

with $b = \frac{1}{16\pi^2}(11 - \frac{2}{3}N_f)$. For fixed point theories these exponents read

$$a_i = \frac{\gamma_i}{2} \quad , \quad s = \ln\frac{Q^2}{Q_0^2} \tag{10}$$

where now the value of the UV finite fixed point $\alpha_s \equiv \alpha^*$, appearing in γ_i, has to be determined by experiment. The NS anomalous
dimension $\gamma_{NS}^n \equiv \gamma_{FF}^F(n)$ is given by[4]

$$\gamma_{FF}^F = \frac{\alpha_s}{2\pi} C_2(R)\left[1 - \frac{2}{n(n+1)} + 4\sum_{j=2}^{n}\frac{1}{j}\right] \tag{11}$$

   In the general case, structure functions (e.g., $F_2^{\mu p}$) receive contributions also from flavor singlet operators. There are two different
types of singlet Wilson operators[4] contributing to the sum in Eq.
(5), one fermionic operator O_F^n and one gluonic operator O_V^n, the
latter being constructed[4] from the fundamental vector fields A_a^μ.
In this case the RG equation becomes a matrix equation according
to the 2x2 singlet anomalous dimension matrix $\hat{\gamma}(n)$, because of the
four possible matrix elements of $O_{F,V}^n$ between the external fermionic
(F) and gluonic (V) states: the operators mix under renormalization
which is usually referred to as "singlet operator mixing". This
singlet mixing will play a crucial role for discriminative tests
of QCD! This singlet matrix reads (in a straightforward notation)

$$\hat{\gamma}(n) = \begin{pmatrix} \gamma_{FF}^F & , & \gamma_{FF}^V \\ \gamma_{VV}^F & , & \gamma_{VV}^V \end{pmatrix}$$

with γ_{FF}^{F} given by Eq. (11) and[4]

$$\gamma_{VV}^{V} = \frac{d_S}{2\pi}\left\{ C_2(G)\left[\frac{1}{3} - \frac{4}{n(n-1)} - \frac{4}{(n+1)(n+2)} + 4\sum_{j=2}^{n}\frac{1}{j}\right] + \frac{4}{3}T(R)\right\}$$

$$\gamma_{VV}^{F} = -\frac{d_S}{2\pi}\frac{4(n^2+n+2)}{n(n+1)(n+2)}T(R) \qquad (12)$$

$$\gamma_{FF}^{V} = -\frac{d_S}{2\pi}\frac{2(n^2+n+2)}{n(n^2-1)}C_2(R)$$

where the group invariants are as follows: for non-abelian vector theories (QCD) we have

$$C_2(G) = 3, \quad C_2(R) = \frac{4}{3}, \quad T(R) = \frac{1}{2}N_f \qquad (13)$$

and for an abelian gluon field theory these quantities simply read

$$C_2(G) = 0, \quad C_2(R) = 1, \quad T(R) = 3N_f. \qquad (14)$$

For all subsequent considerations we shall take $N_f = 4$. Note that only $C_2(G)\delta_{ab} \equiv f_{acd}f_{bcd}$ in γ_{VV}^{V} is a direct measure of the gluon self-interactions (triple gluon vertex) in Eq. (2). Similar expressions for anomalous dimensions can be derived for non-abelian and abelian scalar gluon theories[6,7]. The Q^2 evolution of singlet structure functions is now obtained by diagonalizing the RG matrix equation via $\hat{\gamma} = \gamma_-\hat{P}^- + \gamma_+\hat{P}^+$ with

$$\gamma_{\pm} = \frac{1}{2}\left[\gamma_{FF}^{F} + \gamma_{VV}^{V} \pm \sqrt{(\gamma_{VV}^{V} - \gamma_{FF}^{F})^2 + 4\gamma_{VV}^{F}\gamma_{FF}^{V}}\right] \qquad (15)$$

and where the projection operators are given by

$$\hat{P}^- = \begin{pmatrix} \bar{P}_{11}, & \bar{P}_{12} \\ \bar{P}_{21}, & 1-\bar{P}_{11} \end{pmatrix}$$

and $\hat{P}^+ = 1 - \hat{P}^-$ with

$$\bar{P}_{11} = \frac{\gamma_{FF}^{F} - \gamma_+}{\gamma_- - \gamma_+}, \quad \bar{P}_{21} = \frac{\gamma_{VV}^{F}}{\gamma_- - \gamma_+}, \quad \bar{P}_{12} = \frac{\gamma_{FF}^{V}}{\gamma_- - \gamma_+} \qquad (16)$$

The RG prediction for the Q^2 dependence of a general structure function in Eq. (5) is then obtained to be[5]

$$\langle F(Q^2)\rangle_n = \sum_{i=NS,\pm} \langle F_i(Q_o^2)\rangle_n \, e^{-s\,a_i(n)} \qquad (17)$$

with the RG exponents given by Eqs. (9) and (10), and

$$\langle F_\pm(Q_o^2)\rangle_n = \binom{1-\alpha_n}{\alpha_n} \langle x\Sigma(Q_o^2)\rangle \mp \beta_n \langle xG(Q_o^2)\rangle_n \qquad (18)$$

where for brevity we have defined $\alpha_n \equiv p_{11}^-(n)$ and $\beta_n \equiv p_{21}^-(n)$.
As we shall see α_n will play a crucial role in discriminating be-
tween different field theories. The gluon distribution $G(x,Q^2)$ in
the nucleon is defined by $\langle xG(Q_o^2)\rangle_n \equiv \langle p|O_V^n|p\rangle$ and the
fermionic singlet, defined by $\langle x\Sigma(Q_o^2)\rangle_n \equiv \langle p|O_F^n|p\rangle$, is
just

$$x\Sigma(x,Q^2) \equiv x\sum_q [q(x,Q^2) + \bar{q}(x,Q^2)] \qquad (19)$$

which is directly measured by $F_2^{\nu N}$ (above charm threshold and always
assuming $s = \bar{s}$, $c = \bar{c}$). Using Eq. (18) the singlet contributions to
Eq. (17) can be rewritten in the following convenient form

$$\langle x\Sigma(Q^2)\rangle_n = [\alpha_n \langle x\Sigma(Q_o^2)\rangle_n + \beta_n \langle xG(Q_o^2)\rangle_n] e^{-s\,a_-(n)}$$
$$+ [(1-\alpha_n)\langle x\Sigma(Q_o^2)\rangle_n - \beta_n \langle xG(Q_o^2)\rangle_n] e^{-s\,a_+(n)} \qquad (20a)$$

$$\langle xG(Q^2)\rangle_n = [(1-\alpha_n)\langle xG(Q_o^2)\rangle_n + \frac{\alpha_n(1-\alpha_n)}{\beta_n}\langle x\Sigma(Q_o^2)\rangle_n] e^{-s\,a_-(n)}$$
$$+ [\alpha_n\langle xG(Q_o^2)\rangle_n - \frac{\alpha_n(1-\alpha_n)}{\beta_n}\langle x\Sigma(Q_o^2)\rangle_n] e^{-s\,a_+(n)} \qquad (20b)$$

It should be emphasized that these equations are not independent:
once $\langle x\Sigma(Q_o^2)\rangle_n$ and $\langle xG(Q_o^2)\rangle_n$ are fixed from experiment by
fitting, say, Eq. (20a) to the measured Q^2 dependence of $\langle x\Sigma(Q^2)\rangle_n$,
then Eq. (20b) is trivially satisfied and does not constitute an
independent test of QCD. We shall come back to this point later.

3. COMPARISON WITH EXPERIMENT

The most straightforward and simple, although in many cases not
very stringent and instructive, tests of field theories are to
compare just moments of structure functions with experiment which

are directly predicted by any field theory. Let us start with the theoretically most simple case of non-singlet structure functions.

3.A. Non-Singlet Moments

As we have seen in Eq. (8) the Q^2 evolution of the moments of a non-singlet structure function $F_{NS} = F_2^{ep} - F_2^{en}$, $F_3^{\nu N}$, etc. is governed by only $\underline{one}$ anomalous dimension γ_{FF}^F. Considering first $F_3^{\nu N}$, Eq. (8) tells us that for QCD

$$\langle x F_3^{\nu N}(Q^2) \rangle_n^{-1/a_{NS}(n)} \sim \ln \frac{Q^2}{\Lambda^2} \tag{21}$$

i.e., the $(-1/a_{NS})$ -th power of the n-th moments are expected to lie along straight lines when plotted against $\ln Q^2$ with a common intercept $\ln Q^2 = \ln \Lambda^2$. These predictions have been found to be in very good agreement[8] with the data for $\Lambda \simeq 0.5$ GeV as shown in Fig. 3(a). Similar conclusions have been reached from analyzing[9] the BEBC data but it should be emphasized, however, that these latter results rely heavily on measurements between $Q^2 = 0.6$ and 2 GeV^2 – a region neither appropriate for the parton model nor for the legitimacy of perturbative calculations. Even in the CDHS experiment[8], where $Q^2 \gtrsim 6.5$ GeV^2, ill understood kinematical target mass effects $(\sim x^2 m_N^2/Q^2)$ play a non-negligible role: assuming that these effects can be in part accounted for by Nachtmann moments[10] (which results from the trace-terms in the NS Wilson operator of definite spin), the fitted slopes $\underline{decrease}$ by more than 10 % as shown in Fig. 3(b). The importance of this statement will become clear in a moment.

However, these measured NS moments can be equally well explained, in the presently measured region of Q^2, by conventional fixed point field theories. For an abelian vector gluon theory, Eqs. (9)-(11), (13) and (14) tell us that

$$a_{NS}^{vector} = \frac{25 \alpha^*}{16 \pi} a_{NS}$$

where a_{NS} is the RG exponent of QCD, and thus Eq. (8) predicts

$$\langle x F_3^{\nu N}(Q^2) \rangle_n^{-1/a_{NS}} = C_n(Q_0^2) \left(\frac{Q^2}{Q_0^2} \right)^{\frac{25\alpha^*}{16\pi}} \tag{22}$$

where the unknown normalization constants $C_n(Q_0^2) \equiv \langle x F_3^{\nu N}(Q_0^2) \rangle_n^{-1/a_{NS}}$ have to be fitted to the data at an arbitrary value of $Q^2 = Q_0^2$.[11]

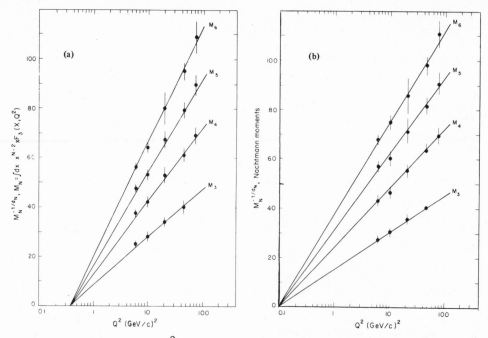

Fig. 3. Fit to the Q^2 dependence of (a) ordinary Cornwall-
Norton moments $M_n \cong \langle x F_3^{\nu N}(Q^2) \rangle_n$ and (b) Nachtmann
moments according to QCD. The figures are taken from
Ref. 8.

A similar power-like behavior in Q^2 is predicted by, for example,
non-Abelian scalar-gluon theories[5-7,11] where

$$a_{NS}^{scalar} = \frac{\alpha^*}{8\pi} \tilde{a}_n \tag{23}$$

with $\tilde{a}_n = \frac{4}{3}[1 - \frac{2}{n(n+1)}]$ which gives for Eq. (8)

$$\langle x F_3^{\nu N}(Q^2) \rangle_n^{-1/\tilde{a}_n} = \tilde{C}_n(Q_0^2)\left(\frac{Q^2}{Q_0^2}\right)^{\frac{\alpha^*}{8\pi}} \tag{24}$$

The predictions of abelian scalar-gluon theories are as in Eq. (24)
with α^* multiplied by a factor of 3/4. From Fig. 4 it can be seen
that the predictions according to Eqs. (22) and (24) are in equally
good agreement[11] with experiment as are the straight line fits in
Fig. 3. Thus non-singlet quantities can only provide us with a
consistency check of a given theory but cannot discriminate between
QCD and other finite fixed point theories of strong interactions
(unless precision measurements can be extended to $Q^2 = 200$ or

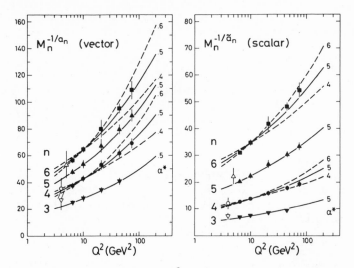

Fig. 4. Comparison of measured[8] moments $M_n \cong \langle x F_3^{\nu N}(Q^2) \rangle_n$
with the predictions[11] of abelian vector-gluon theories,
Eq. (22), and non-abelian scalar-gluon theories, Eq. (24),
for various choices of the fixed point α^*. The low-
statistics data (open circles and triangles) are from
Ref. 9.

300 GeV^2 , as it is evident from Fig. 4). This, however, is not
too surprising since the Q^2 dependence of NS moments is uniquely
determined by just <u>one</u> anomalous dimension and, therefore,
quantities such as $\langle F_{NS}(Q^2) \rangle_n^{-1/a_{NS}}$ are mainly sensitive to
differences in a logarithmic and a power-like behavior in Q^2. This
is in contrast to structure functions which receive also contributions from flavor-<u>singlet</u> Wilson operators, such as F_2, the Q^2 dependence of which is determined by <u>three</u> different anomalous dimensions in Eq. (17): these subtleties of singlet-mixing will play a
crucial role in discriminating between different field theories
which will be discussed in the next Section.

Another theoretically very attractive test of field theories,
which measures ratios of anomalous dimensions directly, is obtained[9]
by comparing the logarithms of two moments $\langle F_{NS} \rangle_n$ and $\langle F_{NS} \rangle_{n'}$
which, according to Eq. (8), should result in straight lines (in
the 1-loop order) with slopes $a_{NS}(n)/a_{NS}(n')$:

$$\frac{d \ln \langle F_{NS} \rangle_n}{d \ln \langle F_{NS} \rangle_{n'}} = \frac{a_{NS}(n)}{a_{NS}(n')} \tag{25}$$

These slopes are obviously independent of Λ and α^*, as well as of

the number of flavors. Moreoever it should be emphasized that
Eq. (25) can discriminate only between vector and scalar gluons,
but not between subtleties such as their abelian or non-abelian
group structure: This is because the only difference between an
abelian and non-abelian structure of the qqg-coupling in vector-
gluon theories is due to the "color charge of a quark" $C_2(R)$ in
Eq. (11) which cancels in the ratio in Eq. (25); and similarly for
scalar-gluon theories. For illustration we compare in Table 1 a
typical prediction for the ratio of anomalous dimensions, according
to Eq. (11), with the measured $F_3^{\nu N}$ -moments[8,9]. As we can see,
the measured slope[12] of ordinary x-moments $\langle x F_3^{\nu N}(Q^2) \rangle_n$ is in
good agreement with the predictions of vector-gluon theories. How-
ever, as in the previous case, target mass effects (Nachtmann
moments) play a non-negligible role, although $Q^2 \gtrsim 6.5$ GeV^2 for the
CDHS experiment[8], which <u>decrease</u> the slopes by more than 10 % as
compared to our ordinary moments. On the other hand the slope predictions of scalar theories are typically 20 % smaller than those
of vector theories. Thus, at present, not even scalar-gluon theories
can be ruled out within 1σ on the basis of this non-singlet moment-
slope test in Eq. (25).

A similar conclusion has been reached by Harari[13] as far as the
slopes of NS moments are concerned. He assumes that $xF_3(x,Q^2)$ is
a slowly varying function in the <u>whole</u> x-region for increasing
values of Q^2 which, by the usual bremsstrahlung effects of field
quanta in any field theory, led him to the ansatz

$$ x F_3(x,Q^2) \simeq x^{\frac{1}{2f(Q^2)}} (1-x)^{3g(Q^2)} \tag{26} $$

with f and g arbitrary but slowly varying functions of Q^2, obeying
$f(Q_0^2) = g(Q_0^2) = 1$ and f', $g' \gtrsim 0$. From g'/f' = 0 and f'/g' = 0 one
easily finds lower and upper bounds, respectively, for the slopes
of the logarithmic moment ratios in Eq. (25) which should be valid
for a general class of field theories. Since, as we have already
discussed, the measurements[8,9] for ordinary and Nachtmann-moment
ratios scatter throughout the whole region allowed by these upper
and lower bounds, Harari[13] concludes that this specific test provides no evidence for QCD. It should be pointed out, however, that
in general Harari's "general" bounds are not general. This comes
about because Harari assumes more than can be tested by n > 2
moments: whereas n > 2 moments are sensitive to the <u>large</u>-x region
only (x $\gtrsim$ 0.3), the ansatz in Eq. (26) assumes the structure function
to be a slowly varying function of x and Q^2 in the <u>whole</u> x-region
$(0 \leq x \leq 1)$. It is well known[5,14] that scalar-gluon theories do
satisfy this latter assumption over the whole x-region, and thus
it is not surprising that the ratios of anomalous dimensions in
Eq. (25) predicted by scalar-gluon theories, using Eq. (23), lie
outside Harari's "upper" and "lower bounds": specifically, the

	experiment		theory	
$\langle x F_3^{\nu N} \rangle_n$		Nachtmann	vector	scalar
$a_{NS}(6)/a_{NS}(4)$	1.34 ± 0.07	1.18 ± 0.09	1.29	1.06

Table 1. Comparison of the theoretical predictions for the
n/n' = 6/4 moment ratio with CDHS measurements.[8,12]

ratios $a_{NS}^{scalar}(n)$ / $a_{NS}^{scalar}(n')$ always lie below Harari's "lower bounds".

Recently, a similar moment analysis has been performed[15,16] using the Fermilab data for $F_2^{\mu p}$ and $F_2^{\mu p} - F_2^{\mu n}$ in addition to the SLAC-MIT ep,n data for large values of x. Besides the leading α_s contributions, the subleading α_s^2 corrections[17] (2-loops in anomalous-dimensions) have also been taken into account. If one naively uses 2-loop corrections to $F_{NS} = F_2^{\mu p} - F_2^{\mu n}$ one expects $d \ln \langle F_{NS} \rangle_6 / d \ln \langle F_{NS} \rangle_4$ to change from 1.29 in the 1-loop order to 1.33 in the 2-loop order[17] (for $Q_0^2 = 4$ GeV^2, $Q^2 = 50$ GeV^2 and $\Lambda = 0.5$ GeV). The observed value turns out to be somewhat larger, 1.6 ± 0.2, but a very good agreement with the measurements of the various structure functions is obtained[15,16] by taking into account n-dependent Λ's, i.e. Λ_n increases for increasing n which is "naturally" explained theoretically[18] by going beyond the leading 1-loop order: typically[16] $\Lambda_4 \approx 0.4$ GeV and $\Lambda_6 \approx 0.5$ GeV.

To summarize, we have seen that present measurements of non-singlet moments cannot discriminate between different field theories of the strong interactions and can only provide us with a (necessary) consistency check of QCD. In order discriminatively to test QCD we therefore must turn to structure functions such as F_2 which contain dominant <u>singlet</u> components. This should prove more rewarding since fixed point theories differ from QCD mainly in their singlet mixing properties, because of their very different gluonic anomalous dimensions[5,6] as for example implied by Eq. (12).

3.B. Singlet Moments

The most important and instructive moment to study is the lowest n = 2 moment of F_2, i.e., the area under $F_2(x,Q^2)$. Recall that the fermionic singlet in Eq. (19) is directly measured in neutrino scattering

$$F_2^{\nu N}(x, Q^2) = x \Sigma (x, Q^2) \quad , \qquad (27)$$

whereas deep inelastic e (or μ) scattering off nucleons measures
in addition also the NS part, as for example

$$F_2^{\mu p}(x, Q^2) = \frac{5}{18} x\Sigma(x, Q^2) + \frac{1}{6} x[u + \bar{u} - d - \bar{d} - s - \bar{s} + c + \bar{c}] \qquad (28)$$

with $u = u(x, Q^2)$ etc., and where the Q^2 dependence of the NS expression in square-brackets is determined solely by a_{NS} as in Eq. (8).
For brevity we will discuss the area under the pure singlet structure
function $F_2^{\nu N}$ although the discussion of $\langle F_2^{\mu p}(Q^2)\rangle_2$ is very
similar[6]. According to Eq. (20a) the Q^2-dependence of the lowest
(n=2) moment of the singlet component is given by

$$\langle x\Sigma(Q^2)\rangle_2 = \alpha_2 + [\langle x\Sigma(Q_0^2)\rangle_2 - \alpha_2] \, e^{-s \, a_+(2)} \qquad (29)$$

where we have used $\alpha_2 = \beta_2$, and $\langle xG\rangle_2 = 1 - \langle x\Sigma\rangle_2$ and $a_-(2) = 0$,
by energy-momentum conservation. This quantity, being the total
fractional momentum carried by the fermionic constituents in the
nucleon, is then directly measured by

$$\int_0^1 F_2^{\nu N}(x, Q^2) \, dx = \langle x\Sigma(Q^2)\rangle_2 \quad . \qquad (30)$$

At moderate $Q^2 \simeq$ 2-4 GeV^2, corresponding to our input Q_0^2, experiment
tells us that[9,19] $\langle x\Sigma(Q_0^2)\rangle_2 \simeq 0.52$ and hence, according
to Eq. (29) and since $a_+(2) = 56/75 > 0$, $\langle x\Sigma(Q^2)\rangle_2$ is an increasing
or decreasing function of Q^2 depending on whether α_2 is larger or
smaller than 1/2, respectively. Substituting the different possible
values of the group invariants of Eqs. (13) and (14) into Eqs. (11)
and (12) and also into the appropriate expressions for anomalous
dimensions of scalar-gluon theories[6], it turns out that $\alpha_2 < 1/2$ only
for QCD where $\alpha_2 = 3/7$. It is a unique feature[5] of all other presently known field theories that $\alpha_2 > 1/2$ (specifically[6] $\alpha_2 = 6/7$,
9/10 and 72/73 for abelian vector-gluon, non-abelian scalar-gluon
and abelian scalar-gluon theories, respectively) which forces
$\int_0^1 F_2(x, Q^2) \, dx$ to increase with Q^2. Since $\int_0^1 F_2(x, Q^2) \, dx$ is
experimentally observed[9,19,20] to decrease with Q^2 (or at most to
be constant), all theories except QCD are already excluded on the
basis of this single qualitative observation! In Fig. 5 we compare
the data for $\int_0^1 F_2(x, Q^2) \, dx$ with the predictions of QCD (solid
curves) and of the abelian vector field theory (dotted curves),
for which we have taken the fixed point α^* to be 0.5, in agreement
with our analysis[11] of NS moments in Fig. 4. The predictions of
scalar-gluon theories are in even worse agreement with the data
since their values for α_2 are always larger than 6/7. Similar conclusions have already been reached some time ago by a more detailed
quantitative analysis[5,14] of $F_2^{e(\mu)N}(x, Q^2)$.

It should be noted that, although for n = 2 we have

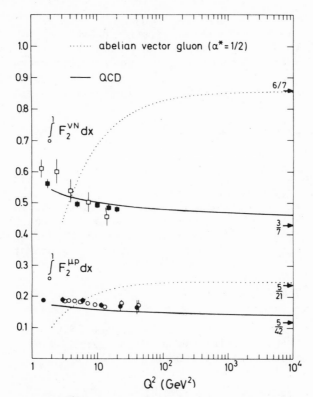

Fig. 5. Comparison[6] of the Q^2 evolution of the area under F_2, predicted by vector gluon theories, with the νN data of Refs. 9 and 19, and with the μp data of Ref. 20.

$$\alpha_2 = \frac{\gamma_{VV}^{V}(2)}{\gamma_{FF}^{F}(2) + \gamma_{VV}^{V}(2)} = \frac{\text{(diagrams)}}{\text{(diagrams)}} \quad , \tag{31}$$

this discriminative test of QCD is <u>not</u> sensitive to the triple gluon coupling since the coefficient of $C_2(G)$ in $\gamma_{VV}^{V}(2)$ vanishes, and therefore the whole contribution to $\gamma_{VV}^{V}(2)$ is due to the term proportional to $T(R)$ in Eq. (12), i.e., to the external wave function renormalization, the vacuum polarization $\sim\!\!O\!\!\sim$ in Eq. (31). Thus, α_2 measures mainly the color charge of quarks, i.e., the quark-gluon coupling in Eq. (1) but not the color charge of gluons, i.e., the Yang-Mills structure of Eq. (2).

Since higher n ($\gtrsim 3$) moments weigh mainly the large x region (x $\gtrsim 0.3$), the study of n > 2 moments of any structure function cannot provide us with additional information on the gluon structure of the theory. This is so because $\gamma_+(n) > \gamma_-(n) = \gamma_{FF}^F(n) - O(1/n^2 \ln n)$ for $n > 2$ and thus always just one anomalous dimension $\gamma_{NS} \equiv \gamma_{FF}^F$ dominates. Therefore, the subtle and very important singlet-mixing properties of the theory, which allow us to study the detailed gluon structure, are only effective for small n, i.e., in the small x-region, not accessible to any moment analysis.

To summarize, recent measurements on $\int_0^1 F_2(x, Q^2)\, dx$ enable us already to eliminate all possible finite fixed point theories by purely <u>qualitative</u> arguments, leaving us with QCD as the only viable theory of the fundamental strong interactions[21]. Since this very discriminative test, as well as any higher $(n > 2)$ moment analysis of F_2, is sensitive only to the quark-gluon coupling of QCD in Eq. (1), we have to resort to the full x- and Q^2-dependence of structure functions (especially in the <u>small</u> x-region, $x \lesssim 0.2$) in order to test and learn about the specific gluon structure of QCD.

3.C. Scaling Violations in $F_2(x,Q^2)$

The most efficient and direct way to obtain the explicit x-dependence of structure functions is to do a numerical Mellin-inversion[22] of the moments predicted by QCD in Eq. (17):

$$F(x,Q^2) = \frac{1}{2\pi i} \int_{c-i\infty}^{c+i\infty} dn\; x^{-n+1} \langle F(Q^2) \rangle_n \qquad (32)$$

As it is apparent from Eq. (20a) we now need, in addition to the[6] input quark distributions (which we have fitted in a standard way to experiment at $Q_0^2 \simeq 4\ \text{GeV}^2$), the gluon distribution $G(x,Q_0^2)$. To check the sensitivity of the predicted scaling violations to the choice of $G(x,Q_0^2)$, we have performed the calculations once with the "standard" gluon distribution $xG(x, Q_0^2 \simeq 4\ \text{GeV}^2) = 2.6(1-x)^5$ and once with $G(x,Q_0^2) = 0$. This latter choice obviously violates the energy momentum sum rule and is intended only as a check on the above mentioned sensitivity to $G(x,Q_0^2)$. Further, to make sure that the results do not sensitively depend on our "standard" input gluon distribution chosen, we have repeated the calculations using a broad gluon $xG(x,Q_0^2) = 0.88(1+9x)(1-x)^4$ as suggested by the Caltech group[23] and which appears to be in better agreement with recent experiments[9,16,19]: within a few percent our predictions in Figs. 6 and 7 (solid curves) remain unchanged. As one can see from Figs. 6 and 7 the scaling violations with the "standard" gluon distribution (full lines) do not differ significantly (i.e., by less than a standard deviation) from the ones with a zero input gluon distribution (dashed lines). A distinction can be made only in the <u>small</u>

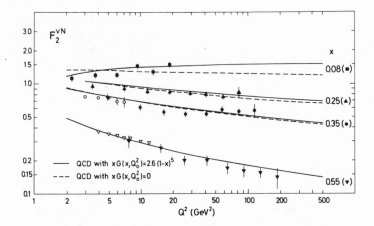

Fig. 6. Predictions[6] of scaling violations according to QCD as
        compared with neutrino data[19] (solid points) and ed data[24]
        (open points) multiplied by 9/5.

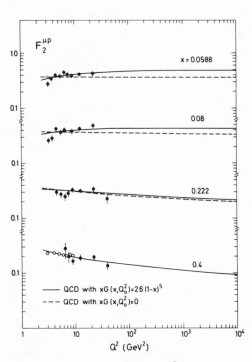

Fig. 7. Comparison of the predictions[6] for scaling violations
        with μp data[20] (solid points) and with ep data[24] (open
        points).

x ($<$ 0.2) region and at higher values of Q^2: future precision
measurements of F_2 for $0.05 \lesssim x \lesssim 0.2$ (heavy quark production should
become important[25] only for smaller values of x) and for Q^2 up to
100-200 GeV^2, say, should prove very useful to pin down the gluon
distribution in the nucleon! Thus any moment analysis of F_2 with
n $\gtrsim$ 3, which is sensitive to the large x-region only, for testing
QCD and determining the gluon distribution[9,15,26] is rendered
meaningless and statistically insignificant.

The fact that the predictions for scaling violations in F_2 are
insensitive to the gluon content of the nucleon for x $>$ 0.1 can be
easily understood from the explicit values of the projection matrix
elements α_n and β_n and how they enter Eq. (20a): From Table 2 we
see that $\alpha_n \approx 1$ and $\beta_n \ll 1$ for n $>$ 2, and therefore only the term proportional to $\alpha_n \langle x\Sigma(Q_0^2)\rangle_n$ in Eq. (20a) will survive, except for
small values of n (small x) where the gluon-terms gradually begin
to contribute. This is, of course, in contrast to the Q^2-evolution
of the gluon distribution itself where the terms proportional to
$\langle xG(Q_0^2)\rangle_n$ in Eq. (20b) always dominate. We shall turn to this point
in the next Section.

Alternatively, instead of using the "heavy artillery" of the renormalization group, the insensitivity of scaling violations to the
detailed shape of the gluon distribution in the nucleon can be understood using the physical more transparent and intuitive language
of Altarelli and Parisi[27]. Here the Q^2 evolution of q(x,Q^2) and
G(x,Q^2) is described by a coupled set of integro-differential equations which are uniquely determined by the parton i $\rightarrow$ parton j decay
probabilities $P_{ji}(x)$. These probabilities can be obtained from the
fundamental interaction vertices of QCD in a probe (beam) independent (!) way[27], which is in contrast to our renormalization group
approach using the light cone expansion. Moreover, the (Mellin)

n	$a_{NS}(n)$	$a_-(n)$	$a_+(n)$	α_n	β_n
2	0.427	0	0.747	0.429	0.429
3	0.667	0.609	1.39	0.925	0.288
4	0.837	0.817	1.85	0.98	0.17
5	0.971	0.960	2.19	0.992	0.119
6	1.08	1.07	2.46	0.996	0.091

Table 2. Values for the renormalization group exponents $a_i(n)$
and for the projection matrix elements $\alpha_n \equiv P_{11}^-(n)$
and $\beta_n \equiv P_{21}^-(n)$ for a four flavor QCD (N_f = 4).

moments of these coupled set of equations are identical to the RG moment-equations (20a) and (20b) by keeping in mind that

$$
\int_0^1 dx \; x^{n-1}
\begin{bmatrix}
P_{qq}(x) \\
2N_f \, P_{qg}(x) \\
P_{gq}(x) \\
P_{gg}(x)
\end{bmatrix}
= -\frac{\pi}{d_s}
\begin{bmatrix}
\gamma_{FF}^{F}(n) \\
\gamma_{VV}^{F}(n) \\
\gamma_{FF}^{V}(n) \\
\gamma_{VV}^{V}(n)
\end{bmatrix}
\tag{33}
$$

Since P_{qg}, which couples $G(x,Q^2)$ to the evolution equation for $q(x,Q^2)$, is small only the term proportional to P_{qq} dominates the effects of scaling violations and therefore the latter are rather insensitive to the gluon content of the nucleon.

For alternative, non-field-theoretic (Regge-like) approaches to scaling violations we refer to Refs. 28 and 29, and to the lecture of G. Preparata. However, these (generalized) vector-meson dominance models can be "trusted" mainly in the small x-region where they might serve to extrapolate[29,30] structure functions down to $Q^2 \simeq 0$ – a region which cannot be reached by perturbative QCD. The power and beauty of explaining scaling violations with field theoretic methods (i.e., radiative corrections in QCD) remains, however, un-challenged in as much as they provide us with a framework for the whole x-region with essentially only one free parameter Λ.

4. MEASURING THE TRIPLE GLUON VERTEX

So far we have seen that all deep inelastic tests strongly favor QCD over any other field theory, and these tests were sensitive only to the quark-gluon coupling and left the non-abelian vertex in Eq.(1) unchallenged. Needless to say that, once the non-Abelian charac-ter of the quark-gluon coupling is established, renormalizability of the theory requires the gluon self-couplings of Eq. (2), i.e., a Yang-Mills gauge-field Lagrangian, which are so essential for asymptotic freedom. This argument in favor of the full QCD Lagrangian is certainly convincing enough to most theorists, it probably is of little – if any – "proof" to most experimentalists. It is therefore also desirable to look for additional measurements which are directly sensitive to the gluon content of the nucleon and thus to the triple-gluon vertex of QCD in order to "see" ex-perimentally the Yang-Mills structure which plays such a prominent role in QCD.

The most direct way to test the triple-gluon vertex directly would be the Q^2 evolution of $G(x,Q^2)$ itself[31] as predicted by the

moments in Eq. (20b). As it has been already discussed, the terms
in Eq. (20b) proportional to $\langle x\Sigma(Q_0^2)\rangle_n$ are suppressed for $n \gtrsim 3$
because of the values of α_n and β_n in Table 2: Thus

$$\langle xG(Q^2)\rangle_n / \langle xG(Q_0^2)\rangle_n \simeq \exp\left(-5\,a_+(n)\right) \qquad \text{with}$$

$a_+(n)$ being critically dependent on $\gamma_{vv}^v(n)$ in Eq. (12) and thus on
the triple-gluon vertex. This sensitivity to the triple-gluon vertex
(the term proportional to $C_2(G)$ in γ_{vv}^v) is demonstrated in Fig. 8
by the difference between the solid and dashed curves, the latter
being the result with $C_2(G) = 0$ in γ_{vv}^v. For these quantitative
predictions we have used[31] the input ratios $\langle x\Sigma(Q_0^2)\rangle_n / \langle xG(Q_0^2)\rangle_n$
determined experimentally[9,19] from the Q^2 variation of F_2-moments
predicted by Eq. (20a). (Similar results hold, of course, also for
the explicit x-dependence, but in order to avoid any ambiguities
on the theoretically ill understood x-dependence of $G(x,Q^2)$ we discuss only moments of structure functions). It should be emphasized
that the predictions of Eq. (20b) for $\langle xG(Q^2)\rangle_n$ cannot serve
as an independent test of QCD if one determines[9,19] the gluon distributions from fitting to the scaling violations of F_2 predicted by
Eq. (20a) since, once $\langle x\Sigma(Q_0^2)\rangle_n$ and $\langle xG(Q_0^2)\rangle_n$ are fixed
by experiment via $\langle x\Sigma(Q^2)\rangle_n$ in Eq. (20a), Eq. (20b) is trivially satisfied. Thus we need an <u>additional, independent</u> determination of $G(x,Q^2)$! This can be achieved[31] by measuring the longitudinal structure function $F_L = F_2 - 2xF_1$ which, to leading order α_S,

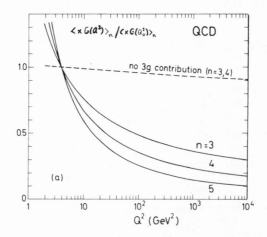

Fig. 8. Predictions[31] for the Q^2 evolution of gluon moments
        according to Eq. (20b). The dashed curve for n = 3,4
        demonstrates the sensitivity of these predictions to the
        gluon self-couplings and corresponds to the contribution
        with the triple-gluon coupling turned off.
        ($C_2(G) = 0$ in γ_{vv}^v), and is similar but slightly larger
        for n = 5.

receives its contribution from

$$F_2(x,Q^2) \qquad\qquad\qquad G(x,Q^2)$$

i.e.,

$$\langle F_L(Q^2)\rangle_n = C_n^q \langle F_2(Q^2)\rangle_n + a\, C_n^g \langle x G(Q^2)\rangle_n \tag{34}$$

with $a = \sum_q e_q^2 = 10/9$ for electroproduction and $a = 4$ for ν and $\bar{\nu}$ scattering on matter, and where the moments of the longitudinal projections of the fundamental parton processes are given by[32]

$$C_n^q = \frac{4\alpha_s(Q^2)}{3\pi\,(n+1)} \quad , \quad C_n^g = \frac{2\alpha_s(Q^2)}{\pi\,(n+1)(n+2)} \quad . \tag{35}$$

From Eq. (34) we see that good data on $\langle F_L(Q^2)\rangle_n$, together with the experimental knowledge of[9,19,20] $\langle F_2(Q^2)\rangle_n$, $n = 2,3\ldots$, can be translated into a reasonable knowledge of the gluon density

$$\langle x G(Q^2)\rangle_n = \frac{\pi(n+1)(n+2)}{2a\,\alpha_s(Q^2)} \langle F_L(Q^2)\rangle_n - \frac{2(n+2)}{3a} \langle F_2(Q^2)\rangle_n \quad . \tag{36}$$

For $n > 2$ moments, measurements for $x \gtrsim 0.3$ should suffice. Furthermore, as can be seen from Fig. 8, measurements in the region $10 \text{ GeV}^2 \lesssim Q^2 \lesssim 100 \text{ GeV}^2$ will be required in order clearly to pin down the triple-gluon coupling; at these large values of Q^2, non-perturbative contributions to F_L can be safely neglected[33] since they are of the order k_T^2/Q^2 or m^2/Q^2, with k_T being the intrinsic transverse momentum of partons and m some typical hadronic mass scale. We are aware of the fact that measurements of F_L, or $R = F_L/F_2$, are exceedingly difficult[34], but feasible[35] in the not too distant future. However, a precision measurement of F_L would be equally fundamental in providing us with a direct and sensitive test of the Yang-Mills structure (gluon self-couplings) of QCD.

For the present status of the measurements of $R \equiv \sigma_L/\sigma_T = F_L/F_2$ we refer to the talk of R. Taylor[36]. Here we only would like to mention that at least the qualitative trend expected by the parton model and the QCD prediction in Eq. (34) (using a "standard" gluon distribution $xG(x,Q_0^2) \approx 2.6(1-x)^5$ as input) is in agreement with the scarce data available up to now. Let us compare these data with the predictions for [33,37] $R(x,Q^2) = R^{\text{intrinsic}} + R^{QCD}$ where the intrinsic part is due to taking into account kinematical target mass effects (intrinsic transverse momenta) and R^{QCD} results from Eq. (34): the SLAC ep measurements ($0.3 \lesssim x \lesssim 0.8$, $3 \text{ GeV}^2 \lesssim Q^2 \lesssim 18 \text{ GeV}^2$) give on the average [36] $R = 0.21 \pm 0.1$ to be compared with $R(0.5, 8 \text{ GeV}^2) = 0.025 + 0.035 = 0.06$, whereas the Fermilab μp experiment ($0.003 < x < 0.1$, $1 \text{ GeV}^2 \lesssim Q^2 \lesssim 30 \text{ GeV}^2$) gives an average value of [15,16] $R = 0.52 \pm 0.35$, which can be compared with the prediction $R(0.02, 15 \text{ GeV}^2) = 0.0 + 0.2 = 0.2$. The expected increasing trend of R for decreasing x seems to be reproduced by the data. To illustrate the expected [33,37] x- and Q^2-dependence of $R(x,Q^2)$ we give a few predictions in Table 3. These values can be increased if one chooses a harder (flatter) gluon distribution than our "standard" choice.

Similar direct tests of the triple-gluon vertex can be obtained by looking [38] for the Q^2 evolution of gluon jets in heavy quarkonium decay, i.e., measuring the gluon decay function $D_g^h(z,Q^2)$ in $e^+e^- \to Q\bar{Q} \to 3g \to h +$ anything of successive 1^3S_1 quarkonium states (e.g., at $Q_T^2 \simeq 100 \text{ GeV}^2$ and, provided toponium exists, at $Q_{t\bar{t}}^2 \simeq 1000 \text{ GeV}^2$)

$Q^2 (\text{GeV}^2)$	x	R	$R^{\text{intrinsic}}$	R^{QCD}
2	0.8	0.1	0.08	0.02
	0.5	0.16	0.1	0.06
	0.2	0.19	0.04	0.15
	0.02	0.53	0.0	0.53
10	0.8	0.025	0.015	0.01
	0.5	0.05	0.02	0.03
	0.2	0.08	0.01	0.07
	0.02	0.23	0.0	0.23
20	0.8	0.014	0.006	0.008
	0.5	0.035	0.01	0.025
	0.2	0.063	0.003	0.06
	0.02	0.18	0.0	0.18
100	0.8	0.007	0.002	0.005
	0.5	0.02	0.002	0.018
	0.2	0.04	0.0	0.04
	0.02	0.12	0.0	0.12

Table 3. Predicted values for [33] $R(x,Q^2) = R^{\text{intrinsic}} + R^{QCD}$ for electroproduction. The predictions at $Q^2 = 2 \text{ GeV}^2$ correspond to naive (Q^2-independent) parton distributions.

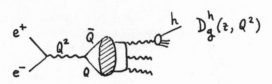

Since D_q^h satisfies[39] a similar renormalization group equation as in Eq. (20b) (or, equivalently, similar Altarelli-Parisi equations as do the distribution functions), the predicted[38] Q^2 evolution is again critically sensitive to the gluon self-couplings and is similar to the predictions in Fig. 8.

Finally it has been suggested[40] to look for T-odd asymmetries in the hadronic decays of heavy quarkonia produced in $e^+ e^-$ collisions, where electrons and positrons are longitudinally polarized with opposite helicities. For Υ production this asymmetry is expected to be as small as[40] 0.3 %, and therefore non-perturbative effects due to final state interactions may totally mask these predictions.

5. DEEP INELASTIC HEAVY QUARK PRODUCTION

Due to the large mass, all heavy quark flavors (c,b,...) in the nucleon are expected[41] to exist only as quantum fluctuations at short distances and the corresponding distribution can be consistently calculated within QCD. Therefore at $Q^2 \gg 4m_Q^2$, heavy quark Q=c,b,... production in deep inelastic electron (muon) nucleon scattering should, in leading order, be adequately described[42,25] by the renormalization group improved virtual Bethe-Heitler process

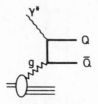

which is proportional to the gluon content $G(x,Q^2)$ of the nucleon. Thus the nucleon should be considered to consist only of the three light quarks u,d,s and of gluons G, with all other heavy quark flavors being produced via these light quark and gluon fields. The total charm contribution to F_2 can then be calculated to be[25] ($b\bar{b}$ production is about two orders of magnitudes smaller than $c\bar{c}$)

$$F_2^{c\bar{c}}(x,Q^2) = \frac{1}{N} \int_{ax}^{y_{max}} \frac{dy}{y} \, y G(y,Q^2) \, f_2^{\gamma^* g \to c\bar{c}}\left(\frac{x}{y},Q^2\right) \qquad (37)$$

with $a = 1 + 4 m_c^2 / Q^2$, $Q^2 \equiv -q^2$, and for total charm production $N \equiv 1$, $y_{max} = 1$, whereas for J/ψ production $N =$ (number of charmonium states) $\simeq 8$, $y_{max} = x(1 + 4m_J^2/Q^2)$; furthermore the fundamental sub-process $\gamma^* g \to c\bar{c}$ gives

$$f_2^{\gamma^* g \to c\bar{c}}(z, Q^2) = \frac{4}{9} \frac{\alpha_s(Q^2)}{\pi} \left\{ v \left[4z^2(1-z) - \frac{z}{2} - \frac{2m_c^2}{Q^2} z^2(1-z) \right] \right.$$
$$\left. + \left[\frac{z}{2} - z^2(1-z) + \frac{2m_c^2}{Q^2} z^2(1-3z) - \frac{4m_c^4}{Q^4} z^3 \right] \ln \frac{1+v}{1-v} \right\} \tag{38}$$

with $v^2 = 1 - 4m_c^2 z/Q^2(1-z)$. From the predictions for "open charm" and J/ψ production according to Eq. (37) as shown in Fig. 9, we observe a very <u>steep</u> and non-negligible contribution for $x \lesssim 0.01$ which increases rapidly not only for decreasing values of x but also for increasing Q^2. For example, fitting the x dependence of the charm predictions at $Q^2 = 4$ GeV^2, the charm sea is expected to behave like (for $m_c = 1.25$ GeV)

$$x c(x, Q^2 \simeq 4 \text{ GeV}^2) = 0.05 (1-x)^{30} \quad . \tag{39}$$

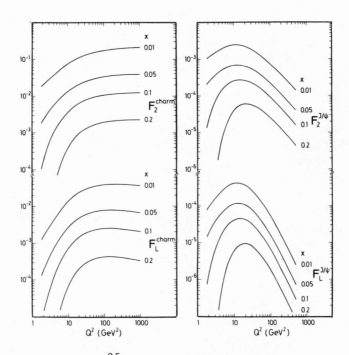

Fig. 9. Predictions[25] for total charm and J/ψ production according to Eq. (37).

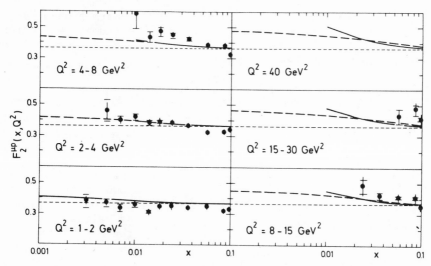

Fig. 10. Comparison of the predictions[25] of total charm production, Eq. (37), with μp data[20]. These predictions
are added to the short-dashed curves which correspond
to the light quark contributions to $F_2^{\mu p}$ (x=0.01)=0.365.
The full QCD predictions of Eq. (37) yield the solid
curves, while the long-dashed curves correspond to
using a naive Q^2-independent gluon distribution in
Eq. (37).

Furthermore, Fig. 9 tells us that, at Q^2 = 10 GeV^2 and x ≃ 0.01,
the total charm and J/Ψ production in F_2 is about 0.08 and 0.003,
respectively, which is about a 20 % contribution to the measured
total value of F_2. To illustrate the size of F_2^{charm}, we compare it
with actual data in Fig. 10 on top of $F_2^{\mu p}$(x ≃ 0.01) = 0.365. It seems
appropriate to conclude that the amount of charm produced by the
process $\gamma^* g \to c\bar{c}$ accounts for <u>all</u> the charm component of $F_2(x,Q^2)$
presently observed. The CERN-EMC[43] and the Berkeley-Princeton[44] μp
experiment at Fermilab should provide us with stringent tests of
this production mechanism of heavy quark flavors.

Furthermore, photoproduction of charm is also expected to proceed via the Bethe-Heitler process $\gamma g \to c\bar{c}$ for which QCD makes
firm predictions[46]. Again, μp precision measurements[43,44] should
be able to discriminate between alternative production mechanisms[47].

In neutrino scattering charm is already produced by the weak
current in zeroth order, in contrast to electroproduction, via the
naive quark-parton process $W^+ s \to c$. Additional (non-leading)
corrections originate[41,48] from the virtual weak Bethe-Heitler
process $W^+ g \to c\bar{s}$. Such calculations, however, depend now critically
on the light s-quark mass chosen because of the appearance of terms
like $\ln (\hat{s} - m_c^2)/m_s^2$; these large logarithms should be absorbed

as usual into the RG improved strange quark distribution s of the
0-th order $W^+ s \to c$ process.

6. NON-LEADING CORRECTIONS

Finally, I would like to comment briefly on various recent
analyses concerning subleading α_s corrections to Wilson coefficients
("finite terms") and 2-loop contributions to anomalous dimensions.
Taking into account these non-leading terms, i.e.,

$$\beta(\alpha_s) = -\beta_0 \alpha_s^2 - \beta_1 \alpha_s^3$$
$$\gamma^n(\alpha_s) = \gamma_0^n \alpha_s + \gamma_1^n \alpha_s^2 \tag{40}$$
$$C_i^n(\alpha_s) = 1 + c_i^n \alpha_s \quad ,$$

and inserting these expansions into the solution (7) of the RG
equation

$$C_i^n(Q^2/\mu^2, \alpha_s(\mu^2)) = C_i^n(1, \alpha_s(Q^2)) \exp\left[-\frac{1}{2} \int_{\alpha_s(\mu^2)}^{\alpha_s(Q^2)} d\alpha \, \frac{\gamma^n(\alpha)}{\beta(\alpha)}\right] \tag{41}$$

where $C_i^n(1, \alpha_s(Q^2)) \equiv C_i^n(\alpha_s(Q^2))$, the Q^2-dependence of moments
of structure functions in Eq. (8) is predicted to be

$$\langle F_i(Q^2) \rangle_n = \langle q_0 \rangle_n \left\{ 1 + \alpha_s(Q^2) \left[c_i^n + \left(\frac{\gamma_1^n}{2\beta_0} - \frac{\gamma_0^n \beta_1}{2\beta_0^2} \right) \right] \right\} e^{-\frac{\gamma_0}{2\beta_0} S} . \tag{42}$$

Here, $\langle q_0 \rangle_n$ denotes the moments of the matrix elements of the
local Wilson operator in Eq. (5) between target states which are
nothing else but appropriate combinations of input quark distributions fitted to experiment, and, because of Eq. (40), the corrected
form of $\alpha_s(Q^2)$ is

$$\frac{1}{\alpha_s(Q^2)} = \beta_0 \ln \frac{Q^2}{\Lambda^2} + \frac{\beta_1}{\beta_0} \ln \frac{\ln Q^2/\Lambda^2}{\ln \mu^2/\Lambda^2} \quad , \tag{43}$$

instead of Eq. (3), with $\Lambda^2 \equiv \mu^2 \exp[-1/\beta_0 \alpha_s(\mu^2)]$ and
$\beta_0 = 4\pi \ell$. It is clear from Eq. (42) that $O(\alpha_s)$ corrections to
the Wilson coefficients C_i^n have to be taken into account, once the
2-loop contributions β_1^n and γ_1^n to β and γ^n are considered, in
order to include consistently all contributions in a given order
of perturbation theory. We recall that only the <u>whole</u> Eq. (41)
corresponds to a physical measurable quantity, whereas the individual quantities $C_i^n(\alpha_s(Q^2))$ and exp[...] depend upon the precise

definition of the Wilson operator (the renormalization prescription): Although the parameters γ_0^n, β_0 and β_1 are gauge and renormalization prescription independent, the quantities c_i^n and γ_1^n depend on the renormalization prescription and on the gauge chosen. Thus c_i^n and γ_1^n must be calculated in the same renormalization scheme in order to obtain a physical, convention independent answer for Eq. (41).

The γ_1^n for non-singlet operators (i.e., for NS structure functions) has been calculated in Ref. 49; comparing these results quantitatively[50] with the data for $F_2^{\mu P}$ for $x \gtrsim 0.4$ showed that the only effect of subleading contributions was a change in Λ by about 20-30% as compared to the value of Λ obtained by fitting the leading order (1-loop) expressions to experiment.

Furthermore, the rather lengthy calculations of γ_1^n for singlet structure functions have recently been completed[17,51] which, together with the $O(\alpha_s)$ contributions[18,52] to the coefficient functions C_i^n, allow now a detailed quantitative comparison of non-leading terms with experiment. As already discussed in Section 3.A, a moment analysis[15,16] of F_2 yields (by using n-dependent[18] Λ's) an equally good agreement with experiment as do the leading 1-loop terms γ_0^n with Λ held fixed. Moreover, studying[53] the explicit x- and Q^2-dependence of $F_2^{e(\mu)P}(x, Q^2)$ in the 2-loop approximation also gives an equally good agreement with experiment as does the 1-loop approximation, provided the scale Λ of the effective coupling constant is reduced from 0.5 GeV to about 0.3 GeV. Therefore, although further detailed analyses are certainly required before we can draw definite conclusions, it appears that non-leading terms do not significantly alter the successful quantitative results based on the 1-loop approximation; this gives us some additional confidence in the usefulness and validity of perturbative lowest order calculations in QCD.

For practical applications a very neat "trick" has been suggested[52] to study "finite" α_s terms in C_i^n without having to take into account the 2-loop contribution γ_1^n explicitly. This is achieved by defining effective Q^2-dependent parton distributions relative to F_2, i.e., by demanding that $F_2(x, Q^2)$ expressed in terms of them should have the same form as in the naive quark model (F_2 is given a special status because it satisfies the Adler sum rule):

$$\langle q(Q^2) \rangle_n \equiv C_2^n(\alpha_s(Q^2)) \langle q_0 \rangle_n \, exp[\cdots] \qquad (44)$$

instead of the usual definition $\langle \tilde{q}(Q^2) \rangle_n \equiv \langle q_0 \rangle_n \, exp[\cdots]$ where $exp[\cdots]$ is our general RG exponent in Eq. (41). Similar definitions[52] apply to singlet (gluon) densities. Because of the inclusion of the coefficient function in the definition (44) of parton densities, the Q^2-evolution equations of these quantities are the same[52] as in Section 2 keeping only the 1-loop anomalous dimensions

γ_0^n, with γ_1^n being suppressed by $O(\alpha_s)$. Thus, once the input parton distributions $q(x,Q_0^2)$ are fitted at $Q^2 = Q_0^2$ to
$F_2(x,Q^2) = x \sum_q [q(x,Q^2) + \bar{q}(x,Q^2)]$, which evolve according to Eqs. (20), say, one can study the effects of finite terms in structure functions (or processes) <u>other</u> than F_2 since

$$\langle F_i(Q^2)\rangle_n = \frac{C_i^n(\alpha_s(Q^2))}{C_2^n(\alpha_s(Q^2))}\; C_2^n(\alpha_s(Q^2))\,\langle q_0\rangle_n\, exp[\cdots]$$

$$\equiv \frac{C_i^n(\alpha_s(Q^2))}{C_2^n(\alpha_s(Q^2))}\,\langle q(Q^2)\rangle_n$$

(45)

which is of course a fully gauge invariant and renormalization prescription independent procedure. Thus, expanding in α_s, we always get <u>differences</u> $c_i^n - c_2^n$ of "finite" terms contributing to F_i with $i \neq 2$:

$$\langle F_3(Q^2)\rangle_n = (-\langle q(Q^2)\rangle_n + \langle \bar{q}(Q^2)\rangle_n)[1 + \alpha_s(Q^2)(c_{3,q}^n - c_{2,q}^n)]\;,\; etc.$$ (46)

where the fermionic Wilson coefficients $c_{i,q}^n$ result from[52]

and gluonic Wilson coefficients $c_{i,G}^n$ are calculated from

The importance of these "finite" $\alpha_s(Q^2)$ corrections for phenomenological applications are obvious. For example, the effect of gluon corrections on sea distributions to be determined from neutrino reactions are enormous[52]:

$$\sigma_{RH}^\nu \sim \frac{1}{2} \int_0^1 dx\, (2x F_1^\nu + x F_3^\nu)$$

(47)

$$= \underbrace{(\bar{U}+\bar{D})(1-0.16\alpha_s)}_{\simeq 0.06} + \underbrace{0.018\alpha_s(U+D+2S)}_{\simeq 0.003\ (small)} - \underbrace{0.106\alpha_s G}_{\simeq 0.02}$$

$$\sigma^{\nu}(y=0) - \sigma^{\nu}(y=1) \sim \int_0^1 dx \, (F_2^{\nu} - xF_1^{\nu} + \tfrac{1}{2}xF_3^{\nu})$$

$$= (\bar{U}+\bar{D})(1-0.018\alpha_s) + 0.16\alpha_s(U+D+2S) + 0.106\alpha_s G \tag{48}$$

$$\underbrace{\phantom{(\bar{U}+\bar{D})(1-0.018\alpha_s)}}_{\simeq 0.06} \quad \underbrace{}_{\simeq 0.025} \quad \underbrace{}_{\simeq 0.02}$$

where $U \equiv \langle xu(Q^2)\rangle_2$, etc. and we have neglected the small contributions from the charm sea. It is already clear from these equations that the gluon corrections cannot be neglected for a precise quantitative determination of sea densities; even quarks play a non-negligible role in determining the deviations from flatness in $\sigma^{\nu}(y)$ as it is evident from Eq. (48).

Similarly, one can and has to check simultaneously, using the same parton distributions determined in deep inelastic reactions, the importance of "finite" terms in other reactions such as Drell-Yan dimuon production $(pp \to \mu^+\mu^- + X)$. Here "finite" α_s terms are obtained[52,54] from the same, but crossed diagrams which yielded the above quantities $c_{i,q}^n$ and $c_{i,G}^n$. The corrected Drell-Yan formula thus obtained is

$$\frac{d\sigma_{DY}^{P_1 P_2 \to \mu^+\mu^- X}}{dQ^2} = \frac{4\pi\alpha^2}{9sQ^2} \int_0^1 \frac{dx_1}{x_1} \int_0^1 \frac{dx_2}{x_2} \Big\{ \Big[\sum_q e_q^2 \, q^{(1)}(x_1,Q^2)\, \bar{q}^{(2)}(x_2,Q^2) + (1\leftrightarrow 2) \Big]$$

$$\times \Big[\delta(1-z) + \alpha_s \,\Theta(1-z)\big(f_q^{DY}(z) - 2f_{2,q}(z)\big) \Big]$$

$$+ \alpha_s \,\Theta(1-z)\Big[\tfrac{1}{x_1} F_2^{(1)}(x_1,Q^2)\, G^{(2)}(x_2,Q^2) + (1\leftrightarrow 2) \Big] \Big[f_G^{DY}(z) - f_{2,G}(z) \Big] \tag{49}$$

with $z = Q^2/x_1 x_2 s$ and where the two terms $f_q^{DY} - 2f_{2,q}$ and $f_G^{DY} - f_{2,G}$, with $c_i^n \equiv \int_0^1 dz\, z^{n-1} f_i(z)$, are given, for example, in Ref. 52. It is now generally agreed[55,56] that the quark-gluon correction in Eq. (49) is small (less than -10%), provided the full Q^2-dependence of the (nonscaling) parton distributions is taken into account! However, the existing analyses[56,57] concerning the size of the α_s correction to $q\bar{q}$ scattering in Eq. (49) yield entirely contradictory results.

ACKNOWLEDGEMENT

Some of the work described here has been done in collaboration with M. Glück; I would like to thank him for several helpful discussions.

REFERENCES AND FOOTNOTES

1. D.J. Gross and F. Wilczek, Phys. Rev. Lett. $\underline{30}$, 1343 (1973);
   Phys. Rev. $\underline{D8}$, 3633 (1973);
   H.D. Politzer, Phys. Rev. Lett. $\underline{30}$, 1346 (1973).

2. S. Coleman and D.J. Gross, Phys. Rev. Lett. $\underline{31}$, 851 (1973).

3. R.A. Brandt and G. Preparata, Nucl. Phys. $\underline{B27}$, 541 (1971);
   Y. Frishman, Phys. Rev. Lett. $\underline{25}$, 966 (1970); Ann. of Phys.
   $\underline{66}$, 373 (1971);
   R. Jackiw, R. van Royen, and G.B. West, Phys. Rev. $\underline{D2}$, 2473
   (1970).

4. D.J. Gross and F. Wilczek, Phys. Rev. $\underline{D9}$, 980 (1974);
   H. Georgi and H.D. Politzer, ibid. $\underline{9}$, 416 (1974).

5. M. Glück and E. Reya, Phys. Rev. $\underline{D16}$, 3242 (1977).

6. M. Glück and E. Reya, DESY 79/13 (1979), to appear in Nucl.
   Phys. B.

7. N. Christ, B. Hasslacher and A.H. Mueller, Phys. Rev. $\underline{D6}$,
   3543 (1972).

8. CDHS coll., J.G.H. de Groot et al., Phys. Lett. $\underline{82B}$, 292 (1979).

9. BEBC coll., P.C. Bosetti et al., Nucl. Phys. $\underline{B142}$, 1 (1978).

10. O. Nachtmann, Nucl. Phys. $\underline{B63}$, 237 (1973);
    S. Wandzura, Nucl. Phys. $\underline{B122}$, 412 (1977).

11. E. Reya, DESY 79/02 (1979), to appear in Phys. Lett. B.

12. For further measured slopes we refer to Refs. 8 and 9 but these
    are not to be construed as giving several independent tests of
    QCD, since the moments are highly correlated with one another
    and not much new information is provided once the result for
    one pair of moments is given.

13. H. Harari, SLAC-PUB-2254 (1979), submitted to Nucl. Phys. B.

14. M. Glück and E. Reya, Phys. Lett. $\underline{69B}$, 77 (1977).

15. H.L. Anderson, these Proceedings.

16. H.L. Anderson et al., A Measurement of the Nucleon Structure
    Functions, University of Chicago preprint, 1979.

17. E.G. Floratos, D.A. Ross, and C.T. Sachrajda, Phys. Lett. 80B, 269 (1979).

18. W.A. Bardeen, A.J. Buras, D.W. Duke, and T. Muta, Phys. Rev. D18, 3998 (1978).

19. CDHS coll., J.G.H. de Groot et al., Phys. Lett. 82B, 456 (1979).

20. B.A. Gordon et al., Phys. Rev. Lett. 41, 615 (1978); H.L. Anderson et al., Phys. Rev. Lett. 40, 1061 (1978).

21. Combined field theoretic models of QCD and fixed point theories, the latter being treated as small perturbations, can, of course, at the present state of the art, not be excluded by presently available experiments. See, for example, K. Watanabe, University of Nagoya preprint DPNU-59-78 (1978).

22. M. Glück and E. Reya, Phys. Rev. D14, 3034 (1976).

23. R.P. Feynman, R.D. Field and G.C. Fox, Phys. Rev. D18, 3320 (1978).

24. E.M. Riordan et al., SLAC-PUB-1634 (1975), unpublished.

25. M. Glück and E. Reya, DESY 79/05 (1979), and Phys. Lett. 83B, 98 (1979).

26. D.W. Duke and R.G. Roberts, Rutherford Lab. preprint RL-79-025, T.238 (1979).

27. G. Altarelli and G. Parisi, Nucl. Phys. B126, 298 (1977); see also K.J. Kim and K. Schilcher, Phys. Rev. D17, 2800 (1978).

28. See, for example, M. Greco in Lepton and Hadron Structure, Erice 1974, p. 262, and references therein; G. Preparata, this Conference.

29. G.J. Gounaris and S.B. Sarantakos, Phys. Rev. D18, 670 (1978).

30. I would like to thank G.J. Gounaris for a discussion on this point.

31. E. Reya, DESY 79/15 (1979).

32. A. Zee, F. Wilczek and S.B. Treiman, Phys. Rev. D10, 2881 (1974); A. de Rújula, H. Georgi and H.D. Politzer, Ann. of Phys. 103, 315 (1977); I. Hinchliffe and C.H. Llewellyn Smith, Nucl. Phys. B128, 93 (1977)

33. M. Glück and E. Reya, Nucl. Phys. B145, 24 (1978).

34. Conventional radiative corrections, especially for antineutrino reactions, might further complicate the matter (A. de Rújula, R. Petronzio and A. Savoy-Navarro, Ref. TH.2593-CERN (1979)).

35. See, for example, E. Gabathuler, XIX International Conference on High Energy Physics, Tokyo, 1978, and private communication concerning the CERN EMC collaboration; CDHS coll., F. Eisele and K. Kleinknecht, private communication.

36. R. Taylor, this Conference.

37. A. de Rújula et al., in Ref. 32.

38. K. Koller, T.F. Walsh and P.M. Zerwas, Phys. Lett. 82B, 263 (1979)

39. T. Uematsu, Kyoto University preprint RIFP-292, June 1977, and submitted to the 1977 International Symposium on Lepton and Photon Interactions, Hamburg; Phys. Lett. 79B, 97 (1978); see also J.F. Owens, Phys. Lett. 76B, 85 (1978).

40. A. de Rújula, R. Petronzio and B. Lautrup, Nucl. Phys. B146, 50 (1978).

41. E. Witten, Nucl. Phys. B104, 445 (1976).

42. M.A. Shifman, A.I. Vainshtein and V.I. Zakharov, Nucl. Phys. B136, 157 (1978).

43. J.J. Aubert, this Conference; H. Stier, this Conference.

44. M. Strovink, this Conference.

45. J.P. Leveille and T. Weiler, Nucl. Phys. B147, 147 (1979).

46. M. Glück and E. Reya, Phys. Lett. 79B, 453 (1978).

47. H. Fritzsch and K.H. Streng, Phys. Lett. 72B, 385 (1978).

48. J. Babcock and D. Sivers, Phys. Rev. D18, 2301 (1978).

49. E.G. Floratos, D.A. Ross and C.T. Sachrajda, Nucl. Phys. B129, 66 (1977); ibid. B139, 545 (1978).

50. A.J. Buras, E.G. Floratos, D.A. Ross and C.T. Sachrajda, Nucl. Phys. B131, 308 (1977).

51. E.G. Floratos, D.A. Ross and C.T. Sachrajda, Ref.TH.2566-
    CERN (1978).

52. G. Altarelli, R.K. Ellis and G. Martinelli, Nucl. Phys. B143,
    521 (1978); ibid. B146, 544 (1978).

53. D.A. Ross, Caltech preprint CALT-68-699 (1979).

54. J. Abad and B. Humpert, Phys. Lett. 78B, 627 (1978); and
    University of Wisconsin preprint COO-881-44 (1978);
    J. Kubar-André and F.E. Paige, Phys. Rev. D19, 221 (1979);
    K. Harada, T. Kaneko and N. Sakai, Ref.TH.2619-CERN (1979),
    and Erratum.

55. A.P. Contogouris and J. Kripfganz, Scale Violations and the
    Quark-Gluon Correction to the Drell-Yan Formalism, McGill
    University preprint (1978).

56. G. Altarelli, R.K. Ellis and G. Martinelli, MIT preprint
    CTP #776 (1979).

57. J. Kripfganz and A.P. Contogouris, The Quark-Quark Correction
    to the Drell-Yan Formalism, McGill University preprint (1979).

A REALISTIC LOOK AT DEEP INELASTIC PHYSICS

G. Preparata

CERN, Geneva, Switzerland
and
Istituto di Fisica dell'Università, Bari, Italy

INTRODUCTION: REALISM VERSUS QUANTUM CHROMODYNAMICS

Let me begin by commenting on the title I have chosen for these talks. It seems to me that, for what we know, *realistic* is an adjective which cannot be attributed to the picture of deep inelastic physics which today dominates the high-energy physics scene, and is based on quantum chromodynamics (QCD)*. And before people get scandalized, and recite the amazing list of striking successes of this theory, let me try to explain why I believe that there is a fundamental lack of *realism* in the theoretical ideas most of my colleagues are now pursuing.

In order not to get entangled in an endless philosophical debate about what physical *reality* is, I would like to stress that by this word I denote a general class of natural phenomena with which we have become familiar enough, and of which we possess descriptive models of such an adequacy that we are able, successfully, to predict what will happen in situations which we have not encountered previously.

Let me illustrate this by some examples. When we apply the ideas of Galileo-Newtonian mechanics to the observation of the motion of a rocket, we are realistically describing the physical situation. In the same way we are realistically describing the behaviour of a hydrogen atom, by putting to work the machinery of quantum mechanics. On the other hand, we are not giving a realistic description of Maxwell phenomena by invoking, as most scientists

---

* Consult most of the current literature on the subject.

at the end of the last century did, the existence of an all per-
vading "ether". Hoping that, although schematically, I have clarified what I mean by a physically realistic theory, I shall now bring
forth the reasons why I do not think that QCD, the way people understand it today, is realistic.

As a Lagrangian quantum field theory QCD is certainly a good
theory. It is renormalizable, it is a gauge theory, it contains
the quantum numbers of colour and flavour indispensable for a theory
of hadronic interactions. *Qualitatively* it is great. But it is in
the *quantitative* aspects that QCD fails, at least for the time being,
thus bringing about a picture of hadronic phenomena which is utterly
unrealistic. Science cannot survive if the qualitative description
does not yield to the quantitative one. Nature's book, said Galilei,
is written in a mathematical language. And he was certainly right.

Nobody can object that a *truly* quantitative QCD theory does not
exist; before we get to that point we must be able to handle the
solutions of a quantum field theory far away from the free field
limit, i.e. in a *non-perturbative* régime. In the fifty years or so
that quantum field theories have been with us, no really general
scheme of attack of non-perturbative problems has been found, thus
rendering this fascinating chapter of quantum mechanics largely a
terra incognita. Strictly speaking, QCD also is a no-man's land.
But quantum electrodynamics (QED), many will object, is not. And
they are right; because Nature was so kind as to give us α, a small
coupling constant, so that perturbation theory could be used to give
an approximate (order by order in α) description of the whole world
of leptons and photons, which has been gratifying us enormously with
very predictive and accurate results. But also QCD -- again many
will argue -- has a small coupling, $\alpha_s(Q^2)$, in the large momentum
transfer region. And it is in this statement, which is at the very
root of present understanding of QCD, that, in my opinion, a grave
lack of *realism* (in the sense clarified above) overshadows the
whole approach. For the asymptotic freedom behaviour of $\alpha_s(Q^2)$
follows only from a *perturbative* analysis of QCD. Once we are in
the perturbative world, a world where quarks and gluons exist as
asymptotic states, there is no known way, at least to me, to get
out of it. We simply remain in it. Note that such a world is a
perfectly possible one, very close to the QED universe; *but it is
not the real world*. The real world is, on the other hand, populated by a multitude of hadronic states, whose spectrum can be described remarkably well by a simple quark model; no convincing
evidence has so far been presented in favour of the existence of a
free, "asymptotic", quark. Thus the QCD theorist today faces the
following dilemma: his calculations become possible only in a
world which is unreal; in the real world, on the other hand, his
calculations are impossible. Unfortunately only a small minority
of QCD theorists show awareness of this paradoxical situation, and

take a cautious attitude towards applying the QCD "predictions" to experimentally observed facts; today the literature is flooded with "rigorous" QCD predictions, and with "remarkable" experimental confirmations of the most diverse QCD calculations. But if one looks at the real theoretical meaning of the present "practical" QCD, one finds the old quark-parton model barely disguised by the subtleties of the renormalization group, which change the well-known parton model results effectively in a very minor way. The old difficulties of interpretation of the parton model remain in all of their gravity; and the knowledge that one is backed up by a fully-fledged Lagrangian field theory is of little consolation: one is working in a perturbative world and stays there. The link between perturbative QCD, populated by quark and gluon quanta, and the *real* world remains enshrouded in the deep mystery of "confinement", precisely in the same way as in the parton model.

To get back to deep inelastic physics, much emphasis has recently been put just on the deviations from simple parton behaviour, predicted by the renormalization group. The patterns of scaling violations, which have been revealed by experiments on lepton-nucleon scattering at large Q^2, have been taken as an almost "final" proof of the validity of QCD. We have heard at this Seminar[1] that there are now a few difficulties in fitting the data with the simple QCD parametrization and that the "Perkins' plots" do not seem to pin down so much asymptotically free quantum field theories against other types of behaviour. We are already at a point where higher order perturbative corrections are invoked to face what begins to appear to be a crisis. Put briefly, confusion seems to be increasing.

I hope that the present-day difficulties of QCD will at least open up new spaces to doubt and to an objective appreciation of what alternatives to the dominant theory there are at our disposal, and possibly to the finding of new ones.

In the following I shall expose to your criticism a set of theoretical ideas I have been developing for the greater part of this decade[2]. Their main motivation is to produce a *realistic* description of the hadronic world, which has a chance to cope not only with deep inelastic phenomena but also with the particles' spectra and their high-energy scattering; and by *realistic* I mean a description which makes use of quarks only in a way which is *from the beginning* compatible with what we know about the *real* physical world, i.e. that quarks differ from other spin-½ particles like leptons in a very fundamental way. On the other hand, the mysterious "confinement" mechanism that is invoked by QCD theorists as a type of *deus ex machina* to save the picture, like the classic scenic stratagem, has the flaw of being an artifice, which comes into the play only at the end.

Thus a realistic picture of hadrons must contain, if not explain, the confinement mechanism from the very start. This is the most distinctive character of the alternative to present-day QCD that I have to offer.

1. A REALISTIC LOOK AT DEEP INELASTIC SCATTERING

1.1 The Basic Ideas of the Massive Quark Model[2],[3]

The massive quark model (MQM) is a theoretical framework suited to describing dynamical degrees of freedom, to which no asymptotic quanta correspond. According to present knowledge quarks are of this type. A full discussion of the physical motivations and of the theoretical implications of the MQM has been given five years ago in lectures I delivered here in Erice[3]. Here I shall only summarize the main points:

  i) The dynamical degrees of freedom which we identify with quarks can be localized only inside finite space-time domains (bags) associated with "colourless" hadrons;

 ii) In the bag-domains the quark degrees of freedom behave like (effective) low-mass hadrons;

iii) Quarks are coupled to electromagnetic (e.m.) and weak currents in a point-like fashion.

Physical processes will be described by diagrams, analogous to quantum field theory diagrams, with the important distinctive feature that, according to (i), the propagators associated with quark lines cannot have any singularity, i.e. must be entire analytic functions which, by (ii), drop fast on the real p^2 axis. Furthermore, (ii) implies that in the appropriate high-energy regions the quark amplitudes exhibit Regge behaviour, with the same Regge trajectories that dominate high-energy hadronic scattering. Finally, (iii) introduces the observed point-like behaviour of quarks, when they are "probed" by weak and e.m. currents.

1.2 MQM Description of Deep Inelastic Scattering

As everybody knows, by deep inelastic scattering (DIS), one means the process (see Fig. 1)

$$J_\mu(q) + N(p) \rightarrow \text{all} \tag{1}$$

where a current J_μ with momentum q scatters off a nucleon of momentum p, to produce a highly inelastic state of mass W^2. By means of the optical theorem, and of the point-like nature of the current-quark

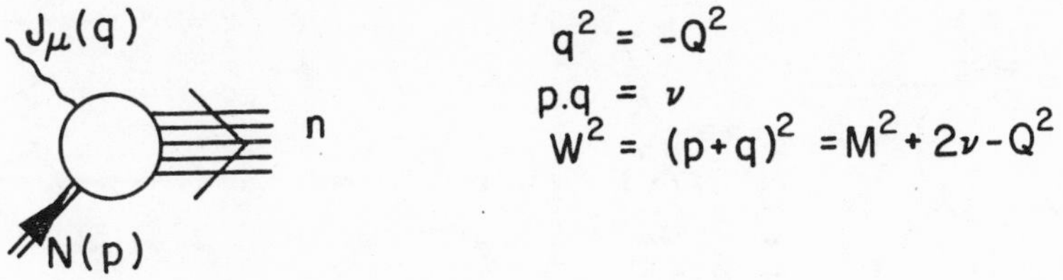

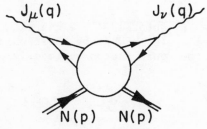

$$q^2 = -Q^2$$
$$p.q = \nu$$
$$W^2 = (p+q)^2 = M^2 + 2\nu - Q^2$$

Fig. 1 DIS and its kinematics

coupling, the MQM diagram one is led to consider is shown in Fig. 2, where the cut denotes the operation of taking the imaginary part. DIS is characterized by the famous Bjorken limit, i.e.

$$\begin{matrix} Q^2 \to \infty \\ \nu \to \infty \end{matrix} \quad \text{with} \quad x = \frac{Q^2}{2\nu} \quad \text{fixed}.$$

The quark amplitude in Fig. 2 can be obtained by integrating over the quark legs a six-point function involving four quark legs and two nucleon legs; and in the Bjorken limit, by (ii) above, we can make a Mueller-Regge analysis of the six-point function which leads to the expansion I have drawn in Fig. 3. This can easily be shown[2,3] to give, in the case of eN scattering:

$$\frac{1}{m^2} \nu W_2^{eN}(x,Q^2) \to \sum_i \sum_F Q_F^2 \times [\, f_{N,F}^i(x) + \bar{f}_{N,F}^i(x)\,] \, (Q^2)^{\alpha_i(0)-1} \qquad (2)$$

where F stands for the flavour index and Q_F denotes the charge of the quark of flavour F. The striking aspect of Eq. (2) is that for $\alpha_i(0)$, i.e. for the Pomeron trajectory, our result coincides with the quark-parton model, without obviously showing its already-mentioned difficulties. Thus we see that in the MQM *Bjorken scaling and Feynman scaling for quark amplitudes have the same origin.*

Fig. 2 The MQM diagram for DIS

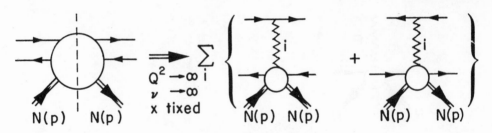

Fig. 3 The Mueller-Regge expansion for the DIS six-point function

Furthermore, for $\alpha_i(0) < 1$ we have non-scaling contributions of a subasymptotic character which, for finite values of Q^2, can give appreciable deviations form Bjorken scaling. From hadronic physics the next-to-leading trajectories (the meson-trajectories, like ρ, A_2, f_0, ...) have $\alpha_M(0) = \frac{1}{2}$; thus Eq. (2) *predicts scaling violating terms behaving like $(Q^2)^{-\frac{1}{2}}$.*

We can make further use of the Mueller-Regge approach to study the limits of quark (antiquark) "fragmentation functions" $f^i_{N,F}(x)\left[\bar{f}^i_{N,F}(x)\right]$ when $x \to 1$ (triple-Regge limit) and $x \to 0$ (double-Regge limit).

It is well known that when $x \to 1$, the Mueller-Regge residue functions $f^i_{N,F}(x)$ and $\bar{f}^i_{N,F}(x)$ are dominated by the triple-Regge graph of Fig. 4, which give the following results:

$$f^i_{N,F}(x) \xrightarrow[x \to 1]{} c^i_{N,F} \, (1-x)^{\alpha_i(0) - 2\alpha_{2q}(0)} , \qquad (3)$$

$$\bar{f}^i_{N,F}(x) \xrightarrow[x \to 1]{} \bar{c}^i_{N,F} \, (1-x)^{\alpha_i(0) - 2\alpha_{4q}(0)} . \qquad (4)$$

Notice that two new trajectories' intercepts have appeared in our analysis, $\alpha_{2q}(0)$ and $\alpha_{4q}(0)$. What is their value? From a similar analysis of the nucleon form factor we readily obtain

$$G_M(Q^2) \xrightarrow[Q^2 \to \infty]{} (Q^2)^{-1 + \alpha_{2q}(0)} , \qquad (5)$$

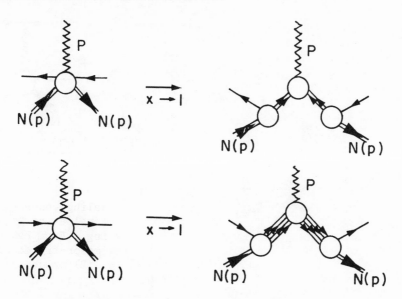

Fig. 4   The triple-Regge limit of the quark-fragmentation functions

which together with (3) is the well-known Drell–Yan–West connection. We know that, to a very good approximation, the magnetic
form factor of the nucleon behaves like $(Q^2)^{-2}$ for large Q^2, thus
we get $\alpha_{2q}(0) = -1$. For trajectories with n quarks exchanged,
dimensional arguments would give

$$\alpha_{nq}(0) = 1 - n \,, \tag{6}$$

an Ansatz that we hope to be able to derive in a more mature version
of the MQM. Thus (3) and (4), by virtue of (6), become

$$f^i_{N,F}(x) \xrightarrow[x \to 1]{} c^i_{N,F} (1-x)^{2+\alpha_i(0)} \; ; \tag{3'}$$

$$\bar{f}^i_{N,F}(x) \xrightarrow[x \to 1]{} \bar{c}^i_{N,F} (1-x)^{6+\alpha_i(0)} \, . \tag{4'}$$

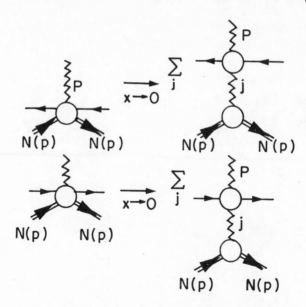

Fig. 5   The double-Regge limit of the quark-fragmentation functions

In the limit $x \to 0$ it is the double-Regge limit (Fig. 5) which determines the behaviour of $f^i_{N,F}(x)$ and $\bar{f}^i_{N,F}(x)$. By a straightforward analysis we get

$$f^i_{N,F}(x) \xrightarrow[x \to 0]{} \sum_j \gamma^{i,j}_{N,F}(x)^{-\alpha_j(0)} = \gamma^{i,P}_{N,F}(x)^{-1} + \gamma^{i,M}_{N,F}(x)^{-1/2} + \ldots \quad (7)$$

$$\bar{f}^i_{N,F}(x) \xrightarrow[x \to 0]{} \sum_j \bar{\gamma}^{i,j}_{N,F}(x)^{-\alpha_j(0)} = \bar{\gamma}^{i,P}_{N,F}(x)^{-1} + \bar{\gamma}^{i,M}_{N,F}(x)^{-1/2} + \ldots \quad (8)$$

which exhibit the characteristic Regge behaviour of DIS structure functions when $\nu \gg Q^2$.

Having pinned down the behaviour of the fragmentation functions for $x \to 1$ and $x \to 0$, we will attempt a description of the complete fragmentation functions by interpolating in the smoothest and simplest way between the two regions. In the next section I shall briefly report on the results of such analysis carried out in collaboration with Castorina and Nardulli[4].   We end this section by giving the MQM general results for DIS structure functions. Up to terms of order $1/Q^2$ we have: (notice that $\bar{T}_2(x,Q^2) = (2x)F_1(x,Q^2)$ is the Callan-Gross relation, a direct consequence of the MQM):

eN scattering

$$F_2^{eN}(x,Q^2) = 2x\,F_1^{eN}(x,Q^2) = \sum_F Q_F^2\, x\,(q_{N,F}(x,Q^2) + \bar{q}_{N,F}(x,Q^2)) \qquad (9)$$

νN scattering

$$F_2^{\nu N}(x,Q^2) = 2x\,F_1^{\nu N}(x,Q^2) = \sum_F (J^\dagger J)_F^F\, x\,(q_{N,F}(x,Q^2) + \bar{q}_{N,F}(x,Q^2)), \qquad (10)$$

$$xF_3^{\nu N}(x,Q^2) = \sum_F (J^\dagger J)_F^F\,(q_{N,F}(x,Q^2) - \bar{q}_{N,F}(x,Q^2)), \qquad (10')$$

where Q_F is the quark change, and J is the weak-interaction $\Delta Q = +1$ matrix in flavour space. $\bar{\nu}N$ scattering is simply obtained from (10) by substituting for J its Hermitian conjugate.

1.3 A model for DIS structure functions[4]

I shall now briefly discuss the model of Ref. 4. Besides the scaling violating terms, due to the exchange of meson trajectories with $\alpha_M(0) = \frac{1}{2}$, which we shall take into account, there are other scaling violations coming from kinematics (quark and target-mass effects) which are $O(1/Q^2)$ and will be neglected; this will therefore limit our analysis to Q^2 values above, say, $Q^2_{min} \simeq 2$ GeV^2. There is, however, a region where it was shown[5] that one can take into account both the kinematical constraint:

$$\nu W_2(Q^2,\nu) \xrightarrow[Q^2 \to 0]{} Q^2 \frac{d}{dQ^2}\,\nu W_2(Q^2,\nu)\Big|_{Q^2=0}, \qquad (11)$$

and the dispersive contribution from nearby vector meson poles, to write (m_ρ^2 is the ρ mass squared)

$$\nu W_2(Q^2,\nu) \xrightarrow[x \to 0]{} \frac{Q^2}{Q^2 + m_\rho^2}\,F_2(0) \qquad (12)$$

which works accurately for all values of Q^2 (see Fig. 6).

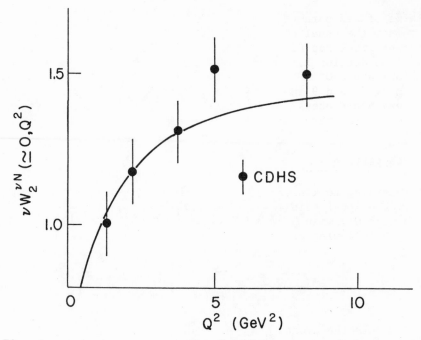

Fig. 6   Comparison of CDHS data[6] for $\nu W_2(x \simeq 0, Q^2)$ with (12)

The model consists in writing the quark (antiquark) distribution functions $q^{(E)}(x,Q^2)[\bar{q}^{(E)}(x,Q^2)]$, for each DIS process E, as a sum of terms of the type $x^\alpha(1-x)^\beta$, where the exponents α and β, as well as their coefficients, are determined by the requirements of (i) giving the correct limits for $x \to 0$ and $x \to 1$, and (ii) producing smooth interpolating functions. Thus we are led to write for a given process E (eN, νN, ...)

$$q^{(E)}(x,Q^2) = \alpha^{(E)}(1-x)^3 + \theta_2^{(E)}\frac{(1-x)^5}{x^{1/2}} +$$

$$+ \frac{\tilde{\theta}^{(E)}}{2}\frac{(1-x)^9}{x}\frac{Q^2}{Q^2+m_\rho^2} + \frac{\delta^{(E)}}{\sqrt{Q^2}}x^{1/2}(1-x)^{5/2} + O\left(\frac{1}{Q^2}\right) \quad (13)$$

and

$$\bar{q}^{(E)}(x,Q^2) = \theta_1^{(E)}\frac{(1-x)^7}{x^{1/2}} + \frac{\tilde{\bar{\theta}}^{(E)}}{2}\frac{(1-x)^9}{x}\frac{Q^2}{Q^2+m_\rho^2} +$$

$$+ \frac{\lambda^{(E)}}{\sqrt{Q^2}}x^{1/2}(1-x)^{13/2} + O\left(\frac{1}{Q^2}\right), \quad (14)$$

in terms of six parameters $\alpha^{(E)}$, $\theta_1^{(E)}$, $\theta_2^{(E)}$, $\tilde{\theta}^{(E)}$, $\delta^{(E)}$, and $\lambda^{(E)}$.
Apart from the usual quark-parton model relationships between the
different processes E, with the help of simple models for $x \to 0$ and
$x \to 1$ (see Section 1.5) behaviours we obtain *all* DIS structure
functions above $Q^2 \simeq 2$ GeV^2, in terms of only six parameters.
Figures 7, 8, and 9 summarize our predictions for the complete set
of observable processes. For a comparison with experiments, which
turns out to be very successful, the reader is invited to consult
Ref. 4.

1.4   The Ratio R = σ_L/σ_T

According to kinematics, neglecting quark masses, the ratio R
between the longitudinal and the transverse cross-sections must
vanish like Q^2, when $Q^2 \to 0$ *. For Q^2 large, and x small, the MQM
makes the following prediction:

$$ R \underset{Q^2 \to \infty}{\longrightarrow} \frac{4 \langle k_T^2 \rangle}{Q^2} , \tag{15} $$

where $\langle k_T^2 \rangle$ is the average quark transverse momentum.

The simplest interpolating function we can imagine, which
vanishes at $Q^2 = 0$ and obeys (15), is

$$ R = \frac{Q^2}{4 \langle k_T^2 \rangle} \frac{1}{\left[1 + \frac{Q^2}{4 \langle k_T^2 \rangle} \right]^2} , \tag{16} $$

whose graphs for different values of $\langle k_T^2 \rangle$ are reported in Fig. 10.
Notice that values of R between 0.1 and 0.2 are predicted up to
$Q^2 = 10$ GeV for reasonable values of $\langle k_T^2 \rangle$.

I would like to conclude this section by stressing that all
available experimental data confirm that the MQM can successfully
describe what we know, without having the difficulties of interpretation and the open ends of the field theoretical paradigms of
our time.

---

* We neglect the divergence of the axial current, which is suppressed
  by the quark masses.

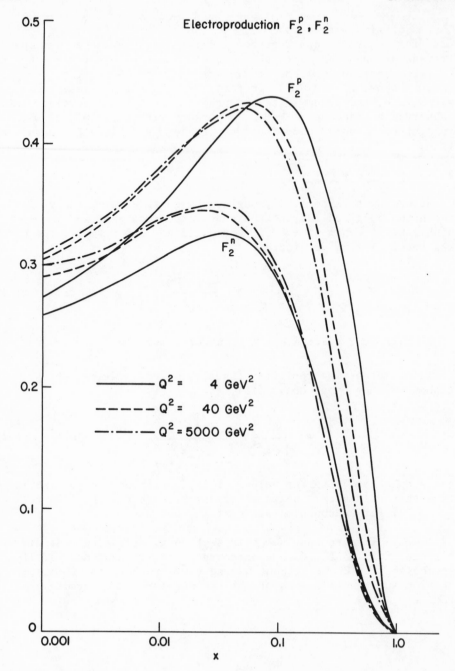

Fig. 7 $F_2(x,Q^2)$ for electron (muon) deep inelastic scattering
         off protons and neutrons, for $Q^2 = 4$, 40, and 5000 GeV^2

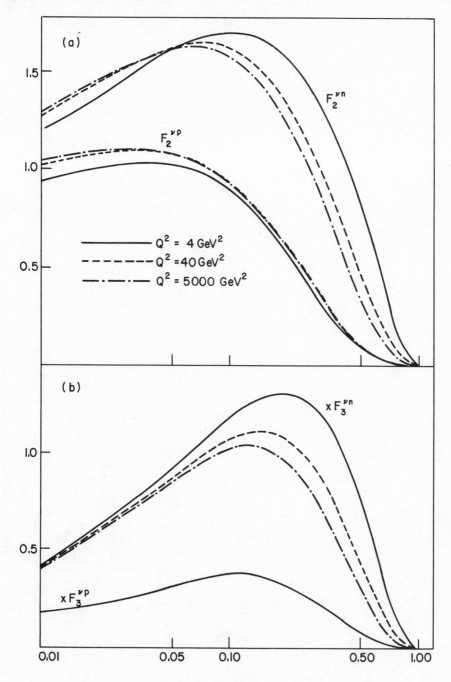

Fig. 8 F_2 (a) and xF_3 (b) structure functions for charged
        currents scattering off protons and neutrons for
        $Q^2 = 4$, 40, and 5000 GeV^2

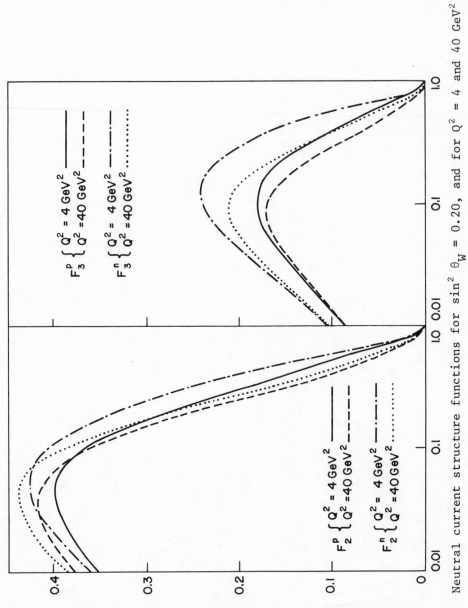

Fig. 9 Neutral current structure functions for $\sin^2 \theta_W = 0.20$, and for $Q^2 = 4$ and 40 GeV^2

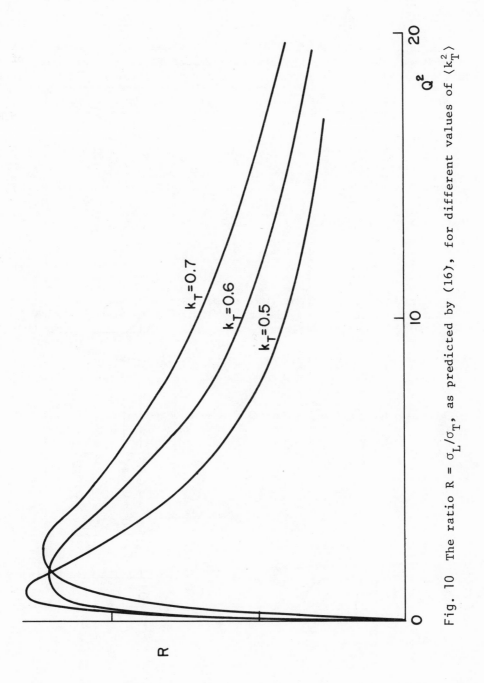

Fig. 10 The ratio $R = \sigma_L/\sigma_T$, as predicted by (16), for different values of $\langle k_T^2 \rangle$

2.   A REALISTIC LOOK AT e^+e^- ANNIHILATION AND FINAL STATES

2.1 A General View of e^+e^- Annihilation in the MQM

A detailed treatment of e^+e^- annihilation in the framework of the MQM was carried out by Gatto and myself in 1973[7]. In this section I shall simply summarize the most important results:

2.1.1 The ratio $R = \sigma(e^+e^- \to \text{hadrons})/\sigma(e^+e^- \to \mu^+\mu^-)$. The MQM diagram of Fig. 11 gives, in the large Q^2 limit, the following result:

$$R \underset{Q^2 \text{ large}}{\longrightarrow} \sum_j c_j \left(\frac{Q^2}{\mu^2}\right)^{\alpha_j(0)-1} = c_P + c_M \left(\frac{\mu^2}{Q^2}\right)^{1/2} + \cdots \tag{17}$$

which scales asymptotically, yielding a value

$$R_\infty = \lambda \sum_F Q_F^2 \tag{18}$$

with λ in general different from three, the coloured quark-free field prediction. As for scaling violations, we predict them to be of $O(1/\sqrt{Q^2})$, with a coefficient

$$c_M \simeq \lambda' \left(\sum_{\substack{F \\ \text{light}}} Q_F\right)^2 = 0 \tag{19}$$

which explains why R seems to tend so precociously to its asymptotic value.

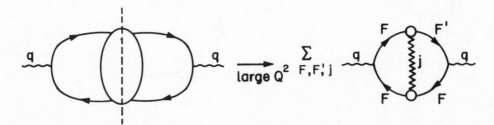

Fig. 11 The MQM diagram for e^+e^- annihilation and its Regge expansion

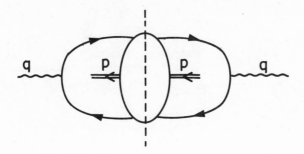

Fig. 12   The MQM diagram for one-particle deep inelastic annihilation

2.1.2 <u>The one-particle inclusive cross-section.</u> Figure 12
shows the relevant MQM diagram. By expanding the six-point function
à la Mueller-Regge in the Bjorken limit $\left[q^2 \to \infty, \; (pq) \to \infty, \; z = 2pq/q^2 \right.$
fixed$\left.\right]$, we obtain

$$\frac{d\sigma}{d\cos\theta \, dz} \to \frac{\pi\alpha^2}{q^2} \; \frac{1+\cos^2\theta}{2} \; z^2 \bar{F}_2^p(z) \, . \tag{20}$$

We notice the typical angular distribution of the single particle p
which follows the $1 + \cos^2 \theta$ law characteristic of the $\mu^+\mu^-$ final
state. Integrating over θ, $d\sigma/dz$ turns out to *scale*, with a scaling
function $\bar{F}_2^p(z)$ whose behaviour for $x \to 1$ and $x \to 0$ are again determined by the triple- and double-Regge limits, in the same way as
for DIS. We thus get:

$$\bar{F}_2^p(z) \xrightarrow[z \to 1]{} (1-z)^{1+2\alpha_{nq}(0)} \tag{21}$$

and

$$\bar{F}_2^p(z) \xrightarrow[z \to 0]{} \sum_j c_j \, (z)^{-2-\alpha_j(0)} \, . \tag{22}$$

When the hadron p is a meson (n = 1) according to (6), $\bar{F}_2^p(z)$ behaves
like $(1 - z)$; for baryons, on the other hand, we have the usual
$(1 - z)^3$ behaviour. Equation (22) implies that when $z \to 0$,
$F_2^p(z) \to z^{-3}$, thus leading to a multiplicity logarithmically increasing with q^2. All such predictions have been remarkably confirmed by experiments done at SPEAR, DORIS, and PETRA[8].

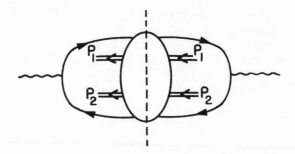

Fig. 13 The MQM diagram for two-particle inclusive e^+e^- annihilation

2.1.3 <u>Two-particle inclusive cross-sections.</u> In order to have
a better idea of the structure of the e^+e^- annihilation final states
we must consider the two-particle correlation functions; the relevant MQM diagram appears in Fig. 13. Performing a familiar
Mueller-Regge analysis, we obtain the following results:

- two-particle correlations are short-range in rapidity like in
  purely hadronic processes;

- relative transverse momentum is limited as in any typical
  hadronic event:

$$\langle k_T \rangle \simeq 300 \; MeV \; ;$$

- this means that event by event we have a two-jet structure;

- the jet axis' angular distribution obeys the $(1 + \cos^2 \theta)$ law.

Since the pioneering experiments in 1975[9], these predictions
of the MQM have found remarkable confirmation in the latest analysis of high-energy e^+e^- annihilation[8].

To conclude this analysis, I would like to stress that the
asymptotic structure of final states predicted by the MQM is the
same as in purely hadronic interactions. The reason for this very
general and powerful connection between hadron interactions and deep
inelastic phenomena is precisely our Regge Ansatz, which gives
quarks, when inside the bag domains, hadronic behaviour.

2.2 <u>Beyond the MQM: the Geometrohadrodynamical Approach</u>[12]

We have seen that the MQM is a very powerful and accurate tool
to describe the multiform aspects of deep inelastic physics from
e^+e^- annihilation to deep inelastic lepton-hadron scattering. How-

ever, the MQM, although *realistic*, is not a *complete* picture of
deep inelastic phenomena. We lack a knowledge of the absolute magnitudes of several "residue functions", so that we cannot predict
absolute cross-section values, but only relationships among them.
A particularly instructive example is the prediction (18) for the
asymptotic e^+e^- annihilation cross-section.

In order to go beyond the MQM, we need a theory which, starting
from the MQM Ansatz (i) (see Section 1.1), describes the hadronic
spectrum whose high-energy behaviour is constrained to obey the
MQM Ansatz (ii), and in which currents, both weak and electromagnetic, are introduced through Ansatz (iii). In the following
I shall provide arguments that the theoretical framework of quark
geometrohadrodynamics (QGD) satisfies the above requirements[12].

2.2.1 The ideas of QGD. A thorough discussion and motivation
of the most important ideas of QGD is contained in the lectures I
delivered at the 1977 Erice Summer School[10]; whoever is eager to
know more about this approach is referred to the School Proceedings.
Here I shall briefly sketch the QGD framework, trying to emphasize
what I regard as the most meaningful points.

First, where does QGD stand *vis-a-vis* the task of understanding
the fascinating aspects of the hadronic world? Figure 14 tries to
answer this question.

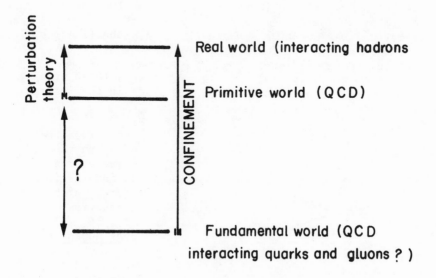

Fig. 14 The QGD strategy

We have, on the one hand, a real world which is revealed by observation and is populated by a multitude of interacting hadrons; on the other hand, we may envisage the existence of a fundamental world, where the flavour and colour degrees of freedom of hadronic matter are present in a simple and beautiful way. It is possible that the latter is in fact the QCD world; no doubt it is simple and beautiful. But the link between this world and the real world is unfortunately missing, depending on the solution of the confinement problem. As I have explained in the Introduction, this missing link is not a mere technicality, but it preempts, at the present time, any realistic appraisal of the hadronic world.

Having no solution to offer to the tremendously difficult problem of confinement, QGD tries to go from the real world to the fundamental world by an intermediate step, the primitive world. In the primitive world we have a multitude of *non-interacting* hadrons -- we shall call them "primitive hadrons" -- whose spectrum is simple and close to the expectations of the "naive quark model": mesons as $q\bar{q}$ states, baryons as qqq systems. It is the contention of the QGD approach that this world is *perturbatively related* to the real world, i.e. that the interactions among these states are reasonably weak, so that a perturbation theory of an appropriate type can be found to go from this world to the real world. There are strong indications, such as the validity of the Zweig rule and the surprising successes of the naive quark model, that this situation may in fact be realized in Nature. Thus it is likely that the so-called strong interactions are indeed fairly weak.

If the QGD programme is successful, i.e. if we can set up a *simple* primitive world from which to conquer the real world through a well-defined perturbative procedure, the step from the primitive to the fundamental world (marked by a question mark in Fig. 14) may turn out to be much easier to perform than the "confinement" step. In any case we would have at our disposal a powerful tool to describe the real hadronic world which has eluded us for the last thirty years.

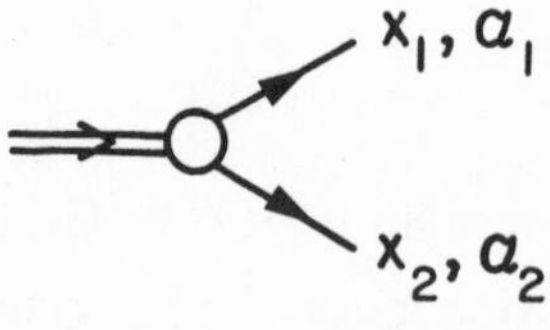

$$= \psi_{\alpha_1}^{\alpha_2} (p; x_1, x_2)$$

Fig. 15   The meson-state wave function

Let me now list the steps that we have taken to build the primitive QGD world and to set up the perturbative procedure to conquer the real hadronic world.

- Introduce simple wave functions (w.f.'s) to describe $q\bar{q}$ and qqq states, the fundamental hadron states (see, for example, Fig. 15).

- Impose the vanishing of the hadronic w.f.'s outside finite space-time domains (bags).

- Inside the bags impose simple free-field equations for the quark fields (i.e. in the space-time variables x,α).

- Characterize the bag domains in very simple terms, giving boundary conditions for the w.f.'s compatible with the free-field equations, and leading to an "internal quark motion" which is as close as possible to the free one (this requirement has been called "maximum freedom").

- Derive from the previous conditions the hadronic spectrum. In the case of mesons made out of light quarks we thus obtain a set of approximately linear and parallel Regge trajectories without odd daughters (see Fig. 16). Remember that in the primitive world these trajectories all have zero width.

As for baryons, QGD provides a simple and surprising picture[11].

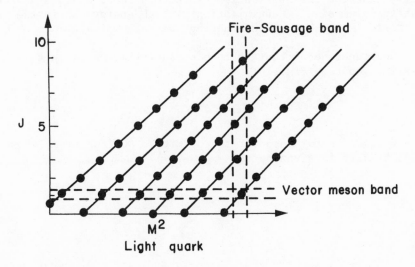

Fig. 16 The J-M^2 plot of light-quark mesonic states according to QCD

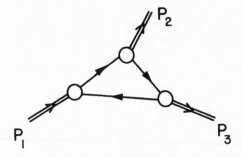

Fig. 17   The three-meson vertex

- Introduce point-like currents coupled to the quark degrees of freedom; and use them to normalize the wave. functions.

  Thus we can calculate, always in a world of non-interacting hadrons, form factors, e^+e^- annihilation (see later), and deep inelastic scattering.

- Introduce couplings among different hadronic states through simple space-time overlaps (see, for example, Fig. 17). This is the first step in constructing the perturbative theory which should produce the real world. The perturbative scheme has not, so far, been analysed in great detail, and represents the most arduous step in the whole QGD approach[*].

  2.2.2 <u>QGD and high-energy e^+e^- annihilation.</u> I shall now briefly describe the application of the QGD ideas discussed above to the problem of e^+e^- annihilation at high energy and in the primitive world, i.e. the non-interacting hadronic world.

  The process contributing to e^+e^- annihilation is in this approximation (see Fig. 18):

$$e^+e^- \longrightarrow \gamma(q) \longrightarrow V_n \tag{23}$$

where V_n is a vector meson of the primitive QGD spectrum, whose mass asymptotically is given by ($R_0^2 = 4$ GeV^{-2})

$$m_n^2 \longrightarrow \frac{4\pi}{R_0^2} n \ . \tag{24}$$

---

[*] In particular, the analysis of the bag "off-shell" behaviour is completely lacking.

Fig. 18   The e^+e^- annihilation in a high-mass vector meson

From the normalized w.f. we can easily compute[10] the transition matrix element $\gamma(q) \to V_n$ and obtain for the ratio R a well-defined sum of δ functions concentrated at the points M_n^2 of the spectrum, given in (24). This unphysical result is obviously due to the zero-width characteristic of the states of the primitive world. Without knowing the detailed structure of the three-meson vertex, we can, however, check[12] that the interaction will give the vector mesons V_n a width Γ_n, such that

$$ m_n \Gamma_n \to \lambda \tag{25} $$

where λ is a universal constant. Now it is only necessary that $\lambda = O(4\pi/R_0^2)$ for us to obtain the result (see Fig. 19)

$$ R_\infty = \left(\frac{4}{\pi}\right) \cdot 3 \sum_F Q_F^2 , \tag{26} $$

i.e. the parton model result is modified by a factor $4/\pi = 1.27$. We cannot yet fully appreciate the meaning of the $4/\pi$ factor, owing to the missing analysis of the effect of interaction corrections.

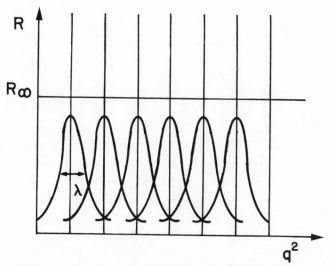

Fig. 19 R_{as} in the primitive and in the real world

We simply notice that (26) is remarkably close to present experimental knowledge[8].

   2.2.3 The fire sausage and the final states. The structure of
the spectrum of the QGD primitive world is such that (see Fig. 16)
at a given mass M^2 there is a set of quasi-degenerate mesonic states
with increasing angular momenta up to a maximum value $\ell_{max} \sim M^2$.
When we switch on the interaction term, the finite range of the
so-called strong interactions leads to an effective decoupling of
all those states which have $\ell > \ell_0 = R_T(M^2)(M/2)$, where $R_T(M^2)$ is
at most logarithmically increasing with the mass of the state. It
is trivial to realize that by an appropriate linear superposition
of the quasi-degenerate states up to the angular momentum value ℓ_0
we can generate coherent states with the cylindrical space-
configuration depicted in Fig. 20, which we may call fire sausages
(FS). Why are FS so important in determining the structure of
final states? The answer can again be given without a detailed
knowledge of the spin structure of the three-meson vertex (Fig. 15).

   Let us in fact estimate the behaviour of the vertex

$$V_n \rightarrow M_1 + M_2 \tag{27}$$

where $M_{1,2}$ are mesons of masses m_1 and m_2. From the structure of
the w.f.'s and of the space-time overlap, it is very easy to show
that:

a)   the by far dominant configuration is when

$$m_1(m_2) \ll m_2(m_1) \tag{28}$$

     with the obvious condition $m_1 + m_2 \leq m_n$;   and

$$m_2(m_1) \ll 0(m_n) \tag{29}$$

b)   if m_2 is the large mass the phases of the overlap integral are
     such that the produced high mass mesonic system M_2 is an FS.

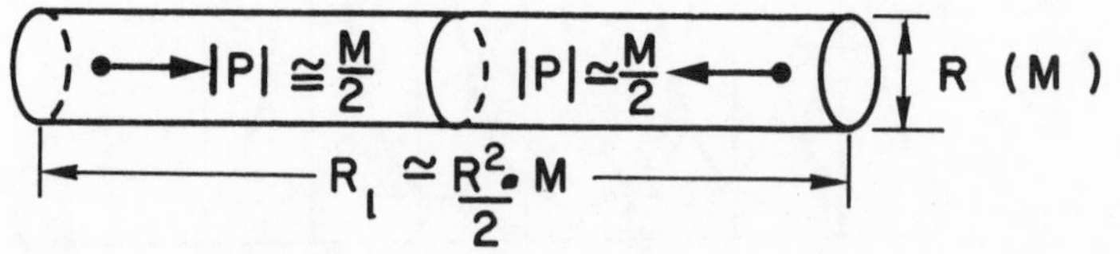

Fig. 20   The space structure of a fire sausage

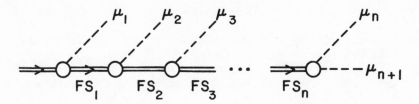

Fig. 21 The cascade decay of the vector meson V_n

Now that V_n has decayed into a small mass meson M_1 and an FS M_2, the decay process will continue until all the high-mass mesons have decayed, thus giving rise to the cascade process of Fig. 21.

With the simplest possible three-meson overlap we can solve the cascade problem and obtain a number of very interesting results:

α) particle production, event by event, takes place along a given direction, with a finite transverse momentum;

β) the one-particle cross-sections scale à la Feynman;

γ) direct production of π, η, and K is suppressed as compared with vector mesons (ρ, ω, ϕ, K^*), which account for the majority of direct production, and are produced transversely polarized;

δ) particle multiplicity increases logarithmically with the vector meson's mass;

ε) particle correlations are short-range in rapidity; and there emerges a cluster structure where clusters are nothing but low-mass resonances.

3.   CONCLUSIONS

Unlike the Introduction the Conclusions will be telegraphic, that is:

i) QCD may be the theory of hadrons but *nobody* knows, so far, how to calculate sensibly with it.

ii) It is indeed possible, for the time being, to embody the surprising simplicity of the real world in a description which has the virtue of being realistic, such as the MQM and its more mature and powerful version of QGD.

I hope I have convinced you that a lot of physics can be conquered starting from simple, but in no way naive, ideas.

REFERENCES

1. See the contributions of H. Anderson and J. Van der Velde to
    this Seminar.
2. The first ideas on the massive quark model (MQM) date back to
    1972, and were first published in G. Preparata, Phys. Rev.
    D 7, 2973 (1973).
3. The most comprehensive description of the MQM is contained in
    G. Preparata, A possible way to look at hadrons; the MQM,
    in Lepton hadron structure (ed. A. Zichichi) (Academic Press,
    NY, 1975), p. 54.
4. P. Castorina, G. Nardulli and G. Preparata, A realistic de-
    scription of deep inelastic structure functions, preprint
    CERN TH 2670 (1979).
5. G. Preparata, Phys. Lett. 36B, 56 (1972).
6. CDHS Collaboration, Inclusive interaction on high-energy neu-
    trinos and antineutrinos in iron, submitted to Z. Phys. C.
7. R. Gatto and G. Preparata, Nucl. Phys. B67, 2973 (1973).
8. For an excellent review of the e^+e^- physics see G. Wolf, A
    review of e^+e^- physics, Lectures at the 1978 Erice Summer
    School (ed. A. Zichichi) (to be published).
9. G. Hanson, e^+e^- production and jet structure at SPEAR, *in*
    Proc. Tbilisi Int. Conf. on High Energy Physics, Tbilisi,
    1976 (JINR, Dubna, 1976), p. Bl.
10. For the most extensive review see G. Preparata, Quark geo-
    metrodynamics: a new approach to hadrons and their in-
    teractions, *in* Proc. of the 1977 Erice Summer School (ed.
    A. Zichichi) (Plenum Press, NY, 1979).
11. G. Preparata and K. Szegö, Nuovo Cimento 47A, 303 (1978).
12. N.S. Craigie and G. Preparata, Nucl. Phys. B102, 497 (1976).